Springer Proceedings in Physics

Volume 257

Indexed by Scopus

The series Springer Proceedings in Physics, founded in 1984, is devoted to timely reports of state-of-the-art developments in physics and related sciences. Typically based on material presented at conferences, workshops and similar scientific meetings, volumes published in this series will constitute a comprehensive up-to-date source of reference on a field or subfield of relevance in contemporary physics. Proposals must include the following:

- name, place and date of the scientific meeting
- a link to the committees (local organization, international advisors etc.)
- scientific description of the meeting
- list of invited/plenary speakers
- an estimate of the planned proceedings book parameters (number of pages/articles, requested number of bulk copies, submission deadline).

More information about this series at http://www.springer.com/series/361

Rajeev K. Puri · Joerg Aichelin ·
Sakshi Gautam · Rohit Kumar
Editors

Advances in Nuclear Physics

Structure and Reactions

Springer

Editors
Rajeev K. Puri
Department of Physics
Panjab University
Chandigarh, India

Joerg Aichelin
SUBATECH, UMR 6457
Université de Nantes
Nantes, France

Sakshi Gautam
Department of Physics
Panjab University
Chandigarh, India

Rohit Kumar
Department of Physics
Panjab University
Chandigarh, India

ISSN 0930-8989 ISSN 1867-4941 (electronic)
Springer Proceedings in Physics
ISBN 978-981-15-9061-0 ISBN 978-981-15-9062-7 (eBook)
https://doi.org/10.1007/978-981-15-9062-7

This Springer imprint is published by the registered company Springer Nature Singapore Pte Ltd.
The registered company address is: 152 Beach Road, #21-01/04 Gateway East, Singapore 189721, Singapore

Foreword

It is with deep satisfaction that I write this foreword to the proceedings of Indo-French workshop on "Multifragmentation, Collective flow and Sub-Threshold Particle Production in Heavy-Ion Reactions." The workshop was aimed to bring together the researchers and academicians from Europe and India who work in the field of Nuclear Physics and provide them with a platform where the exchange of ideas could take place. The workshop was organized at Department of Physics, Panjab University Chandigarh which encouraged the interaction of young undergraduate/postgraduate and research students with the more established academic community and motivated them to pursue Science/Research as a career option. I hope that this tradition of organizing such workshops/conferences at Academic Institute will continue. The papers presented at the workshop contributed the most recent scientific knowledge in the field of Nuclear Structure and Nuclear Reactions. This volume will furnish the researchers in the field with an excellent reference book, providing papers on recent trends in nuclear structure and on the vast phenomenon in nuclear reactions. I also hope that this volume will be an impetus to stimulate further research in all these areas.

I thank all participants and authors for their contribution.

Prof. Raj Kumar
Vice Chancellor
Panjab University
Chandigarh, India

Preface

This book has grown out of the lectures delivered in Indo-French Workshop on *Multifragmentation, Collective flow and Sub-Threshold Particle Production in Heavy-Ion Collisions* which was held at the Department of Physics, Panjab University Chandigarh, India, during February 04–06, 2019. The aim of the workshop was to provide a platform for discussing recent developments in nuclear physics research at intermediate energies carried out in Eastern and Western worlds.

One of the unique and valuable dimensions to the workshop is the way that this event has brought together all the expertise and educators from Europe to discuss recent research going on at their part of the world and thus bridging the gap between the boundaries. Besides formidable Indian speakers, we had invitees from France and Germany which houses the major research facilities of nuclear physics, both in experimental as well as theoretical fronts. In addition, young researchers in the field had also been given the chance to present their work before the expertise and thus gaining wonderful insight about their work. The workshop spanned over 3 days and 10 sessions and we could provide ample time to speakers to present their respective topic at length. Collectively, the research discussed at the workshop contributed to our knowledge and basic understanding of various reaction mechanisms which take place in heavy-ion collisions. This knowledge will lead to further progress in the understanding various interactions that govern the dynamics at sub-nucleonic level, ultimately, may shed light on the fate of the Universe. The workshop aimed at creating a forum for further discussions on the thrust areas of intermediate energy nuclear physics and thus a call for papers was addressed to all the invites of the event. Thus, we are happy to bring out the contributed book compiling all the lectures/talks delivered at the workshop and this volume provides an opportunity for the readers to engage with a selection of referred papers that were presented during the event. The volume contains twenty chapters, each corresponding to one contributed paper by each invitee. All the contributed papers span over one or the other topics of the workshop; multifragmentation, collective flow and sub-threshold particle production. Out of total 18 papers, a *few* belong to experimental studies, while the rest deal with findings obtained using one or the other available theoretical models. The details are listed in the table of contents.

We hope that this volume of lectures will be helpful to the community working in the field of Heavy-Ion Physics. We shall highly appreciate receiving valuable comments and suggestions from the users.

Nantes, France — Joerg Aichelin
Chandigarh, India — Rajeev K. Puri
Chandigarh, India — Sakshi Gautam
Chandigarh, India — Rohit Kumar

Acknowledgments

We gratefully acknowledge all the participants of the workshop who have contributed to the volume. We, organizers, also express our gratitude to every person involved in the workshop, for their unwavering commitment. We are highly thankful to Prof. Raj Kumar, Vice-Chancellor, Panjab University, and Prof. Navdeep Goyal and Devinder Mehta, Chairperson, Department of Physics and Dean Sciences, Panjab University, for their support during the event. The successful completion of workshop could finally lead us to bring this volume of delivered lectures/talks. Lastly, we are most indebted for the generous support given by Indo-French Center for the Promotion of Advanced Research (IFCPAR/CEFIPRA), Department of Science & Technology & Renewable Energy, Chandigarh Administration and Centre of Advanced Study in Physics, Panjab University Chandigarh.

Contents

Editors and Contributors

About the Editors

Prof. Rajeev K. Puri is working as Professor of Physics at Panjab University, Chandigarh, India. He obtained his Ph.D. from Panjab University and then was a post-doctoral fellow at the Institut fur Theoretische Physik, Universitaet Tubingen, Germany and at the Laboratoire de Physique Nucleaire, Universite de Nantes, France. He has around 400 publications in journals and conferences to his credit with citations close to 6500. He has also guided about 25 Ph.D. students. He has received prestigious awards like Scientist of the Year; Himachal Excellence Awards, S.N. Satyamurthy Memorial Award and Young Scientist Research award. He is an editor, member of editorial board, referee of about 47 journals. He is also on the board of academic/advisory bodies of about dozen noted universities throughout India. Currently, he is also holding a position of University coordinator of RUSA (National Higher Education Mission), Panjab University.

Prof. Joerg Aichelin is presently serving as Distinguished Scientist (class exceptionnelle) at the University of Nantes, France. He received his Ph.D. from the University of Heidelberg, and did his post-doctoral research at the Michigan State University, East Lansing, USA. He is a pioneer of theoretical nuclear physics at intermediate/high energies and his proposed n-body dynamical approach to identify fragment production in heavy-ion collisions is a benchmark in the field, which is widely used in the field of heavy-ion physics. He has numerous awards and honors to his credit like Fellow of Institute of Physics, London (2004); Member of Academia Europe (2010–), Spokesperson of a European Network of 20 European theoretical laboratories (2003–), Member of the Program Advisory Committee of the Gesellschaft fuer Schwerionenforschung (GSI) Darmstadt (2004–09), President of the Committee for financing the nuclear/high energy theory in France (2012–16) and many more. He has 239 publications in various reputed journals with over 8010 citations.

Dr. Sakshi Gautam is currently serving as Assistant Professor at the Department of Physics, Panjab University, Chandigarh, India. She obtained her B.Sc. (Hons. School Physics), M.Sc. (Hons. School Physics) as well as Ph.D. from Panjab University. She was a Gold medalist both at the bachelor's and master's level. Her major areas of research are theoretical nuclear physics at intermediate energies and computational physics. She has around 80 publications in journals and conferences of International/national repute. She has successfully completed three research projects till date as principal/co-investigator and delivered numerous talks at symposia/workshops/conferences. She is in the reviewers panel of seven international journals. She was awarded the SERC-Fast Track Fellowship for Young Scientists in 2014.

Dr. Rohit Kumar obtained his Ph.D. in theoretical nuclear physics from Panjab University, Chandigarh, India. He is currently working as a Research Associate at the Physics Department, Panjab University. His primary research interest is to study the fragment production and its associated dynamics at intermediate energy heavy-ion reactions. He has a number of research papers to his credit and has delivered talks in various conferences in India and abroad.

Contributors

Madan M. Aggarwal Department of Physics, Panjab University, Chandigarh, India

Jörg Aichelin SUBATECH, UMR 6457, IMT Atlantique, IN2P3/CNRS, Université de Nantes, Nantes, France;
Frankfurt Institute for Advanced Studies, Frankfurt, Germany

Preeti Bansal Department of Physics, Panjab University, Chandigarh, India;
Department of Physics, Chandigarh University, Mohali, Punjab, India

B. Behera School of Physics, Sambalpur University, Burla, Odisha, India

E. Bonnet SUBATECH, IMTA, CNRS-IN2P3, Nantes, NU, France

E. L. Bratkovskaya GSI Helmholtzzentrum für Schwerionenforschung GmbH, Darmstadtø, Germany;
Institut für Theoretische Physik, Goethe-Universität Frankfurt am Main, Frankfurt, Germany

W. Cassing Institut für Theoretische Physik, Justus-Liebig-Universität Giessen, Giessen, Germany

G. Chaudhuri Variable Energy Cyclotron Centre, Kolkata, India

S. Das Gupta Physics Department, McGill University, Montréal, Canada

P. Das Variable Energy Cyclotron Centre, Kolkata, India

Sakshi Gautam Department of Physics, Panjab University, Chandigarh, India

F. Gulminelli LPC, CNRS/EnsiCaen et Universite, Caen, France

Christoph Hartnack SUBATECH, UMR 6457, IMT Atlantique, IN2P3/CNRS, Université de Nantes, 4 rue Alfred Kastler, Nantes, France

V. Kireyeu Joint Institute for Nuclear Research, Dubna, Moscow region, Russia

V. Kolesnikov Joint Institute for Nuclear Research, Dubna, Moscow region, Russia

Lokesh Kumar Department of Physics, Panjab University, Chandigarh, India

Rohit Kumar Department of Physics, Panjab University, Chandigarh, India

Arnaud Le Févre GSI Helmholtzzentrum für Schwerionenforschung GmbH, Darmstadt, Germany

Yvonne Leifels GSI Helmholtzzentrum für Schwerionenforschung GmbH, Darmstadt, Germany

Ruchi Mahajan Department of Physics, Panjab University, Chandigarh, India

S. Mallik Variable Energy Cyclotron Centre, Kolkata, India

P. Moreau Institut für Theoretische Physik, Goethe-Universität Frankfurt am Main, Frankfurt, Germany

L. Oliva GSI Helmholtzzentrum für Schwerionenforschung GmbH, Darmstadt, Germany;
Institut für Theoretische Physik, Goethe-Universität Frankfurt am Main, Frankfurt, Germany

S. K. Patra Institute of Physics, Bhubaneswar, India;
Homi Bhabha National Institute, Mumbai, India

Rajeev K. Puri Department of Physics, Panjab University, Chandigarh, India

Abdul Quddus Department of Physics, Aligarh Muslim University, Aligarh, India

T. R. Routray School of Physics, Sambalpur University, Burla, Odisha, India

K. P. Santhosh School of Pure and Applied Physics, Kannur University, Payyanur, Kerala, India

Arun Sharma Department of Physics, GDC Billawar, Jammu, India

Sakshi Sharma Department of Physics, Panjab University, Chandigarh, India

O. E. Soloveva Institut für Theoretische Physik, Goethe-Universität Frankfurt am Main, Frankfurt, Germany

T. Song GSI Helmholtzzentrum für Schwerionenforschung GmbH, Darmstadt, Germany

S. Sood Department of Physics, Panjab University, Chandigarh, India

Sucheta Department of Physics, Panjab University, Chandigarh, India

X. Viñas Department de Física Quàntica i Astrofísica and Institut de Ciències del Cosmos (ICCUB), Facultat de Física, Universitat de Barcelona, Barcelona, Spain

V. Voronyuk Joint Institute for Nuclear Research, Dubna, Moscow region, Russia

M. Winn SUBATECH, Université de Nantes, Nantes cedex 3, France

Abbreviations

ALADIN	A Large Acceptance DIpole Magnet
ALICE	A Large Ion Collider Experiment
AMD	Anti-Symmetric Molecular Dynamics
AMPT	A Multi-Phase Transport
AMR	Anti-Magnetic Rotation
ANURIB	Advanced National facility for Unstable Rare-Ion Beams
ATOF	ALADIN Time-Of-Flight
BES	Beam Energy Scan
BNL	Brookhaven National Laboratory
BUU	Boltzmann–Uehling–Uhlenbeck
BW	Blast Wave
BWMF	Bethe-Weizsäcker Mass Formula
CBUU	Coupled-channel Boltzmann Uehling Uhlenbeck
CDFM	Coherent Density Functional Model
CERN	Conseil Européen pour la Recherche Nucléaire
CFL	Color Flavor Locked
CGC	Color Glass Condensate
ChPT	Chiral Perturbation Theory
CM	Center-of-Mass
CPPM	Coulomb and Proximity Potential Model
CTM	Canonical Thermodynamical Model
CYM	Classical Yang Mills
DEFR	Difference of Elliptic Flow
DQPM	Dynamical Quasi-Particle Model
EOS	Equation of State
E-RMF	Effective Mean-Field Theory motivated my RMF
E1	Electric Dipole
E2	Electric Quadrupole
FAIR	Facility for Antiproton and Ion Research
fKLN-CGC	factorized Kharzeev, Lavin and Nardi Color Glass Condensate

FMD	Fermionic Molecular Dynamics
FRS	Fragment Separator
FTRMF	Field Theoretical Relativistic Mean Field
H	Hard
HF	Hatree-Fock
HIC	Heavy-Ion Collisions
HM	Hard Momentum
HSD	Hadron String Dynamics
HRIB	Holifield Radiation Ion Beam
IBUU	Isospin-dependent Boltzmann–Uehling–Uhlenbeck
ImQMD	Improved Quantum Molecular Dynamics
IP-Sat	Impact Parameter Saturation model
IQMD	Isospin Quantum Molecular Dynamics
KB	Kadanoff-Baym
KRATTA	Krakow Tripple Telescope Array
LAND	Large Array Neutron Detector
LHC	Large Hadron Collider
LNL	Lagnaro National Lab
lQCD	Lagrangian Quantum Chromo Dynamics
MCG	Monte Carlo Glauber
MCG-Q	Monte Carlo Glauber model with Quarks as participants
MCG-N	Monte Carlo Glauber model with Nucleons as participants
MGLDM	Modified Generalized Liquid Drop Model
MST	Minimum Spanning Tree
NCQ	Number of Constituent Quarks
NEOS	Nuclear Equation of Stat
NICA	Nuclotron-based Ion Collider fAcility
NM	Nuclear Matter
NN	Nucleon-Nucleon
NSCL	National Superconducting Cyclotron Laboratory
NSE	Nuclear Statistical Equilibrium
PAC	Principal Axis Cranking
PHQMD	Parton-Hadron-Quantum Molecular Dynamics Model
PHSD	Parton-Hadron String Dynamics
PLF	Projectile-Like Fragment
QCD	Quantum Chromo Dynamics
QGP	Quark–Gluon Plasma
QMD	Quantum Molecular Dynamics
QQPM	Quark Quasi-Particle Model
RHIC	Relativistic Heavy-Ion Collisions
RMF	Relativistic Mean Field
S	Soft
SACA	Simulated Annealing Clusterization Algorithm
SHE	Super Heavy Element
SM	Soft Momentum

SNM	Symmetric Nuclear Matter
SPS	Super Proton Synchrotron
STAR	Solenoidal Tracker at RHIC
TOF	Time of Flight
TüQMD	Tübingen Version of the QMD
UrQMD	Ultra-relativistic Quantum Molecular Dynamics
VUU	Vlasov Uehling Uhlenbeck

Chapter 1
Evolution of Cluster Production with Fragmentation Degree

E. Bonnet

Abstract We present in this work the experimental production of clusters in the ^{58}Ni+ ^{58}Ni reactions at incident beam energies between 32 and 90 MeV/nucleon collected with the INDRA apparatus. We focus on the evolution of their contribution with respect to the final fragmentation degree given by the charge of the biggest fragment (Z_{max}) of each event. Using the mass fraction observable, we look at probabilities of nucleons to be free or bound in clusters and how the different species evolve with fragmentation degree and with the incident energy of the collision. We highlight the specific role of the ^{4}He clusters in the whole range of fragmentation degree.

1.1 Introduction

Understanding the production of clusters (namely, nucleons ended together) produced in Heavy-ion collisions (HIC) is a very stimulating question. Bringing experimental constraints on that topic should help to constrain both the modelization of the equation of state (EOS) of non- homogeneous nuclear matter (NM) and also the treatment of the N-body correlations in transport models.

Many experimental signatures of phase transition in the Fermi energy domain can be found in literature [1]. According to that, cluster production should be affected depending on the region of the phase diagram we are looking at. For example, the spinodal decomposition as the driven process of the phase transition has been recently confirmed [2]. From that, we should observe the correlated specific behavior for clusters. In the canonical description of the liquid–gas phase transition, the charge of the biggest fragment, the Z_{max} observable, has been linked to the order parameter of this transition [3] and then can be used to sample the phase diagram. Using it, we can draw easily a continuous evolution between evaporation to vaporization passing through multifragmentation.

E. Bonnet (✉)
SUBATECH, IMTA, CNRS-IN2P3, Nantes, NU, France
e-mail: eric.bonnet@subatech.in2p3.fr

R. K. Puri et al. (eds.), *Advances in Nuclear Physics*, Springer Proceedings in Physics 257, https://doi.org/10.1007/978-981-15-9062-7_1

In this work, we will address the contribution of the different clusters to the produced partitions according to the Z_{max} observable. As we deal with finite systems, we have to consider carefully the trivial conservation of the total number of nucleons. This a specific point concerning HIC respect to NM, the finite number of nucleons involved in the collision process has to be considered. This means that depending on the fragmentation degree associated with different excitation/dissipation/violence, the available nucleons to build clusters is not the same. This picture of available nucleons to build cluster points out also the question of time sequence during the whole process leading to the final partitions. In that sense, the use of Z_{max} has two advantages: it allows to sample the phase diagram and it allows to consider set of final partitions where available nucleons to be shared in clusters are a posteriori equivalent. From the experimental point of view, the charge of the biggest fragment is a straight forward and robust observable.

1.2 Methodology and Experimental Details

The methodology is as following: in the ^{58}Ni+^{58}Ni reactions collected with INDRA, we look at evolution of the contributions of the different species: ^{1,2,3}H, ^{3,4}He and the heavier clusters are gathered in the $A > 4$ family. We choose the ^{58}Ni+^{58}Ni system because it is a rather light system and in this way, we minimize experimental bias over the whole range of fragmentation degree. To draw the cluster contributions, we introduce the mass fraction (X, (1.1)). First, we look at the effect of different beam energies (from 32 to 90 MeV/nucleon) to evaluate the effect of an increase of the energy deposition in the system. Then we introduce the excitation energy (E^*, (1.2)) to apply an additional sorting and to draw a general picture of the different cluster contributions under Z_{max} and E^* constraints.

$$X^{(i)} = m_i A_i/(A_{tot} - A_{max}), \tag{1.1}$$

$$E^* = \sum_{i=1}^{M_{Z\geq 1}} (\epsilon_k^{(i)} + \delta^{(i)}) + M_n(< \epsilon_k^n > + \delta^n) - \delta_{ini}. \tag{1.2}$$

The INDRA apparatus [4] allows the detection and charge identification of all charged products coming from a collision. In addition, thanks to Cesium Iodide scintillators, the isotopic identification is achieved for elements up to Be. The ^{58}Ni+^{58}Ni data presented in this analysis have already used to probe isospin diffusion in semi-peripheral collisions [5] and to measure fusion cross sections for light systems with a significant contribution at 32 and 40 MeV/nucleon [6]. A detailed description of the experiment and the INDRA apparatus can be found in these related publications. In the following, we focus on the forward part of each event because a complete isotopic identification up to ^{10}Be is achieved; the detection efficiency is almost inde-

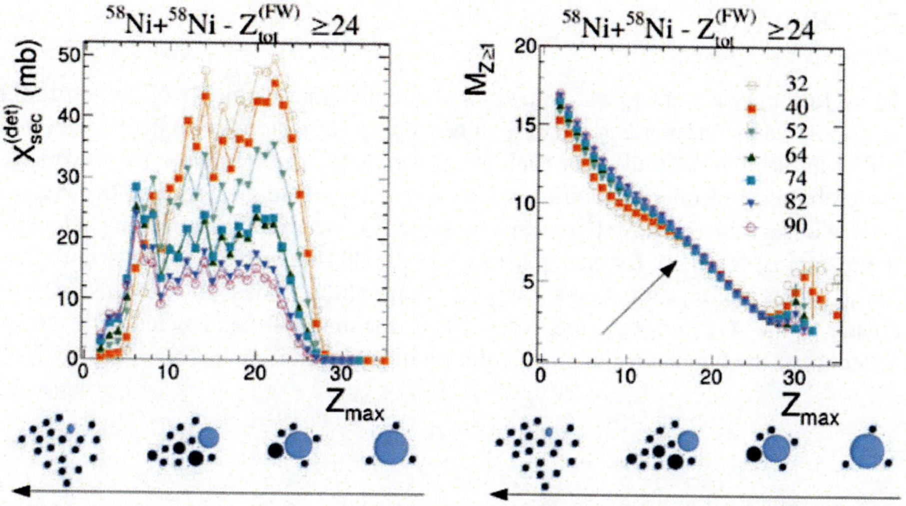

Fig. 1.1 Left panel: evolution of the Z_{max} observable distributions; the Y-scale corresponds to absolute measured cross sections in mb, see [5] for details. Right panel: mean evolution of the total multiplicity of charged produced ($M_{Z\geq1}$) with the Z_{max} observable. The different markers and colors stand for seven beam energies used in the analysis. The cartoon below the panels illustrates how the partitions evolve with fragmentation degree. The black arrow indicates the region where we observe differences in the trends when looking at one beam energy to another, see the test for details

pendent of the reaction mechanism minimizing possible experimental bias. Finally, as ^{58}Ni+^{58}Ni is a symmetric system, it gives, on average, a good description of what happens for the whole system. As we choose Z_{max} as the sorting observable, we have to be sure that if this is present in the selected partitions or not. That's why we keep only events with a missing charge less than 5 compared to the ^{58}Ni charge ($Z_{tot}^{(FW)} \geq 24$). With this condition, we ensure that the Z_{max} fragment is part of the event on the whole range of fragmentation degree.

In Fig. 1.1, the left panel shows distributions of the Z_{max} observable for the selected events, while the right panel shows the mean evolution of total charged multiplicity ($M_{Z\geq1}$) with Z_{max}. The different markers and colors stand for the seven beam energies. The Z_{max} distributions show the population of the whole range of fragmentation with a significant cross section. Increasing beam energies, we observe the expected continuous evolution from evaporation to vaporization events. The mean evolution of total charged multiplicity follows at first order, the conservation of nucleons: more the partition is fragmented, more the number of fragments are important. Nevertheless, we observe already interesting behavior in the intermediate values of Z_{max} for the lowest beam energies: after a linear increase of $M_{Z\geq1}$ in the evaporation regime, the trend is dampened before it goes back in the vaporization regime. This intermediate regime is softned when beam energy becomes higher with an overall correlation more close to a monotonic increase. It is coherent with a scenario where the partitions are produced under constraints in the multifragmentation regime.

1.3 Results

To go further, we want to see behaviors of the different elements of the partitions. To do so, we distinguish the ^{1,2,3}H isotopes, the ^{3,4}He isotopes and the heavier ones gather in the $A > 4$ family. For each, we compute the mass fraction ($X^{(i)}$) which is the probability of one nucleon to belong to the (i) cluster. As we use the Z_{max} as sorting observable we have to remove from the X observable the nucleons belonging to the biggest fragment for each partition. The (1.1) shows the explicit formula used in the following: m_i and A_i are the multiplicity and the mass of the considered (i) cluster, while A_{tot} and A_{max} are, respectively, the mass of the detected partition and the mass of the biggest fragment. Evolution of multiplicities can be found in [7].

1.3.1 Z_{max} Sorting

In Fig. 1.2, the mean evolution of the mass fraction with the Z_{max} are shown for the different clusters understudied. The free protons (^1H) behave differently compared to the other clusters: starting from the evaporation regime, we observe a continuous decrease in their contribution to the final partitions. For ^{2,3}H and ^{3}He clusters, we observe a similar sequence already seen for the $M_{Z\geq 1}$ mean evolution in the Fig. 1.1 but with this representation: an increase in the evaporation and vaporization regime which surrounded a plateau or even a decrease in the multifragmentation regime. For ^{4}He clusters, it is even more clear because the overall shapes are almost not affected by the change in beam energies. On the contrary, for the lightest clusters, there is a clear hierarchy with an increase of their contributions with the beam energy and on the whole range of fragmentation degree. It has to be noted that ^{4}He contribution to the final partitions is dominant whatever the regime. If we look at, now, simultaneously to the evolution of ^{4}He and $A > 4$ contributions, we see that the increase of the ^{4}He contribution surrounded the bell shape of $A > 4$ which sign the multifragmentation window. During the passage between evaporation and multifragmentation, cluster production is frozen until the exit to vaporization. One possible explanation of this experimental fact is that the cost to build fragments and their associated surfaces make the production of additional clusters impossible in terms of energy available in the system. This is especially true for the lowest beam energies (32 and 40 MeV/nucleon) where multifragmentation may occur at the threshold. Indeed, we see that, when looking at the highest beam energies, the production of ^{1,2,3}H and ^{3}He is made possible again due to additional energy deposited in the system. The main effect of increasing the average deposited energy in the system is then an increase of the final light species at the expense of the heavier ones. In this case, fragment partitions are produced excited and go through secondary de-excitation and produce these light clusters. Concerning the vaporization regime, we observe at the beam energy

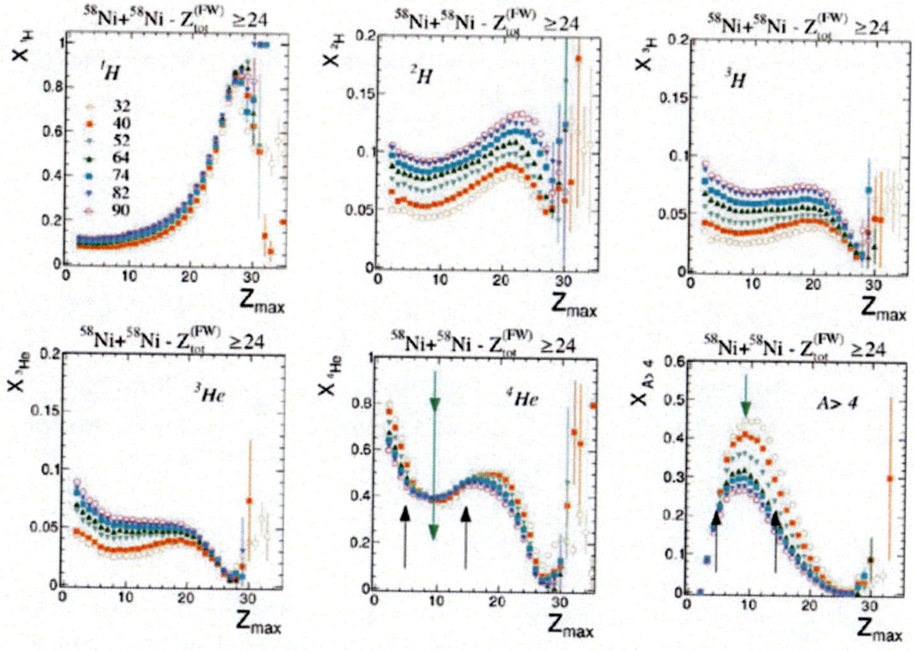

Fig. 1.2 From left to right and top to bottom, mean evolution of the mass fraction with the Z_{max} observable for the following species: $^1H, ^2H, ^3H, ^3He$, and ^4He isotopes and charged products with mass greater than 4 ($A > 4$).The different markers and colors stand for the seven beam energies used in the analysis. The black arrows indicate the entrance and exit of the multifragmentation window, while the green one indicates the maximum production of $A > 4$ fragments

of 32 MeV/nucleon, almost pure ^{4}He events with a continuous decrease in their contribution when beam energy increases.

1.3.2 Z_{max} and E^* Sorting

The previous results were integrated on the whole possible range of dissipation reached during the reaction. We introduce now the excitation energy to bring an additional constraint on the studied partitions. To estimated this excitation energy (E^*), we perform a calorimetry procedure on an event-by-event basis. This procedure is based on the mass and kinetic energies balance and is summarized by the (1.2). The main hypothesis concerns the neutron contribution which is estimated through a fermi gas assumption and the N/Z conservation. Detailed information on the calorimetry can be found in [8]. In Fig. 1.3, we draw the evolution of the mass fraction with, this time, an additional constraint on the excitation energy. Only partitions with $E^* = 7.5 \pm 0.5$ MeV/A are kept in the analysis. We only show the highest beam energies (from 64 to 90 MeV/A) where the statistics are sufficient after this E^*

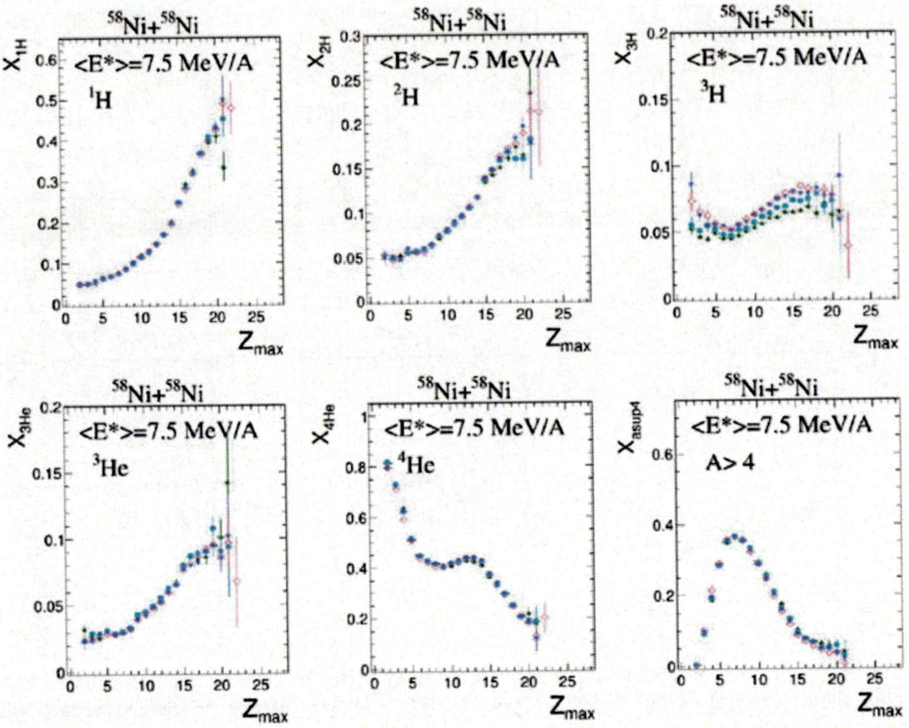

Fig. 1.3 Same as Fig. 1.2, but only the four highest beam energies are shown (from 64 to 90 MeV/A). Only partitions with excitation energy $E^* = 7.5$ MeV/A are kept. See text for details

selection. All beam energies fall on the same curves for all species except for the ^3H clusters which have a small dispersion. This shows that, using the experimental observables Z_{max} and E^*, we can fix the properties of the partitions whatever the beam energy. It is a good indication of the capability to select given regions of the phase diagram. Now, comparing to the previous trends of the Fig. 1.1, we see that ^{2}H and ^{3}He evolutions have changed. Their mass fractions continuously decrease with decreasing Z_{max} values. They join the behavior of free protons. At the opposite, ^{4}He and $A > 4$ evolutions are not affected by the excitation energy selection: the ^{4}He contribution keeps its previous sequence from evaporation to vaporization and still framed the contribution of the heavier clusters. This peculiar remanent shape is very interesting and shows that, for condition close to a microcanonical sampling, we observe the coexistence of evaporation, multifragmentation, and vaporization processes.

1.4 Conclusion

We have looked at the contribution of species which composed final partitions produced in the ^{58}Ni+^{58}Ni reactions. Using Z_{max}, the charge of the biggest fragment of each event, as sorting observable, we have sample classes of events from evaporation to vaporization. We distinguish different behaviors depending on the species understudied (^{1,2,3}H and ^{3,4}He). Looking at the mass fraction, we observe that the $A > 4$ contributions are framed by the lighter clusters contribution. The ^{4}He contribution is predominant whatever the underlying mechanism leading to final observed partitions. The use of several beam energies allow to evidence the effect of cluster production when multifragmentation occurs: at low beam energies, the energetic cost to produce fragments freeze the production of light clusters. This contribution starts to increase again for highest beam energies where the partitions may be produced sufficiently excited to go through secondary decays. Using excitation energy E^* as a second sorting observable, we show that cluster contributions are fully determined whatever the beam energies. We observe also that the ^{4}He clusters are a good candidate to track the exploration of the phase diagram during the reaction process. These results are coherent with a statistical description of fragment production which at first order is conditioned by the excitation energy and the volume of the system as it is foreseen in the freeze-Out picture of the (micro-)canonical statistical models.

References

1. B. Borderie, J.D. Franklanf, Liquid-Gas phase transition in nuclei. Prog. Part. Nucl. Phys. **105**, 82–138 (2019) (and references therein)
2. B. Borderie et al., Phase transition dynamics for hot nuclei. Phys. Lett. B **782**, 291–296 (2018)
3. E. Bonnet et al., Bimodal behavior of the heaviest fragment distribution in projectile fragmentation. Phys. Rev. Lett. **103**, 072701 (2009)
4. J. Pouthas et al., INDRA, A 4 π charged product detection array at GANIL, Nucl. Instrum. Methods Phys. Res. A **357**, 418–442 (1995); J. Pouthas et al., The electronics of the INDRA 4π detection array, Nucl. Instrum. Methods Phys. Res. A **369**, 222–247 (1996)
5. E. Galichet et al., Isospin diffusion in Ni^{58} -induced reactions at intermediate energies. I. Experimental results. Phys. Rev. C **79**, 064614 (2009)
6. P. Lautesse et al., Evolution of the fusion cross- section for light systems at intermediate energies. Eur. Phys. J. A **27**, 349–357 (2006)
7. E. Bonnet, Evolution of cluster production with fragmentation degree. Nuovo Cimento della Societa Italiana di Fisica C **41**, 183 (2018)
8. E. Bonnet et al., Fragment properties of fragmenting heavy nuclei produced in central and semi-peripheral collisions. Nucl. Phys. A **816**, 1–18 (2009)

Chapter 2
New Signatures of Phase Transition from Models of Nuclear Multifragmentation

G. Chaudhuri, S. Mallik, P. Das, and S. Das Gupta

Abstract The study of liquid–gas phase transition in heavy-ion collisions has generated a lot of interest among the nuclear physicists in recent years. In heavy-ion collisions, there is no direct way of measuring the state variables like entropy, pressure, energy, and hence, unambiguous characterization of phase transition becomes difficult. This work proposes new signatures of phase transition that can be extracted from the observables which are easily accessible in experiments. It is observed that the temperature dependence of the first-order derivative of the order parameters in nuclear liquid–gas phase transition exhibit similar behavior as that of the variation of specific heat at constant volume C_v which is an established signature of first-order phase transition. This motivates us to propose these derivatives as confirmatory signals of liquid–gas phase transition. The measurement of these signals is easily feasible in most experiments as compared to the other signatures like specific heat, caloric curve, or bimodality. Total multiplicity, size of the largest cluster are some of the order parameters which have been studied. Statistical models based on canonical ensemble and lattice gas model has been used for the study. This temperature where the peak appears is designated to be the transition temperature and the effect of certain parameters on this has also been examined. The multiplicity derivative signature proposed in this work has been further confirmed by other theoretical models as well as in the experimental study.

G. Chaudhuri (✉) · S. Mallik · P. Das
Variable Energy Cyclotron Centre, 1/AF Bidhan Nagar, Kolkata 700064, India
e-mail: gargi@vecc.gov.in

S. Das Gupta
Physics Department, McGill University, Montréal H3A 2T8, Canada

R. K. Puri et al. (eds.), *Advances in Nuclear Physics*, Springer Proceedings in Physics 257, https://doi.org/10.1007/978-981-15-9062-7_2

2.1 Introduction

The phenomenon of liquid–gas phase transition occurring in heavy ion collisions at intermediate energies is a subject of contemporary interest [1–8]. The nature of nucleon–nucleon strong interaction potential, which is an attractive one with a repulsive core is very similar to the van der Waals potential [4] except for the magnitude. This type of interaction explains the phenomenon of phase transition in ordinary liquid and hence similar observation is very much expected in nuclear systems also. In ordinary liquids, when it is heated, the temperature rises till the boiling point is reached after which it remains constant until the whole amount of liquid is converted to gas. Similarly in order to observe phase transition in the nuclear system, one has to pump energy to the system and the only possible way is by means of nucleus-nucleus collision. In very high energy heavy-ion collision at high density and temperature, hadronic matter transforms to the Quark–Gluon Plasma (QGP) phase. At the intermediate energy regime, nuclear multifragmentation is the dominant mechanism which can be related to a liquid gas kind of transition at sub-saturation nuclear density. Theoretical models of multifragmentation predict the existence of phase transition in infinite nuclear matter. Experimental signatures also indicate the change of state and this can be interpreted as finite-size counterpart of the first-order phase transition in nuclear matter. Different signatures of this transition have been studied extensively both theoretically as well as experimentally [4, 5, 7–9]. The variation of excitation energy and specific heat with temperature are two well-studied signatures theoretically in order to detect the first-order phase transition [10–12].

Phase transition is usually characterized by the specific behavior of state variables like pressure, density, energy, entropy, etc. [13, 14]. The order of phase transition, according to Ehrenfest is determined by the lowest order derivative of free energy that shows a discontinuity. In heavy-ion collisions, there is no direct way of accessing these state variables and hence unambiguous detection of phase transition becomes difficult. The present work is motivated by this limitation and aims at looking for signatures of phase transition that can be extracted from the observables which are easily accessible in experiments. Ideally, phase transition exists in the thermodynamic limit and for a first-order one, entropy should have a finite discontinuity and specific heat a divergence at the phase transition temperature. In finite nuclei, the discontinuity or divergence is replaced by sudden jump or maxima. The variation of total multiplicity or size of the largest cluster (Z_{max}) with temperature is very much similar to that of entropy or excitation energy (caloric curve) with temperature and this can be seen from Fig. 2.1. Hence, the first-order derivative of these observables with temperature is expected to behave in a similar way as those of entropy or energy. This observation led to the investigation of the nature of the derivatives of these multifragmentation observables which can be easily measured in experiments. Encouraging results have been obtained from this study and it has been observed that first-order derivative of the order parameters related to the total multiplicity, largest cluster size (produced in heavy-ion collisions) exhibit similar behavior as that of the variation of specific heat at constant volume C_v which is an established signature of first-order phase

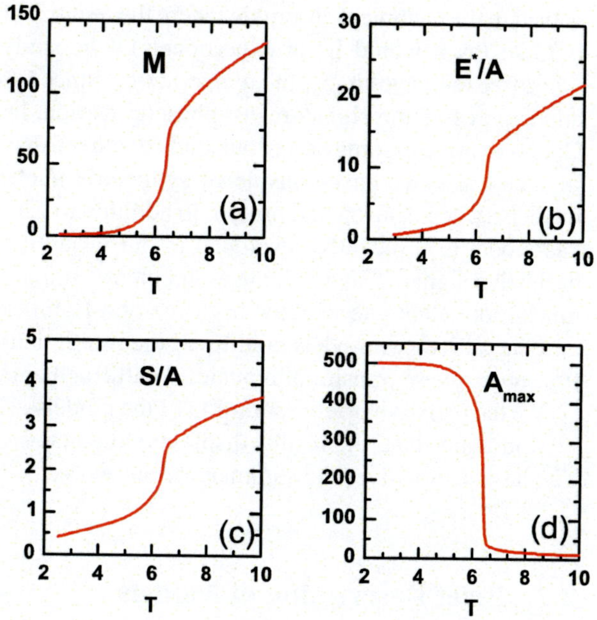

Fig. 2.1 Variation of (**a**) Total Multiplicity M (**b**) excitation energy E^*/A (MeV/nucleon), (**c**) entropy per nucleon S/A (**d**) average size of the largest cluster A_{max} with temperature T (MeV) for the fragments produced in the fragmentation of an ideal one-component system of size A = 500

transition. This motivates us to propose these derivatives of total multiplicity, largest cluster size [15–18] as confirmatory signals of liquid–gas phase transition. Another observable we have proposed here is related to the difference (normalized) between the sizes of the first and second largest clusters which also serve as an order parameter for phase transition in nuclear fragmentation and has been studied experimentally too [19, 20]. The derivatives of all these peaks at the same temperature as specific heat and hence can confirm the phase transition in the fragmentation process. The measurement of these signals are easily feasible in most experiments as compared to the other signatures like specific heat, caloric curve, or bimodality. This temperature where the peak appears is designated to be the transition temperature and the effect of certain parameters on this has also been examined.

We have mainly used a statistical model based on the canonical ensemble which is better known by the Canonical Thermodynamical Model (CTM) [21] in order to study the fragmentation of nuclei. In such models of nuclear disassembly, it is assumed that because of multiple nucleon–nucleon collisions a statistical equilibrium is reached and disintegration pattern is solely decided by the statistical weights in the available phase space. The temperature rises and the system expands from normal density and composites are formed on the way to disassembly as a result of density fluctuation. As the system reaches between three and six times the normal volume, the interactions between composites become unimportant (except for the long-range Coulomb interaction) and one can do a statistical equilibrium calculation to obtain the yields of composites at a volume called the freeze-out volume. This model can be implemented in different statistical ensembles (microcanonical, canonical and grand canonical [1, 3, 21]. In our calculation, the partitioning into available channels is solved in the canonical ensemble where the number of particles in the nuclear system is finite (as it would be in experiments). The study is done for different nuclear sizes, freeze-out volumes, and temperatures. Since Coulomb interaction is long range and suppresses the signatures of phase transition, hence, we have switched off the Coulomb force in some part of our study in order to have a better idea of the signatures. In such cases, we have considered symmetric nuclear matter and no distinction is made between neutron and proton. In addition to CTM, we have also used the lattice gas model [22] recently developed in our group in order to study the multiplicity derivative signal. This model uses geometry similar to the percolation model but is much more elaborate with the insertion of a Hamiltonian. Both the thermodynamic and the lattice gas models confirmed the multiplicity derivative as the signature of first-order phase transition in nuclear multifragmentation.

We have given a brief description of the models used in our calculation in the next section. After that, the results displaying the new signatures are proposed in detail. The last section gives the summary of our work.

2.2 Brief Description of Models

2.2.1 *The Canonical Thermodynamical Model*

In this section, we describe briefly the canonical thermodynamical model which is briefly designated as CTM. We assume that a system with A_0 nucleons and Z_0 protons at temperature T has expanded to a higher than normal volume and the partitioning into different composites can be calculated according to the rules of equilibrium statistical mechanics. In a canonical model, the partitioning is done such that all partitions have the correct A_0, Z_0 (equivalently N_0, Z_0). Details of the implementation of the canonical model can be found elsewhere [21]; here, we give the essentials necessary to follow the present work.

The canonical partition function is given by

$$Q_{N_0,Z_0} = \sum \prod \frac{\omega_{I,J}^{n_{I,J}}}{n_{I,J}!}. \tag{2.1}$$

Here, the sum is over all possible channels of breakup (the number of such channels is enormous) which satisfy $N_0 = \sum I \times n_{I,J}$ and $Z_0 = \sum J \times n_{I,J}$; $\omega_{I,J}$ is the partition function of one composite with neutron number I and proton number J, respectively, and $n_{I,J}$ is the number of this composite in the given channel. The one-body partition function $\omega_{I,J}$ is a product of two parts: one arising from the translational motion of the composite and another from the intrinsic partition function of the composite:

$$\omega_{I,J} = \frac{V_f}{h^3}(2\pi mT)^{3/2}A^{3/2} \times z_{I,J}(int). \tag{2.2}$$

Here, $A = I + J$ is the mass number of the composite and V_f is the volume available for translational motion; V_f will be less than V, the volume to which the system has expanded at breakup. We use $V_f = V - V_0$, where V_0 is the normal volume of nucleus with Z_0 protons and N_0 neutrons. In this calculation, we have used a fairly typical value of $V = 6V_0$.

The probability of a given channel $P(\mathbf{n}_{I,J}) \equiv P(n_{0,1}, n_{1,0}, n_{1,1}......n_{I,J}.......)$ is given by

$$P(\mathbf{n}_{I,J}) = \frac{1}{Q_{N_0,Z_0}} \prod \frac{\omega_{I,J}^{n_{I,J}}}{n_{I,J}!}. \tag{2.3}$$

The average number of composites with I neutrons and J protons is seen easily from the above equation to be

$$\langle n_{I,J} \rangle = \omega_{I,J} \frac{Q_{N_0-I,Z_0-J}}{Q_{N_0,Z_0}}. \tag{2.4}$$

The constraints $N_0 = \sum I \times n_{I,J}$ and $Z_0 = \sum J \times n_{I,J}$ can be used to obtain different looking but equivalent recursion relations for partition functions

$$Q_{N_0,Z_0} = \frac{1}{N_0} \sum_{I,J} I\omega_{I,J} Q_{N_0-I,Z_0-J}. \tag{2.5}$$

These recursion relations allow one to calculate Q_{N_0,Z_0}

We list now the properties of the composites used in this work. The proton and the neutron are fundamental building blocks, thus $z_{1,0}(int) = z_{0,1}(int) = 2$, where 2 takes care of the spin degeneracy. For deuteron, triton, ^{3}He, and ^{4}He, we use $z_{I,J}(int) = (2s_{I,J} + 1)\exp(-\beta E_{I,J}(gr))$, where $\beta = 1/T$, $E_{I,J}(gr)$ is the ground state energy of the composite and $(2s_{I,J} + 1)$ is the experimental spin degeneracy of the ground state. Excited states for these very low mass nuclei are not included. For mass number $A = 5$ and greater, we use the liquid drop formula. For nuclei in isolation, this reads ($A = I + J$)

$$z_{I,J}(int) = \exp \frac{1}{T}[W_0 A - \sigma(T)A^{2/3} - \kappa \frac{J^2}{A^{1/3}} - C_s \frac{(I-J)^2}{A} + \frac{T^2 A}{\epsilon_0}]. \tag{2.6}$$

The derivation of this equation is given in several places [3, 21], so we will not repeat the arguments here. The expression includes the volume energy, the temperature-dependent surface energy, the Coulomb energy, and the symmetry energy. The term $\frac{T^2 A}{\epsilon_0}$ represents contribution from excited states since the composites are at a non-zero temperature.

We also have to state which nuclei are included in computing Q_{N_0,Z_0} (17). For I, J, (the neutron and the proton number) we include a ridge along the line of stability. The liquid drop formula above also gives neutron and proton drip lines and the results shown here include all nuclei within the boundaries.

The long-range Coulomb interaction between different composites can be included in an approximation called the Wigner–Seitz approximation. We incorporate this following the scheme set up in [3].

2.2.2 *The Evaporation Code*

The statistical multifragmentation model described above calculates the properties of the collision averaged system that can be approximated by an equilibrium ensemble. Ideally, one would like to measure the properties of excited primary fragments after emission in order to extract information about the collisions and compare directly with the equilibrium predictions of the model. However, the time scale of a nuclear reaction(10^{-20} s) is much shorter than the time scale for particle detection (10^{-9} s). Before reaching the detectors, most fragments decay to stable isotopes in their ground states. Thus before any model simulations can be compared to experimental data, it is indispensable to have a model that simulates sequential decays. A Monte Carlo technique is employed to follow all decay chains until the resulting products are unable to undergo further decay. For the purposes of the sequential decay calculations, the excited primary fragments generated by the statistical model calculations are taken as the compound nucleus input to the evaporation code. Hence, every primary fragment is decayed as a separate event.

We consider the deexcitation of a primary fragment of mass A, charge Z, and temperature T. The successive particle emission from the hot primary fragments is assumed to be the basic deexcitation mechanism. For each event of the primary breakup simulation, the entire chain of evaporation and secondary breakup events is Monte Carlo simulated. The standard Weisskopf evaporation scheme is used to take into account evaporation of nucleons, d, t, ^3He, and α. The decays of particle stable excited states via gamma rays were also taken into account for the sequential decay process and for the calculation of the final ground-state yields. We have also considered fission as a deexcitation channel though for the nuclei of mass < 100 its role will be quite insignificant. The process of light particle emission from a compound nucleus is governed by the emission width Γ_ν at which a particle of type ν is emitted. The different equations for calculation of particle, gamma, and fission widths are given in details in [23] and we will skip them here. Once the emission widths are known, it is required to establish the emission algorithm which decides whether a particle is being emitted from the compound nucleus. This is done [24] by first calculating the ratio $x = \tau/\tau_{tot}$, where $\tau_{tot} = \hbar/\Gamma_{tot}$, $\Gamma_{tot} = \sum_\nu \Gamma_\nu$ and $\nu = n, p, d, t, ^3He, \alpha, \gamma$ or fission and then performing Monte Carlo sampling from a uniformly distributed set of random numbers. In the case that a particle is emitted, the type of the emitted particle is next decided by a Monte Carlo selection with the weights Γ_ν/Γ_{tot} (partial widths). The energy of the emitted particle is then obtained by another Monte Carlo sampling of its energy spectrum. The energy, mass, and charge of the nucleus is adjusted after each emission. This procedure is followed for each of the primary fragment produced at a fixed temperature and then repeated over a large ensemble and the observables are calculated from the ensemble averages. The number and type of particles emitted and the final decay product in each event are registered and are taken into account properly keeping in mind the overall charge and baryon number conservation.

2.2.3 Lattice Gas Model

The lattice gas model is considerably more complicated than the percolation model [22] but expositions of the model exist [8, 25, 26] and we refer to [26] for details. Let $A = N + Z$ be the number of nucleons in the system that dissociates. We consider D^3 cubic boxes where each cubic box has volume $(1.0/0.16) fm^3$. D^3 is larger than A (they have the same value in the bond percolation model). Here, $D^3/A = V_f/V_0$ where V_0 is the normal volume of a nucleus with A nucleons and V_f is the freeze-out volume where partitioning of nucleons into clusters is computed. For nuclear forces, one adopts nearest neighbor interactions. Following normal practice, we use neutron–proton interactions $v_{np} = -\ 5.33$ MeV and set $v_{nn} = v_{pp} = 0.0$. Coulomb interaction between protons is included. Each cube can contain 1 or 0 nucleon. There is a very large number of configurations that are possible (a configuration designates which cubes are occupied by neutrons, which by protons and which are empty; we sometimes call a configuration an event). Each configuration has an energy. If a temperature is specified, the occupation probability of each configuration is proportional to its energy: P∝exp(-E/T). This is achieved by Monte Carlo sampling using the Metropolis algorithm.

Calculation of clusters needs further work. Once an event is chosen we ascribe to each nucleon a momentum. Momentum of each nucleon is picked by Monte Carlo sampling of a Maxwell–Boltzmann distribution for the prescribed temperature T. Two neighboring nucleons are part of the same cluster if $\mathbf{P}_r^2/2\mu + \epsilon < 0$ where ϵ is v_{np} or v_{nn} or v_{pp}. Here, $\mathbf{P}_r$ is the relative momentum of the two nucleons and μ is the reduced mass. If nucleon i is bound with nucleon j and j with k then i, j, k are part of the same cluster. At each temperature, we calculate 50,000 events to obtain average energy $< E >$ and average multiplicity n_a (where a is the mass number of the cluster) of all clusters. A cluster with one nucleon is a monomer, one with two nucleons is a dimer, and so on. The total multiplicity is $M = \sum n_a$ and $\sum a n_a = A$, where $A = N + Z$ is the mass number of the dissociating system.

2.3 Results

The variation of total multiplicity with temperature for fragmenting systems as calculated by CTM is very similar to that of entropy. This motivated us to look for the derivative of multiplicity which is expected to behave similarly as the derivative of entropy w.r.t temperature which is nothing but the specific heat at constant volume C_V. Unlike the entropy, one can measure the total multiplicity $M = \sum M_a$ (a being the mass number of the composites) with 4π detectors in the laboratory. In CTM the derivative of M with T as a function of T is seen to have a maximum. Figure 2.2a, c shows the total multiplicity for fragmenting system having proton number $(Z) = 82$ and neutron number $(N) = 126$ and and its derivative dM/dT ((b) and (d)). Results for both real nuclei and the one for one kind of particles have been displayed in order to emphasize the effects of Coulomb interaction. The rise and the peak are much sharper in absence of Coulomb interaction clearly indicating the role of the long-range interaction in suppressing the signatures of phase transition. The features become less sharp as in $Z = 28$ and $N = 30$, as the system size decreases (Fig. 2.3).

In the next two Figs. 2.4 and 2.5, we compare dM/dT and C_V for the two same systems as used in Figs. 2.2 and 2.3. We also consider the situation where the Coulomb is switched off. The peak in dM/dT coincides with the maximum of specific heat at constant volume C_v as a function of temperature for all the cases. It's an established fact that specific heat at constant volume peaks at the transition temperature and this is a signature of first-order phase transition. Hence, based on our results as presented in Figs. 2.4 and 2.5, we conclude that dM/dT can be a signature of phase transition and the advantage is it gives an exact value of the transition temperature where the maximum of dM/dT occurs. Next, we calculate the entropy since it's well known

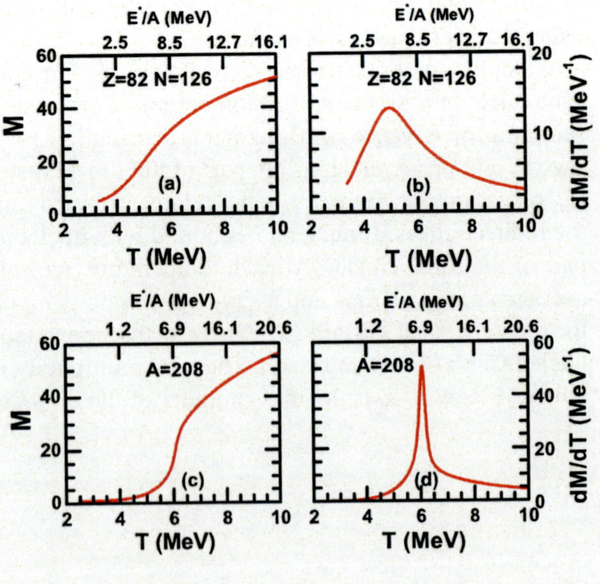

Fig. 2.2 Variation of multiplicity M (**a** and **c**) and dM/dT (**b** and **d**) with temperature (bottom x axes) and excitation per nucleon (top x axes) from the CTM calculation for fragmenting systems having Z = 82 and N = 126 (**a** and **b**). (**c** and **d**) represent the same but for a hypothetical system of one kind of particle with no Coulomb interaction but the same mass number (A = 208). $E^* = E - E_0$, where E_0 is the ground-state energy of the dissociating system in the liquid drop model whose parameters are given in [21]

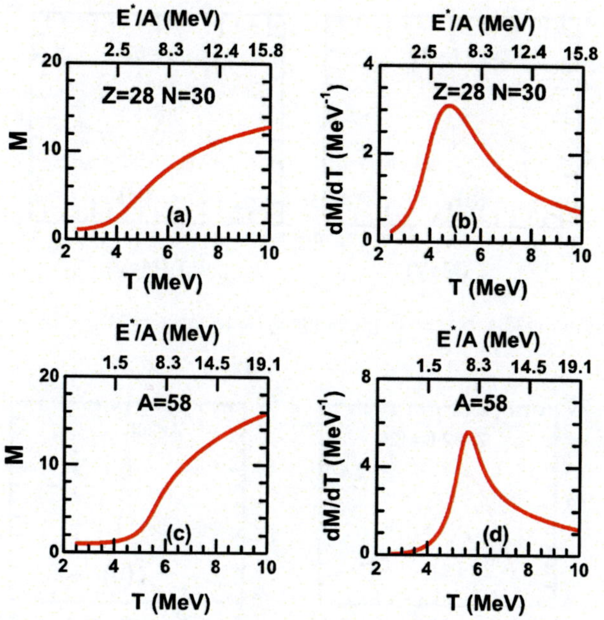

Fig. 2.3 Same as Fig. 2.2 but the fragmenting systems are Z = 28 and N = 30 (**a** and **b**) and A = 58 (**c** and **d**)

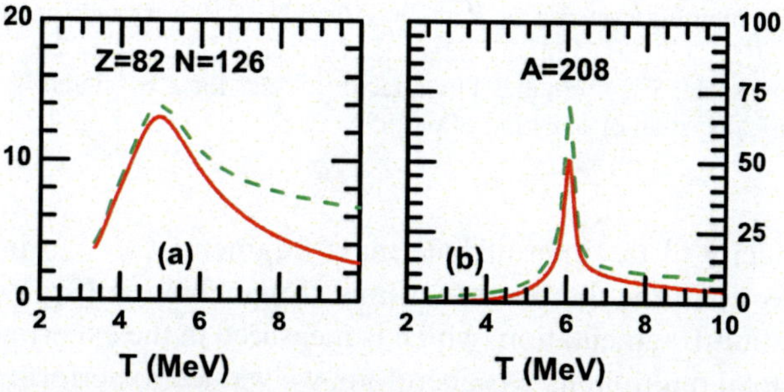

Fig. 2.4 Variation of dM/dT (red solid lines) and C_v (green-dashed lines) with temperature from CTM for fragmenting systems having Z = 82 and N = 126 (**a**) and for hypothetical systems of one kind of particle with no Coulomb interaction of mass number A = 208 (**b**). To draw dM/dT and C_v in the same scale, Cv is normalized by a factor of 1/50

that it shows a sharp rise near the transition temperature. We have compared the temperature variation of dM/dT and the entropy for the fragmenting system Z = 82, N = 126, and also for an ideal condition, neglecting the Coulomb interaction. This is displayed in the next plot Fig. 2.6. It is evident that, for both the cases, the entropy changes rapidly in that regime of temperature scale where dM/dT exhibits a maximum. The presence of Coulomb interaction in a real system, only, smears the rise of entropy and the peak in dM/dT. This behavior, further, establishes that dM/dT is indeed a signature of phase transition.

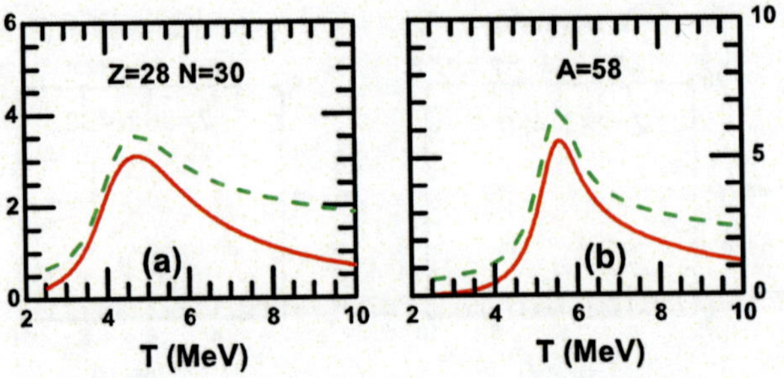

Fig. 2.5 Same as in Fig. 2.4, but the fragmenting systems are Z = 28 and N = 30 (**a**) and A = 58 (**b**)

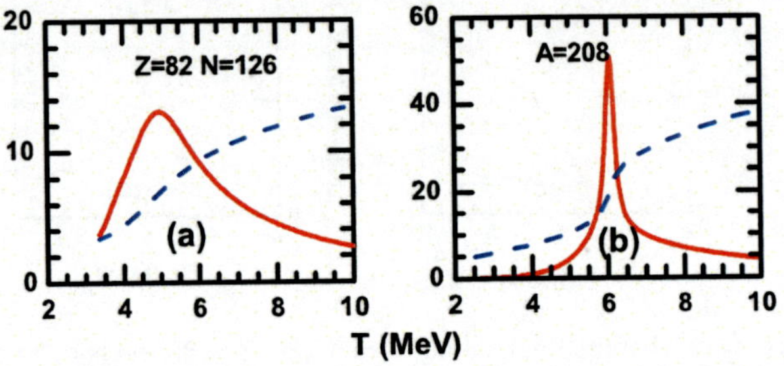

Fig. 2.6 Variation of entropy (blue-dashed lines) and dM/dT (red solid lines) with temperature from CTM for fragmenting systems having Z = 82and N = 126 (**a**) and for hypothetical system of one kind of particle with no Coulomb interaction of mass number A = 208 (**b**). To draw S and dM/dT in the same scale, S is normalized by a factor of 1/20 for Z = 82 and N = 126 system and 1/50 for hypothetical system of one kind of particle

The multiplicity of the intermediate mass fragments (M_{IMF}) in heavy-ion collisions strongly confirms the process of multifragmentation [3]. It is an important observable of multifragmentation, which is measured in the experiment, sometimes, instead of the total multiplicity M. Therefore, we wanted to perform a similar test on the derivative of M_{IMF}. We have plotted the variation of M_{IMF} and its temperature derivative with temperature for the system Z = 82, N = 126 in Fig. 2.7, and compared dM_{IMF}/dT with C_V. M_{IMF} and dM_{IMF}/dT display a similar behavior as that of the total multiplicity and its derivative except for the fact that the peak position of its derivative does not coincide with that of C_V. This is expected because the calculation of C_V involves all the fragments irrespective of their mass or charge, but in M_{IMF}, only selected fragments are included.

Last but not least, we would like to study the effect of secondary decay on the excited fragments formed after multifragmentation. In a heavy-ion collision, when a nucleus breaks up through the process of nuclear multifragmentation, the resulting composites are called primary fragments. The primary fragments are excited in general and lose excitation through sequential two-body decay, and thus change the total multiplicity. The final cold fragments, called secondary fragments, are detected

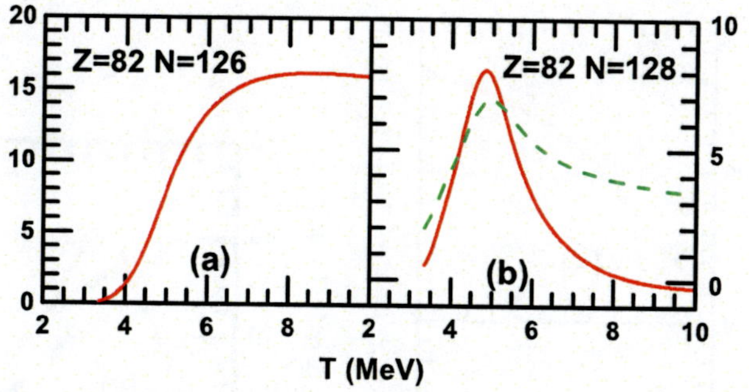

Fig. 2.7 Variation of intermediate mass fragment(IMF) multiplicity M_{IMF} (**a**) and first-order derivative of IMF multiplicity dM_{IMF}/dT (b) with temperature from CTM calculation for fragmenting systems having Z = 82 and N = 126. Variation of C_v with temperature (T) is shown by green-dashed line in (**b**). To draw dM_{IMF}/dT and C_v in the same scale, C_v is normalized by a factor of 1/100

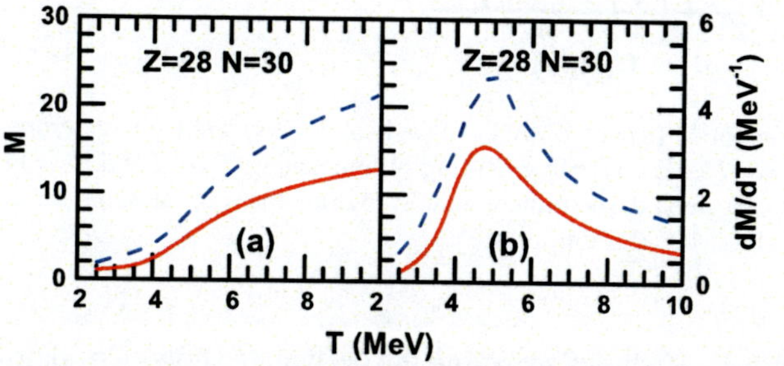

Fig. 2.8 Effect of secondary decay on M (**a**) and dM/dT (**b**) for fragmenting systems having Z = 28 and N = 30. Red solid lines show the results after the multifragmentation stage (calculated from CTM), whereas blue-dashed lines represent the results after secondary decay of the excited fragments

in the laboratory. The fragments that we are dealing with in our study (using CTM), are primary fragments. The secondary decay may affect the total multiplicity in such a way that might change the behavior of multiplicity discussed above. As we are interested in the experimental signature, we investigate the effect of the secondary decay in our calculation and do the same study with the multiplicity of the secondary fragments.

We have plotted the multiplicities of the primary and the secondary fragments and their derivatives in Fig. 2.8. It is apparent that the effect of secondary decay does not alter our previous observation. Moreover, it enhances the signals, the total multiplicity changes more rapidly, and the peak in dM/dT is sharper in case of the secondary fragments. Thus, the maxima of multiplicity derivative can be extracted successfully through experiments with an unaltered transition temperature.

In order to further test multiplicity derivative as a possible signature for first-order phase transition, we have carried out the investigation using the lattice gas Model[22]. This is shown in Fig. 2.9a, b. We have plotted M and its derivative against

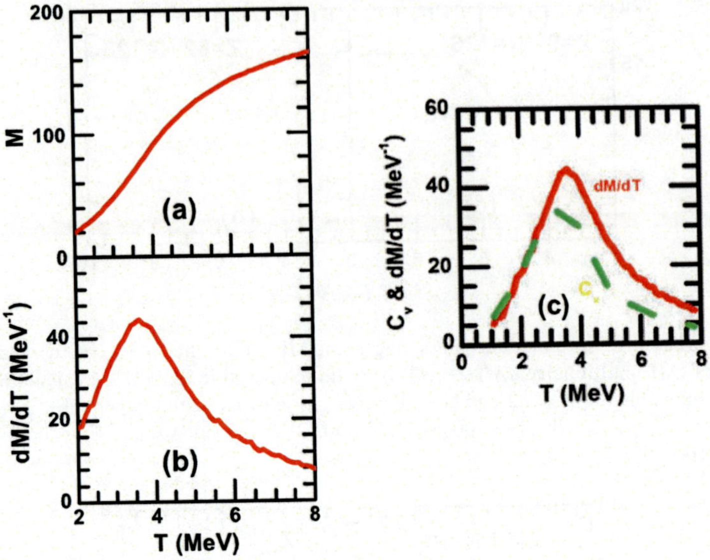

Fig. 2.9 Variation of M (**a**) and dM/dT (**b**) (red solid lines) and C_v with temperature from lattice gas model at $D = 8$ (see text) for fragmenting system having $Z = 82$ and $N = 126$. (**c**) dM/dT (red solid lines) and C_v (green dashed lines) with T; to draw them in the same scale, C_v is normalized by a factor of $1/10$; dM/dT is unit of MeV^{-1}

the temperature T. M shows a rise and the derivative shows a peak as expected. Plots of dM/dT and $d < E > /dT$ are shown in Fig. 2.9c. C_v goes through a maximum at some temperature which is a hallmark of first-order phase transition and this occurs at the same temperature, where dM/dT maximizes. This is remarkably similar to results from CTM corroborating the evidence that the appearance of a maximum in dM/dT is indicative of a first-order phase transition.

Our proposed signal of multiplicity derivative dM/dT was tested and verified in different statistical and dynamical models like the statistical multifragmentation model (SMM) [28, 29], Quantum Molecular Dynamics (QMD) model [30], and Nuclear statistical Equilibrium (NSE) model [31]. Our theoretical proposition of this signal got further support when it was experimentally verified recently and tested using three different reactions $^{40}Ar +^{58} Ni$, $^{40}Ar +^{27} Al$, and $^{40}Ar +^{48} Ti$ at 47 Mev/n [32].

The average size of the largest cluster $\langle A_{max}\rangle$ formed in the fragmentation of the excited nuclei acts as an order parameter for first-order phase transition. The variable a_2 which is a measure of the difference between the average size of the first ($\langle A_{max}\rangle$) and the second ($\langle A_{max-1}\rangle$) largest cluster sizes divided by the sum of these two ($a_2 = \frac{\langle A_{max}\rangle - \langle A_{max-1}\rangle}{\langle A_{max}\rangle + \langle A_{max-1}\rangle}$) also has similar behavior as that of $\langle A_{max}\rangle$. So this observable which is measured in some experiments can also act as an order parameter. The analytical expressions leading to the calculation of the average size of first and second largest clusters can be found in [33].

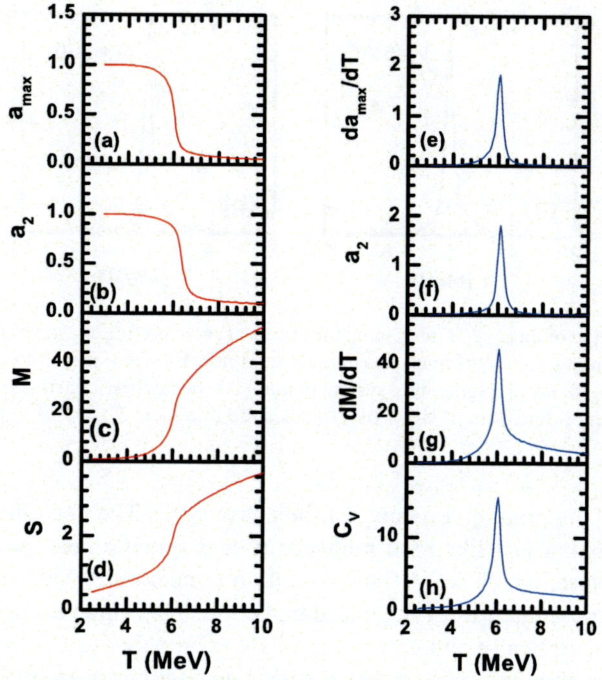

Fig. 2.10 Variation of (a) $\mathbf{a}_{max}$, (b) $\mathbf{a}_2$, (c) M, (d) S, (e) -$d\mathbf{a}_{max}/dT$, (f) -$d\mathbf{a}_2/dT$, (g) dM/dT, and (h) C_v with temperature for fragmenting system of mass A = 200

Now, we will concentrate on these observables in order to study their variation with temperature. We consider an ideal system of A = 200 identical nucleons with no Coulomb force acting between them in order have a better idea of these proposed signatures. Left panels of Fig. 2.10a–d display the variations of the four variables, the normalized size of the average largest cluster $\mathbf{a}_{max}$ ($\mathbf{a}_{max} = \frac{\langle A_{max} \rangle}{A}$), $\mathbf{a}_2$, total multiplicity M and entropy per particle (S/A) with temperature. $\mathbf{a}_{max}$ and $\mathbf{a}_2$ are almost constant and assume a value $\approx$ 1 up to approximately 5 MeV, in the temperature scale. This implies that in this temperature range, the size of the largest fragment produced is almost the same as the size of the fragmenting source. Around T = 6 MeV, both of them fall sharply to a very low value near 0, which indicate the entire system fragments into the light mass nuclei. After that, they remain almost unchanged. These observables, clearly, give a sharp transition near T = 6 MeV and therefore behave as an order parameter of the nuclear phase transition. Now, the last two panels ((c) and (d)) in the left of Fig. 2.10 show the variation of the total multiplicity and entropy per nucleon with temperature. $\mathbf{a}_{max}$ and $\mathbf{a}_2$ display similar behavior as that of the multiplicity and the entropy; the sudden jump (or fall) of these four variables occur almost at the same temperature around 6 MeV. This similarity motivates us to investigate the behavior of the derivatives of $\mathbf{a}_{max}$ and $\mathbf{a}_2$. In the right panel of Fig. 2.10, temperature derivatives of all the four quantities are plotted as a function of temperature. In the right bottom panel Fig. 2.10h, we have plotted C_V, which is

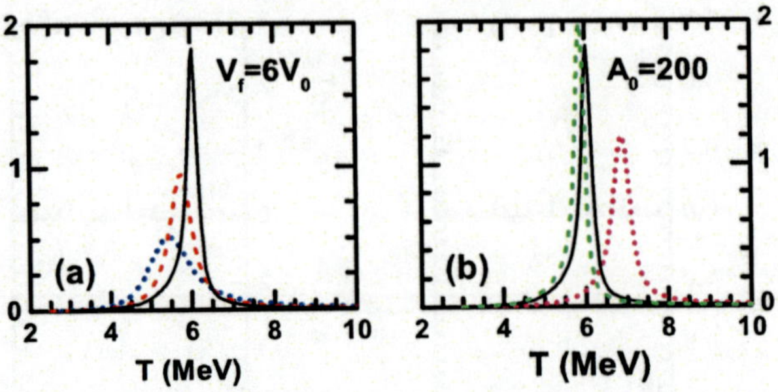

Fig. 2.11 Variation of $d\mathbf{a}_{max}/dT$ with temperature (**a**) at constant freeze-out volume $V_f = 6V_0$ but for three fragmenting system of mass 50 (blue-dotted line), 100 (red-dashed line), and 200 (black solid line) and (**b**) for same fragmenting system of mass 200 but at three constant freeze-out volumes $V_f = 2V_0$ (magenta-dotted line), $V_f = 6V_0$ (black solid line), and $V_f = 8V_0$ (green-dashed line)

related to the temperature derivative of the entropy (S). The derivatives of $\mathbf{a}_{max}$ and $\mathbf{a}_2$ exhibit maxima just like total multiplicity and specific heat, and almost at the same temperature, which we call the transition temperature. This establishes these two variables as signatures of the phase transition. This signature is much easier to access both theoretically and experimentally as compared to the bimodality in the probability distribution of the largest cluster. The later has been used so far in order to detect the existence of phase transition in nuclear multifragmentation but to detect two peaks(bimodality) of equal height in a distribution at a particular temperature (or excitation energy) is far more a difficult job than to simply calculate the derivative in its size with temperature or excitation energy. We strongly believe that this new proposed signature related to the largest cluster size will definitely provide a great impetus to the study of liquid–gas phase transition in heavy-ion collisions.

Next, we have examined how the transition temperature varies with the source size and the freeze-out volume. We have plotted the variation of d$\mathbf{a}_{max}$/dT with T for three different fragmenting systems of size A = 50, 100, 200 at a fixed freeze-out volume $V_f = 6V_0$ in Fig. 2.11a, and the same for three freeze-out volume V_f=3V_0, 4V_0, 8V_0 with fixed source A = 200 in Fig. 2.11b. We see that the peaks are sharper for the more massive source and the higher freeze-out volume. The position of the peak is observed to shift to the higher temperature region for the bigger source size, and the lower temperature side for the greater freeze-out volume. This implies that the smaller system fragments more easily at a lower transition temperature as compared to its bigger counterparts. The peak also becomes sharper for bigger sources which once again proves that phase transition signals are enhanced in larger systems. For freeze-out volume, the result that we have obtained is expected since higher freeze-out volume (lower density) will favor the disintegration of the nucleus, resulting in lower transition temperature.

At the end, we have plotted the transition temperatures as a function of system size at fixed freeze-out volume (left panel (a)), and as a function of freeze-out volume for a fixed system (right panel (b)) in Fig. 2.12. In each panel, four different sets of transition temperatures are plotted. Those sets are obtained from the position of the

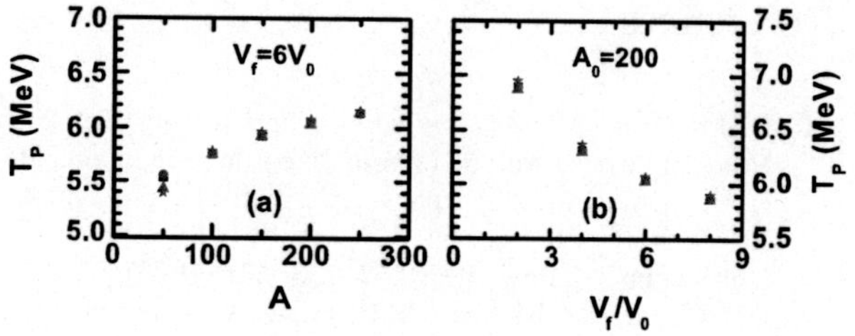

Fig. 2.12 Dependence of the peak position of -$d\mathbf{a}_{max}/dT$, -$d\mathbf{a}_2/dT$, dM/dT, and C_v on fragmenting system size (**a**) and freeze-out volume (**b**)

maxima in $d\mathbf{a}_{max}/dT$, $d\mathbf{a}_2/dT$, dM/dT, and C_V. The transition temperatures obtained from all the four observables give consistent results. Small differences between them can be attributed to the finiteness of the fragmenting system.

2.4 Summary

This work introduces some new signatures of nuclear liquid–gas phase transition which can be measured easily and more accurately in experiments. The observables chosen were the total multiplicity, average largest cluster size a_{max}, and a normalized variable a_2 which assume distinctly different values in liquid and the gas phases thus serving as order parameters of the transition. The variation of these observables with temperature is very much similar to that of entropy or excitation energy and hence the temperature derivatives behave as specific heat at constant volume (C_v) which is an established signature of phase transition. Transition temperature can be identified from the position of the maxima of these derivatives analogous to that of C_v. Multiplicity derivative (dM/dT) serves as a robust signal with very good performance even in the presence of Coulomb interaction which is long range and thereby suppresses signatures of phase transition. This signature not only persists but gets enhanced after secondary decay of the primary hot fragments and thus can be easily detected in experiments which measure total multiplicity. This signal was proposed using the canonical thermodynamical model and was later confirmed by us using the lattice gas model. This was very recently verified in other theoretical models as well as in experiment. The other two observables concerning the derivatives of the largest and second largest clusters also peak at the same temperature as that of C_v and can be considered as signals of first-order phase transition in one component system switching off the Coulomb interaction. It is sometimes easier to measure the size of the largest cluster than to count the total multiplicity covering all the fragments produced in the experiment. The effect of source size and freeze-out volume on the transition temperature is also studied using the one-component model. The extension of this study for real nuclei will be a part of our future work.

References

1. D.H.E. Gross, *Microcanonical Thermodynamics: Phase Transitions in Finite Systems*. Lecture Notes in Physics, vol. 66 (World Scientific, Singapore, 2001)
2. P.J. Siemens, Liquid-gas phase transition in nuclear matter. Nature **305**, 410–412 (1983)
3. J.P. Bondorf, A.S. Botvina, A.S. Iljinov, I.N. Mishustin, K. Sneppen, Statistical multifragmentation of nuclei. Phys. Rep. **257**, 133–221 (1995)
4. S.D. Gupta, A.Z. Mekjian, M.B. Tsang, *Advances in Nuclear Physics*, vol. 26, 89, eds. by J.W. Negele, E. Vogt (Plenum Publishers, New York, 2001).
5. B. Borderie, M.F. Rivet, Nuclear multifragmentation and phase transition for hot nuclei. Prog. Part. Nucl. Phys. **61**, 551–601 (2008)
6. B. Borderie, J.D. Frankland, Liquid-Gas phase transition in nuclei. Prog. Part. Nucl. Phys. (2019). https://doi.org/10.1016/j.ppnp.2018.12.002
7. D.H.E. Gross, Multifragmentation, link between fission and the liquid-gas phase-transition. Prog. Part. Nucl. Phys. **30**, 155–164 (1993)
8. P. Chomaz et al., Nuclear spinodal fragmentation. Phys. Rep. **389**, 263–440 (2004)
9. S. Mallik, S.D. Gupta, G. Chaudhuri, Event simulations in a transport model for intermediate energy heavy ion collisions: applications to multiplicity distributions. Phys. Rev. C **91**, 034616 (2015)
10. J.P. Bondorf, R. Donangelo, I.N. Mishustin, H. Schulz, Statistical multifragmentation of nuclei: (II). Application of the model to finite nuclei disassembly. Nucl. Phys. A **444**, 460–476 (1985)
11. J. Pochodzalla et al., Probing the nuclear liquid-gas phase transition. Phys. Rev. Lett. **75**, 1040–1044 (1995)
12. C.B. Das, S.D. Gupta, A.Z. Mekjian, Specific heat at constant volume in the thermodynamic model. Phys. Rev. C **68**, 031601 (R) (2003)
13. F. Reif, *Fundamentals of Statistical and Thermal Physics* (McGraw Hill, Newyork, 1965)
14. R.K. Pathria, *Statistical Mechanics* (Pergamon Press)
15. F. Gulminelli, P. Chomaz, Distribution of the largest fragment in the lattice gas model. Phys. Rev. C **71**, 054607 (2005)
16. B. Krishnamachari, J. McLean, B. Cooper, J. Sethna, Gibbs-Thomson formula for small island sizes: corrections for high vapor densities. Phys. Rev. B **54**, 8899–8907 (1996)
17. M. Pleimling, W. Selke, Droplets in the coexistence region of the two-dimensional Ising model. J. Phys. A **33**, L199 (2000)
18. G. Chaudhuri, S.D. Gupta, Properties of the largest fragment in multifragmentation: a canonical thermodynamic calculation. Phys. Rev C **75**, 034603 (2007)
19. A. Le Fevre et al., Bimodality: a sign of critical behavior in nuclear reactions. Phys. Rev. Lett. **100**, 042701 (2008)
20. A. Le Fevre et al., Bimodality: a general feature of heavy ion reactions. Phys. Rev. C **80**, 044615 (2009)
21. C.B. Das, S.D. Gupta, W.G. Lynch, A.Z. Mekjian, M.B. Tsang, The thermodynamic model for nuclear multifragmentation. Phys. Rep. **406**, 1–47 (2005)
22. S.D. Gupta, S. Mallik, G. Chaudhuri, Further studies of the multiplicity derivative in models of heavy ion collision at intermediate energies as a probe for phase transitions. Phys. Rev. C **97**, 044605 (2018)
23. G. Chaudhuri, S. Mallik, Effect of secondary decay on isoscaling: results from the canonical thermodynamical model. Nucl. Phys. A **849**, 190–202 (2011)
24. G. Chaudhuri, PhD thesis (Chapter IV), arXiv:nucl-th/0411005
25. J. Pan, S.Das Gupta, M. Grant, First-order phase transition in intermediate-energy heavy ion collisions. Phys. Rev. Lett. **80**, 1182–1185 (1998)
26. S.K. Samaddar, S.D. Gupta, Nuclear fragmentation characteristics from isotopic spin dependent lattice-gas model. Phys. Rev. C **61**, 034610 (2000)
27. S. Mallik, G. Chaudhuri, P. Das, S.D. Gupta, Multiplicity derivative: a new signature of a first-order phase transition in intermediate-energy heavy-ion collisions. Phys. Rev. C **95**, 061601(R) (2017)

28. W. Lin et al., Optical model potential analysis of $\bar{n}$ A and n A interactions Phys. Rev. C **97**, 054615 (2018)
29. W. Lin et al., Solidarity of signal of measures for the liquid-gas phase transition in the statistical multifragmentation model. Phys. Rev. C **99**, 054616 (2019)
30. H.L. Liu et al., Finite-size scaling phenomenon of nuclear liquid-gas phase transition probes. Phys. Rev. C **99**, 054614 (2019)
31. R. Bakeer, W. Awad, H.R. Jaqaman, Clusterisation in low and intermediate energy nuclear matter in the modified nuclear statistical equilibrium model. J. Phys. G **46**, 024105 (2019)
32. R. Wada et al., Experimental liquid-gas phase transition signals and reaction dynamics. Phys. Rev. C **99**, 024616 (2019)
33. P. Das, S. Mallik, G. Chaudhuri, First-order derivative of cluster size as a new signature of phase transition in heavy ion collisions at intermediate energies. Phys. Lett. B **783**, 364–367 (2018)

Chapter 3
Statistical and Dynamical Bimodality in Multifragmentation Reactions

S. Mallik, G. Chaudhuri, F. Gulminelli, and S. Das Gupta

Abstract The bimodal behavior of the order parameter is studied in the framework of Boltzmann–Uehling–Uhlenbeck (BUU) transport model. In order to do that, simplified yet accurate method of BUU model is used which allow calculation of fluctuations in systems much larger than what was considered feasible in a well-known and already existing model. It is observed that depending on the projectile energy and centrality of the reaction, both entrance channel and exit channel effects can be at the origin of the experimentally observed bimodal behavior. Both dynamical and statistical bimodality mechanisms are associated with the theoretical model to different time scales of the reaction, and to different energy regimes.

3.1 Introduction

The bimodal behavior of the order parameter is an important signature of first order phase transition [1–3]. Phase transitions occur in very large systems, but in nuclear physics, practical theoretical calculations (and heavy-ion reaction experiments) need to be done with finite systems.

The largest cluster is an important order parameter for studying nuclear liquid–gas phase transition in intermediate energy heavy-ion reactions. For an infinite system, the normalized order parameter is 1 in the liquid phase and it suddenly drops to 0 when the system crosses the transition temperature. From, statistical point of view, bimodality is a result of the singularity (infinite system) being replaced by the smearing (finite system). At low temperature, the largest cluster probability peaks at the liquid side, whereas at high temperature, it is limited to the gas side only. But in a small range of intermediate temperature, one can expect a double-humped distri-

S. Mallik (✉) · G. Chaudhuri
Variable Energy Cyclotron Centre, 1/AF Bidhan Nagar, Kolkata 700064, India
e-mail: swagato@vecc.gov.in

F. Gulminelli
LPC, CNRS/EnsiCaen et Universite, Caen IN2P3, France

S. Das Gupta
Physics Department, McGill University, Montréal H3A 2T8, Canada

R. K. Puri et al. (eds.), *Advances in Nuclear Physics*, Springer Proceedings in Physics 257, https://doi.org/10.1007/978-981-15-9062-7_3

bution (hence the name bimodality). Different theoretical studies (from lattice gas model and statistical models) [4–10] as well as experimental observations [11–13] confirm this kind of signature of thermal phase transition in heavy-ion reactions.

On the other side, some other theoretical calculations [14–17] as well as experimental measurements [18–20] conclude that, a memory of the entrance channel is clearly present and thermal equilibrium is not achieved. The signal was interpreted in these studies as a dynamical bifurcation of reaction mechanism, induced by fluctuations of the collision rate, which leads to fluctuations of the collective momentum distribution as expected in complex nonlinear dynamical systems.

Therefore, the origin of the experimentally observed bimodality is still not clear completely. In the previous dynamical approaches used [14–17] to study the bimodality phenomenon, the collision final state was determined by the semiclassical one-body transport equation itself, considering simulations evolving until asymptotic times. However, these approaches lack the necessary correlations to properly treat fragment formation in the exit channel, even if they are known to very well describe the entrance channel of heavy-ion reactions at intermediate energy. For this reason, to have a quantitative reproduction of experimental data, the secondary decay of the dynamically formed primary fragments is typically treated in two-step calculations, coupling the transport dynamics to a statistical model (or "afterburner"). As the primary interest is phase transition in nuclear matter due to the nuclear force alone, most theoretical models have considered symmetric nuclear matter where the Coulomb force is switched off [21, 22]. Here, we follow the same practice.

In order to study the dynamical stage for phase transition, one needs to simulate collisions between fairly large nuclei. In order to do that, a simplified yet accurate method of transport model based on Boltzmann–Uehling–Uhlenbeck (BUU) equation is developed recently [23] which allows calculation of fluctuations in systems much larger than what was considered feasible in a well-known and already existing model [27]. For studying the de-excitation phase, Canonical Thermodynamical Model (CTM) is used.

From this theoretical study, it is observed that, depending on the incident energy and impact parameter of the reaction, dynamical as well as statistical bimodality mechanisms can appear, meaning that the different scenario proposed in the literature are both potentially observable in heavy-ion data. Specifically, fluctuations in the stopping dynamics in central collisions lead to different reaction mechanisms that can coexist in the sample characterized by a well-defined value of the impact parameter. This gives rise to a bimodal behavior of the largest cluster probability distribution that can survive to the secondary de-excitation if the deposited energy is low enough, which happens at incident energies around the Fermi energy domain (40 MeV/nucleon). At higher incident energies (100 MeV/nucleon), focusing on binary mid-peripheral reactions, the fluctuations in the energy deposition leads to an excitation energy distribution for the quasi-spectator source which is close to the liquid–gas phase transition range. For these events, local equilibrium is achieved and a thermal bimodality is observed in agreement with statistical expectations.

This article is structured as follows. The modifications in the fluctuation included BUU model which is essential for studying liquid–gas phase transition is introduced in Sect. 3.2, the coupling conditions between the dynamical and statistica treatment are explained in Sect. 3.3, the results concerning the different conditions of dynamical and statistical bimodal behavior are described in Sects. 3.4 and 3.5 respectively, and finally, summary is presented in Sect. 3.6.

3.2 Improvement in BUU Model with Fluctuation

The BUU transport model calculation [23–26] for heavy-ion collisions starts with two nuclei in their respective ground states approaching each other with specified velocities and impact parameters. The ground-state energies and densities of the projectile (mass number A_p) and target (mass number A_t) nuclei are constructed using the Thomas–Fermi approximation [25]. The Thomas-Fermi phase-space distribution is then sampled using Monte Carlo technique by choosing test particles (we use $N_{test} = 100$ for each nucleon) with appropriate positions and momenta. As the projectile and target nuclei propagate in time, the test particles move in a mean-field and occasionally suffer two-body collisions, with probability determined by the nucleon–nucleon scattering cross section, provided the final state of the collision is not blocked by the Pauli principle. The mean-field propagation is done using the lattice Hamiltonian method which conserves energy and momentum very accurately [29]. The mean-field potential is given by

$$U(\rho) = A\left\{\frac{\rho}{\rho_0}\right\} + B\left\{\frac{\rho}{\rho_0}\right\}^{\sigma} + \frac{C}{\rho_0^{2/3}}\nabla_r^2\left\{\frac{\rho(\mathbf{r})}{\rho_0}\right\}, \tag{3.1}$$

where the first two terms represent zero-range Skyrme interaction and the derivative term does not affect nuclear matter properties, but in a finite system, it produces quite realistic diffused surfaces and liquid drop binding energies. This can be archived for $A = -2230.0$ MeV fm^3, $B = 2577.85$ MeV fm$^{7/2}$, $\sigma = 7/6$, $\rho_0 = 0.16$ fm^{-3}, and $c = -6.5$ MeV fm$^{5/2}$ [29]. Two-body collisions are calculated as given in Appendix B of [24], except that pion channels are closed, as there will not be any pion production in this energy regime.

To explain clustering in heavy-ion reaction, one needs an event-by-event computation in transport calculation. Bauer et. al. proposed the following method [27]. Due to collision between projectile nucleus of mass A_p and target nucleus of mass A_t, for each event, two-body collisions are checked between $(A_p + A_t)N_{test}$ test particles. Test particle cross sections are reduced to σ_{nn}/N_{test}; the collisions are further reduced by a factor N_{test}, but if a collision happens between two test particles i and j, then not only these two change momenta, but in addition, $N_{test} - 1$ test particles closest to i in phase space suffer the same momentum change as i; also $N_{test} - 1$ test particles closest to j in phase space are given the same momentum change as j. Physically, this corresponds to nucleons colliding. For conserving energy and momentum simultaneously, one can define $\langle \mathbf{p_i} \rangle = \frac{\sum \mathbf{p_i}}{N_{test}}$; similarly $< \mathbf{p_j} >$. One then considers a collision between $< \mathbf{p_i} >$ and $< \mathbf{p_j} >$ and obtain a $\mathbf{\Delta p}$ for $< \mathbf{p_i} >$ and $-\mathbf{\Delta p}$ for $< \mathbf{p_j} >$. This $\mathbf{\Delta p}$ is added to all $\mathbf{p_i}$'s and $-\mathbf{\Delta p}$ to all $\mathbf{p_j}$'s. This conserves both energy and momentum as one progresses in time. For the second event, new Monte Carlo sampling of A_p on A_t will be started at time zero, similarly for event 3, event 4, etc. The calculation of the collision part becomes very time-consuming and, for this case within each time step, two-body collision is need to be checked between $(A_p + A_t)N_{test}$ test particles. Now, to study nuclear liquid–gas phase transition one needs to simulate collisions between fairly large nuclei. Therefore, it is very difficult to handle this operation with the existing model. Hence, one has to modify the transport model so that it can be used for fairly large nuclei.

To overcome this problem, the fluctuation added BUU method is modified in the following way. N_{test} Monte Carlo simulations of A_p nucleons with positions and momenta and N_{test} simulations of A_t nucleons with positions and momenta are to be done as before. As in cascade calculation [24, 26, 28] for nucleon–nucleon collisions 1 on 1'(event1), 2 on 2'(event2) etc are considered with cross section σ_{nn}. For event 1, within each time step, nn collisions only between 1 and 1' (i.e., between first $(A_p + A_t)$ test particles) will be considered. The collision is checked for Pauli blocking. If a collision between i and j in event 1 is allowed, [27] is to be followed and $N_{test} - 1$ test particles closest to i are to be picked and the same momentum change $\mathbf{\Delta p}$ of them as ascribed to i is to be given. Similarly, $N_{test} - 1$ test particles closest to j are to be selected and these are to be ascribed the momentum change $-\mathbf{\Delta p}$, the same as suffered by j. As a function of time, this is continued till event 1 is over. For Vlasov propagation, all test particles are utilized. For event 2, one has to return to time $t = 0$, the original situation (or a new Monte Carlo sampling for the original nuclei), follow the above procedure but consider nn collisions only between 2 and 2' {i.e., between $(A_p + A_t) + 1$ to $2(A_p + A_t)$ test particles}. This can be repeated for as many events as one needs to build up enough statistics. Finally, to identify fragments, two test particles are considered as the part of the same cluster if the distance between them is less than or equal to 2 fm [17, 30].

Before applying the modified method in phase transition study, at first, one has to check whether modified prescription results are comparable with the existing BUU prescription or not. For this, simulations have to be done by using both the methods for a central collision reaction of symmetric system ^{40}Ca+^{40}Ca. Figure 3.1 shows the comparison of mass distribution obtained from existing and modified BUU prescription for beam energies 25 and 50 MeV/nucleon at $t = 200$ fm/c. The results obtained from two methods are similar because (a) the number of collisions in an event is statistically the same. (b) In the original formulation, the objects that collided were picked from a fine-grain sampling of phase-space density. In the modified method, these are picked from a coarse grain sampling of the same phase-space density. But many events are needed, so statistically, it should not matter. (c) Characteristics of scattering are the same and (d) The same Vlasov propagation is used.

The advantage of this method over the existing method is that here, for one event, nn collisions need to be considered between $(A_p + A_t)$ test particles, whereas in the existing method, collisions need to be checked between $(A_p + A_t)N_{test}$ test particles.

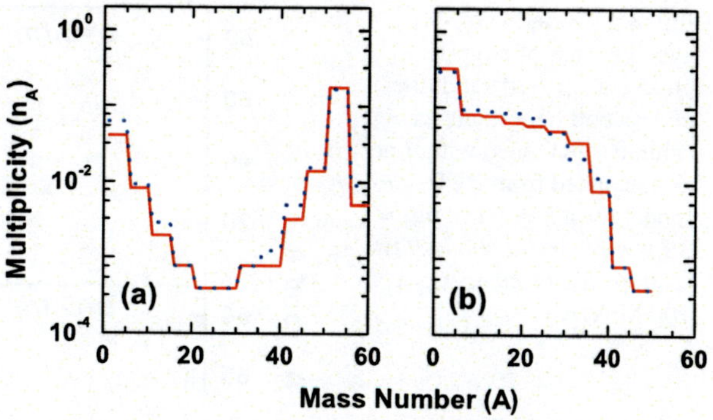

Fig. 3.1 Comparison of mass distribution calculated according to the existing (blue-dotted lines) and the modified (red solid lines) BUU prescription. The average value of 5 mass units are shown. The cases are for central collision of mass 40 on mass 40 for two different beam energies **a** 25 and **b** 50 MeV/nucleon calculated at $t = 200$ fm/c

Hence, in the modified calculation, the total number of combinations for two-body collision is reduced by a factor of $1/N_{test}^2$. Since typically N_{test} is of the order of 100 this is a huge saving in computation and has allowed us to treat reactions at different projectile energies and impact parameter described in the next sections. One bonus of this prescription is that one sees some common ground between the BUU approach and the "quantum molecular dynamics" approach [31].

At the end of the transport calculation, i.e., at freeze-out stage, we get different clusters of a finite number of test particles with known position and momenta. By knowing the number of test particles present in the cluster, one can get the mass, and by knowing the position and momenta of these test particles, one can calculate the potential and kinetic energies, respectively. By adding kinetic and potential energy, the excited state energy of the cluster can be obtained. However, to know excitation, one needs to calculate the ground-state energy also. This is done by applying the Thomas–Fermi method for a spherical (ground state) nucleus having a mass equal to the cluster mass. Knowing PLF mass and its excitation, the freeze-out temperature as well as decay of excited clusters are calculated by using the canonical thermodynamic model CTM [32] which is described in Chap. 2.

3.3 Identification of Freeze-Out

Our first aim is to identify the freeze-out time when one can safely stop transport calculation and switch over to the statistical model. In order to do that, we have investigated the time dependence of (i) mass of the largest and second largest clusters and (ii) isotopy of momentum distribution in the largest and second largest clusters from BUU calculation. Indeed in the binary collisions, we consider the largest and second largest clusters are always the residues of projectile and target. The first signal can therefore help us to determine the time when the projectile and target are completely separated, while the second one will point to the attainment of thermalization of these residues.

The dependence of average size of the largest and the second largest clusters with time for symmetric system $^{40}Ca + ^{40}Ca$ at projectile beam energy 100 MeV/nucleon

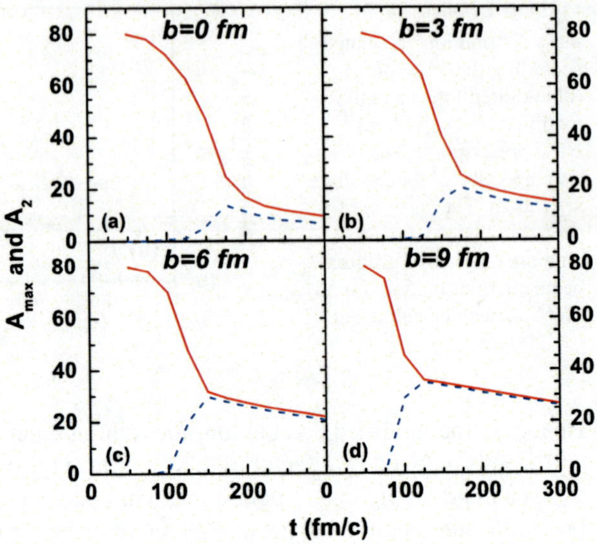

Fig. 3.2 Variation of average mass of largest cluster A_{max} (red solid lines) and second largest cluster A_2 (blue-dashed lines) with time as calculated from BUU model for **a** $b = 0$ fm, **b** $b = 3$ fm, **c** $b = 6$ fm, **d** $b = 9$ fm at projectile beam energy 100 MeV/nucleon

but four different impact parameters ranging from central to peripheral collisions is displayed in Fig. 3.2. The nature of variation is almost identical in each impact parameter. At $t = 50$ fm/c, there was just one system comprising of both projectile and the target nuclei, hence the size of the largest cluster (A_{max}) is close to 80 and the second largest (A_2) is close to 0. With the progress of time, the size of the largest cluster decreases gradually as the system fragments as well as there is evaporation of light clusters and nucleons. For central collision, the size of the participant zone is maximum which results in faster disintegration, therefore, the rate of decrease of largest cluster size is also maximum, while with the increase of impact parameter, participant size decreases which gradually reduces the rate of decrease in the largest cluster size. The size of the second largest starts from zero gradually increases as the target and projectile crosses each other and reaches a maximum when they are completely separated and then again decreases because of secondary decay, and settles to a final value. The evolution of the largest and that of the second largest cluster is pretty similar after the second largest cluster reaches its maximum and the evolution coincides for the most peripheral collisions. This is only because we are dealing with an identical size of projectile and target, and would change if one considers an asymmetric entrance channel.

The time evolution of the average isotropy of the momentum distribution (I) of the largest and second largest clusters is shown in Fig. 3.3. This observable indicates the thermalization of the system, and hence the ideal time to switch over from the dynamical model to the statistical model. This is defined through the following equations. Let the beam direction is along z axis and, for a given event, out of total $(A_p + A_t)N_{test}$ test particles, only N test particles form a cluster, i.e., the mass of the cluster is N/N_{test}. The average momentum of the cluster along $k = x$, y, and z directions can be calculated from the relation

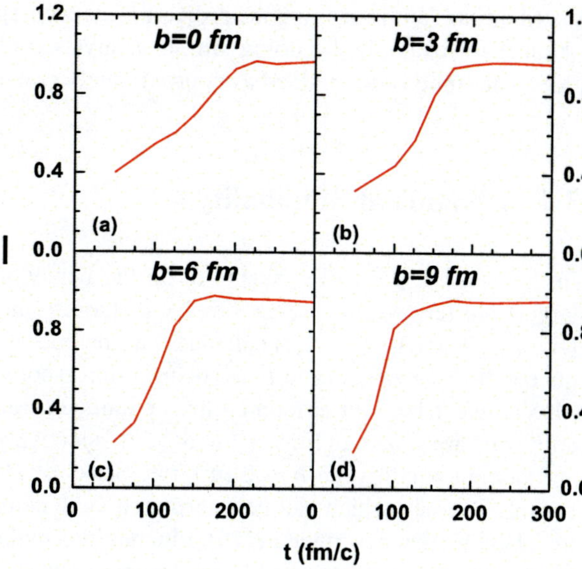

Fig. 3.3 Variation of isotropy of momentum distribution (I) of the largest cluster with time calculated from BUU model for **a** $b = 0$ fm, **b** $b = 3$ fm, **c** $b = 6$ fm, **d** $b = 9$ fm at projectile beam energy 100 MeV/nucleon

$$P_k = \frac{1}{N}\sum_{i=1}^{N} p_{k_i}, \quad (3.2)$$

where p_{k_i} is the k component of momentum of the i-th test particle. The isotropy in momentum distribution can be defined as

$$I = \frac{\frac{1}{N}\sum_{i=1}^{N}(p_{x_i} - P_x)^2 + \frac{1}{N}\sum_{i=1}^{N}(p_{y_i} - P_y)^2}{2 \times \frac{1}{N}\sum_{i=1}^{N}(p_{z_i} - P_z)^2}. \quad (3.3)$$

The quantity is defined such that it is less than 1 when the system is not fully thermalized and still there are some test particles having significant momentum in the beam direction. This will reduce the isotropy. Initially, during the overlapping stage of the projectile and target nuclei, the isotropy is less than unity. With the increase of time, it gradually increases and finally becomes unity when complete thermalization is achieved. Comparing Figs. 3.2 and 3.3, it can be concluded that $I \approx 1$ is archived almost at the same time when the second largest cluster size is also maximum. This freeze-out time varies from about $t_f = 150$ fm/c for most peripheral collision to about $t_f = 200$ fm/c for central collision. For simplicity, we have stopped the dynamical calculation at $t = 175$ fm/c for all impact parameters. Accounting for the precise impact, parameter dependence of the freeze-out time would only marginally affect the distributions shown in this article, and would not affect any of our conclusions which are essentially based on the qualitative properties of the distributions. With changing the projectile beam energy, the freeze-out time will also change, for example, by doing similar analysis for 40 MeV per nucleon reaction freeze-out time is determined as approximately $t = 400$ fm/c.

3.4 Dynamical Bimodality

In this section, we will concentrate on the behavior of probability distribution of largest cluster (A_{max}) and asymmetry of largest and second largest clusters ($a_2 = (A_{max} - A_2)/(A_{max} + A_2)$) calculated at the end of transport simulation at freeze-out condition. $P(A_{max})$ and $P(a_2)$ distribution at constant projectile beam energy 100 MeV/nucleon but four different impact parameters ranging from central to peripheral collisions are shown in Figs. 3.4 and 3.5, respectively.

For central collision ($b = 0$ fm), two peaks of $P(A_{max})$ as well as $P(a_2)$ distributions are seen which can be interpreted as dynamical bimodality very similar to the phenomenon described in [15]. **Fluctuations in the collision rates lead to fluctuations in the momentum distribution, that is, in the degree of stopping of the reaction. We have fixed a mass cut of** $A_{cut} = 37$ (corresponds to the minimum between the two peaks at $b = 0$ fm to distinguish the two event classes as it corresponds to the minimum between the two peaks. Fragments with $A_{max} \geq A_{cut}$ represent stopped events having nearly zero z-component (beam direction) of momentum and scattered isotropically in the center of mass frame, whereas fragments with $A_{max} < A_{cut}$ represent crossed events having high z-component of momentum and scattered either in the forward direction (projectile-like fragments) or backward direction (target-like fragments). This is shown in Fig. 3.6. For non-central collisions, due to lesser

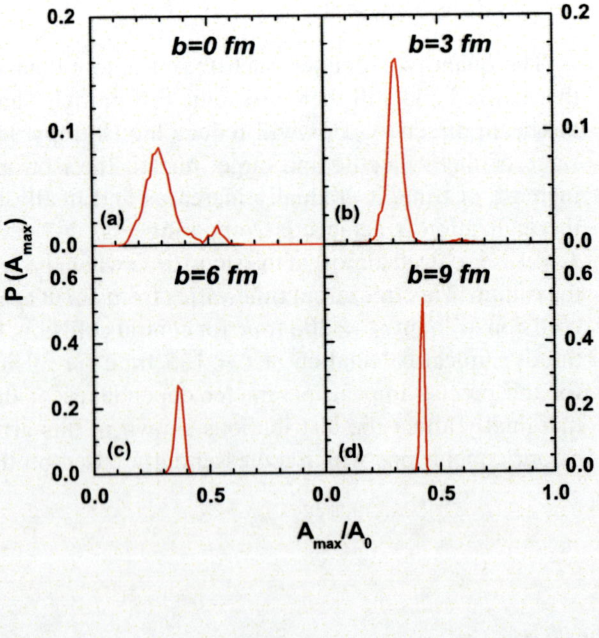

Fig. 3.4 Largest cluster probability distribution $P(A_{max})$ at freeze-out stage ($t = 175$ fm/c) for constant projectile beam energy 100 MeV/nucleon but for four different impact parameters **a** $b = 0$ fm, **b** $b = 3$ fm, **c** $b = 6$ fm, **d** $b = 9$ fm calculated from BUU model

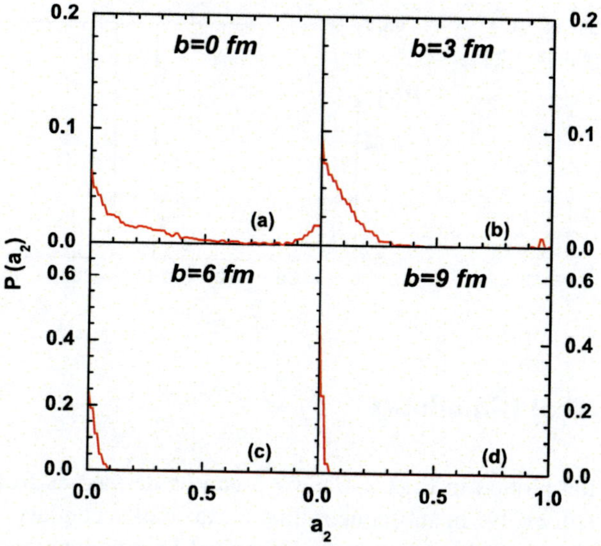

Fig. 3.5 Probability distribution of normalized mass asymmetry of two largest masses $P(\mathrm{a}_2)$ at freeze-out stage ($t = 175$ fm/c) for constant projectile beam energy 100 MeV/nucleon but for our different impact parameters **a** $b = 0$ fm, **b** $b = 3$ fm, **c** $b = 6$ fm, **d** $b = 9$ fm calculated from BUU model

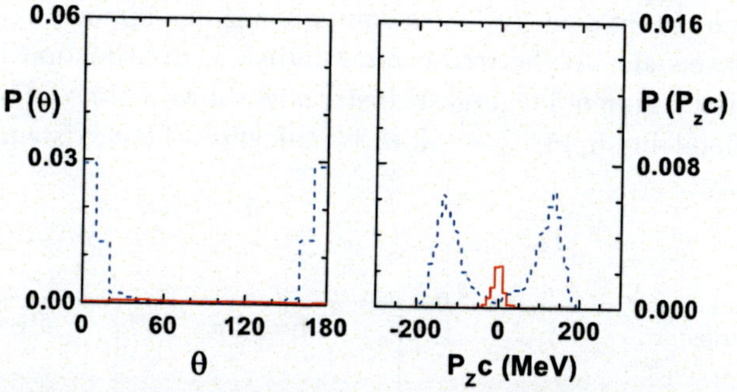

Fig. 3.6 Largest cluster scattering angle (left panel) and momentum (right panel) probability distribution for $A_{max} \geq 37$ (red lines) and $A_{max} < 37$ (blue lines) for central collisions ($b = 0$ fm) at projectile beam energy 100 MeV/nucleon calculated from BUU model at freeze-out time $t = 175$ fm/c. The average value of 10 degrees and 10 MeV are shown for angle and momentum, respectively

overlapping of the projectile and target, almost all are crossed-events and only liquid phase is present. Similar calculations for $b = 3$ fm are shown in Fig. 3.7.

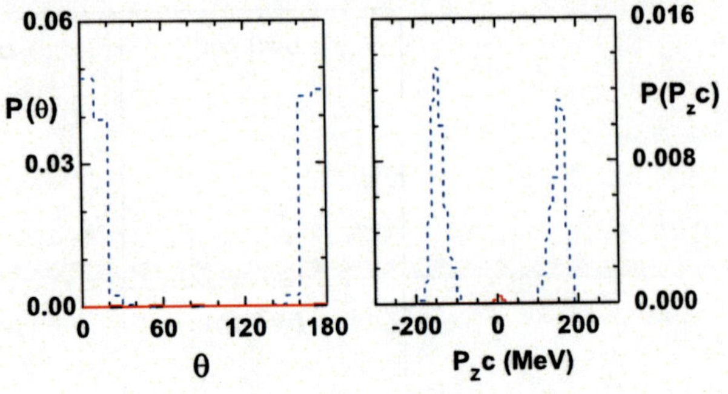

Fig. 3.7 Same as Fig. 3.6 except the impact parameter is 3 fm

3.5 Statistical Bimodality

The distribution plotted in Figs. 3.4 and 3.5 can be defined as freeze-out distribution and can still evolve in subsequent time because of secondary decay which has been calculated by switching over to the Canonical Thermodynamical Model (CTM) calculations from the transport one. In Fig. 3.8, we have plotted the probability distribution of the largest cluster for these four impact parameters. The ones at b = 0 fm are structureless and typical of multifragmentation reactions: the average excitation energy is so high in this case that both fully stopped and incompletely stopped events undergo multiple decays. As a consequence, the bimodality signal observed in Fig. 3.4 disappears. At the mid-central collision, the situation is reversed. The probability distribution of the largest cluster now shows a bimodal behavior (but not dynamical bimodality in Fig. 3.4) which is indicative of the existence of two phases simultaneously.

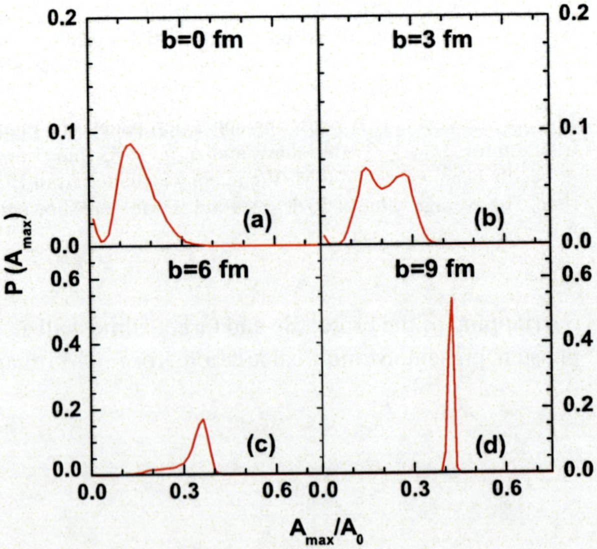

Fig. 3.8 Largest cluster probability distribution $P(A_{max})$ after secondary decay for constant projectile beam energy 100 MeV/nucleon but for four different impact parameters **a** $b = 0$ fm, **b** $b = 3$ fm, **c** $b = 6$ fm, **d** $b = 9$ fm calculated from BUU model

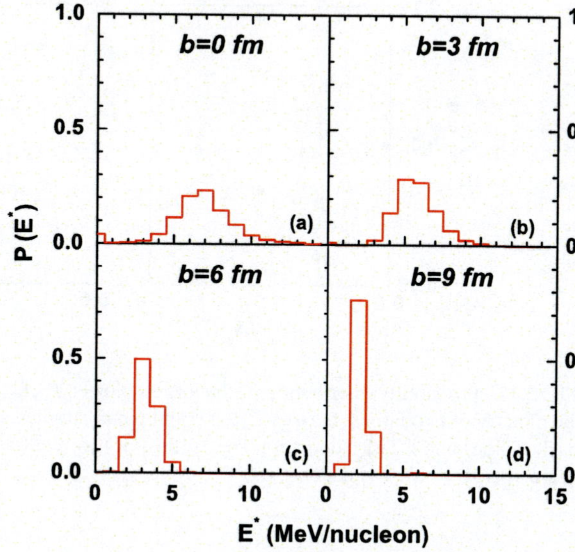

Fig. 3.9 Excitation (E^*) probability distribution for the largest and second largest clusters at constant projectile beam energy 100 MeV/nucleon but for four different impact parameters **a** $b = 0$ fm, **b** $b = 3$ fm, **c** $b = 6$ fm, **d** $b = 9$ fm

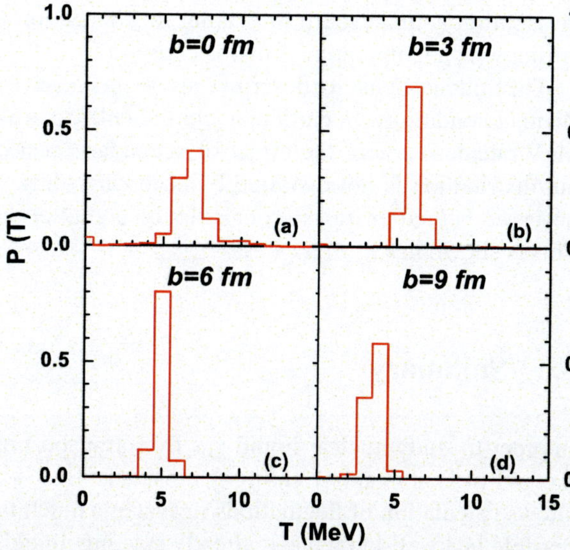

Fig. 3.10 Temperature (T) probability distribution for the largest and second largest clusters at constant projectile beam energy 100 MeV/nucleon but for four different impact parameters **a** $b = 0$ fm, **b** $b = 3$ fm, **c** $b = 6$ fm, **d** $b = 9$ fm

The indication of thermal phase transition will be more clear if one concentrate on the excitation energy (Fig. 3.9) and temperature (Fig. 3.10) spectrum of the largest and second largest clusters with varying centrality of the reaction. The step size selected for displaying these distributions are 1 MeV/nucleon for the excitation energy and 1 MeV/nucleon for the temperature. For the excitation energy in central collision, there is a small peak at low excitation that corresponds to the second largest cluster of the stopped event which is small in size and has less excitation. Using these excitation energies from the transport code as input to the statistical model code, temperature

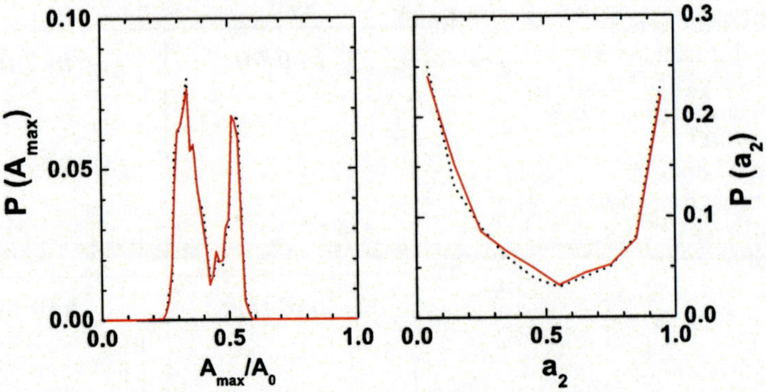

Fig. 3.11 Probability distribution of the largest cluster $P(A_{max})$ (left panel) and normalized mass asymmetry of two largest masses $P(\mathrm{a}_2)$ (right panel) studied after BUU model calculation (black-dashed line) and CTM calculation (red solid line) for central collisions ($b = 0$ fm) at projectile beam energy 40 MeV/nucleon

is obtained. At $b = 3$ fm, the excitation spectrum is broadened but the temperate distribution is quite sharp indicating its connection to the thermal phase transition during which temperature remains constant.

The bimodal behavior described above, however, strongly depends on the entrance channel conditions. In particular, central collisions at lower bombarding energy (40 MeV/nucleon) where the effect of secondary decay is less, therefore, the freeze-out distribution is not distorted by secondary decay and bimodal behavior can be observed both after transport calculation and after the statistical model calculation. This is shown in Fig. 3.11.

3.6 Summary

In order to study nuclear liquid gas from transport model, a simplified yet accurate method of Boltzmann–Uehling–Uhlenbeck (BUU) model is developed [23] which allows calculation of fluctuations in systems much larger than what was considered feasible in a well-known and already existing model.

We have analyzed the bimodal behavior for the symmetric system $^{40}Ca +^{40} Ca$ with varying centrality of the reaction as well as bombarding energies, as predicted by a two-step model. The entrance channel dynamics is described by the BUU transport equation, which is coupled to the statistical CTM decay model at the time of local equilibration of the primary fragments produced in the collision.

Based on the combined theoretical simulation, it is observed that depending on the incident energy and impact parameter of the reaction, both entrance channel and exit channel effects can be at the origin of the experimentally observed bimodal behavior. Specifically, fluctuations in the reaction mechanism induced by fluctuations in the collision rate, as well as thermal bimodality directly linked to the nuclear liquid–

gas phase transition are observed in simulations [33]. These results indicate that heavy-ion reactions at intermediate energies can be used in the laboratory to study dynamical as well as statistical bimodalities

References

1. D.H.E. Gross, *Microcanonical Thermodynamics: Phase Transitions in Finite Systems*. Lecture Notes in Physics, vol. 66 (World Scientific, Singapore, 2001).
2. F. Gulminelli, Phase coexistence in nuclei. Ann. Phys. Fr. **29**(6), 1–121 (2004)
3. S.D. Gupta, A.Z. Mekjian, M.B. Tsang, *Advances in Nuclear Physics*, vol. 26, 89, eds. by J.W. Negele, E. Vogt (Plenum Publishers, New York, 2001)
4. P. Chomaz, F. Gulminelli, First-order phase transitions: equivalence between bimodalities and the Yang-Lee theorem. Phys. A **330**, 451–458 (2003)
5. F. Gulminelli, J.M. Carmona, P. Chomaz, J. Richert, S. Jimenez, V. Regnard, Transient back-bending behavior in the Ising model with fixed magnetization. Phys. Rev. E **68**, 026119 (2003)
6. P. Chomaz, V. Duflot, F. Gulminelli, Caloric curves and energy fluctuations in the microcanonical liquid-gas phase transition. Phys. Rev. Lett. **85**, 3587–3590 (2000)
7. F. Gulminelli, P. Chomaz, Distribution of the largest fragment in the lattice gas model. Phys. Rev. C **71**, 054607 (2005)
8. M. Kastner, M. Pleimling, Microcanonical phase diagrams of short-range ferromagnets. Phys. Rev. Lett **102**, 240604 (2009)
9. G. Lehaut et al., Phase diagram of the charged lattice-gas model with two types of particles. Phys. Rev. E **81**, 051104 (2010)
10. G. Chaudhuri, S.D. Gupta, Properties of the largest fragment in multifragmentation: a canonical thermodynamic calculation. Phys. Rev. C **75**, 034603 (2007)
11. M. Bruno et al., Signals of bimodality in the fragmentation of Au quasi-projectiles. Nucl. Phys. A **807**, 48–60 (2008)
12. E. Bonnet et al., Bimodal behavior of the heaviest fragment distribution in projectile fragmentation. Phys. Rev. Lett. **103**, 072701 (2009)
13. B. Borderie et al., The prominent role of the heaviest fragment in multfragmentation and phase-transition for hot nuclei. Int. J. Mod. Phys. E **19**, 1523–1533 (2010)
14. K. Zbiri et al., Transition from participant to spectator fragmentation in Au+Au reactions between 60 and 150 AMeV. Phys. Rev. C **75**, 034612 (2007)
15. A. Le Fevre et al., Bimodality: a sign of critical behavior in nuclear reactions. Phys. Rev. Lett. **100**, 042701 (2008)
16. A. Le Fevre et al., Bimodality: a general feature of heavy ion reactions. Phys. Rev. C **80**, 044615 (2009)
17. S. Mallik, S.D. Gupta, G. Chaudhuri, Bimodality emerges from transport model calculations of heavy ion collisions at intermediate energy. Phys. Rev. C **93**, 041603(R) (2016)
18. M. Pichon et al., Bimodality: a possible experimental signature of the liquid-gas phase transition of nuclear matter. Nucl. Phys. A **779**, 267–296 (2006)
19. P. Lautesse et al., Experimental overview of Ni+Ni collisions at 32 MeV/nucleon: discriminant analysis and duality in the decay modes of a fusionlike system. Phys. Rev. C **71**, 034602 (2005)
20. O. Lopez, M.F. Rivet, Bimodalities: a survey of experimental data and models. Eur. Phys. J. A **30**, 263–274 (2006)
21. S. Das Gupta, A.Z. Mekjian, Phase transition in a statistical model for nuclear multifragmentation. Phys. Rev C **57**, 1361–1365 (1998)
22. K.A. Bugaev, M.I. Gorenstein, I.N. Mishustin, W. Greiner, Exactly soluble model for nuclear liquid-gas phase transition. Phys. Rev. C **62**, 044320 (2000)

23. S. Mallik, S.D. Gupta, G. Chaudhuri, Event simulations in a transport model for intermediate energy heavy ion collisions: applications to multiplicity distributions. Phys. Rev. C **91**, 034616 (2015)
24. G.F. Bertsch, S.D. Gupta, A guide to microscopic models for intermediate energy heavy ion collisions. Phys. Rep. **160**, 189–233 (1988)
25. S.D. Gupta, S. Mallik, G. Chaudhuri, Heavy ion reaction at intermediate energies: Theoretical Models, To be published from World Scientific Publishers
26. S. Mallik, G. Chaudhuri, S.D. Gupta, Hybrid model for studying nuclear multifragmentation around the Fermi energy domain: the case of central collisions of Xe on Sn. Phys. Rev. C **91**, 044614 (2015)
27. W. Bauer, G.F. Bertsch, S.D. Gupta, Fluctuations and clustering in heavy-ion collisions. Phys. Rev. Lett. **58**, 863–866 (1987)
28. S. Mallik, S.D. Gupta, G. Chaudhuri, Estimates for temperature in projectile-like fragments in geometric and transport models. Phys. Rev. C **89**, 044614 (2014)
29. R.J. Lenk, V.R. Pandharipande, Nuclear mean field dynamics in the lattice Hamiltonian Vlasov method. Phys. Rev. C **39**, 2242–2249 (1989)
30. S. Mallik, G. Chaudhuri, F. Gulminelli, Sensitivity of the evaporation residue observables to the symmetry energy. Phys. Rev. C **100**, 024611 (2019)
31. J. Aichelin, Quantum molecular dynamics-a dynamical microscopic n-body approach to investigate fragment formation and the nuclear equation of state in heavy ion collisions. Phys. Rep. **202**, 233–360 (1991)
32. C.B. Das, S.D. Gupta, W.G. Lynch, A.Z. Mekjian, M.B. Tsang, The thermodynamic model for nuclear multifragmentation. Phys. Rep. **406**, 1–47 (2005)
33. S. Mallik, G. Chaudhuri, F. Gulminelli, Dynamical and statistical bimodality in nuclear fragmentation. Phys. Rev. C **97**, 024606 (2018)

Chapter 4
Study of Isospin Effects in Heavy-Ion Collisions at Intermediate Energies Using Isospin-Dependent Quantum Molecular Dynamics Model

Arun Sharma, Rohit Kumar, and Rajeev K. Puri

Abstract We present results, calculated using the microscopic framework of isospin-dependent quantum molecular dynamics (IQMD) model for some of the asymmetric reactions. We show that isospin effects via symmetry potential and isospin-dependent nucleon–nucleon cross section along with high Fermi momentum in IQMD model can handle the difficulties of quantum molecular dynamics approach. Further, we show that isospin effects via Coulomb forces influence the onset of multifragmentation.

4.1 Introduction

In addition to several radioactive ion beam facilities (RIBs) that already exist, considerable efforts have been made to install or plan new generation radioactive ion beam facilities. The well-renowned facilities like National Superconducting Cyclotron Laboratory (NSCL) at Michigan State University in the United States of America (USA) [1], Legnaro National Lab (LNL) at Padua in Italy [2], Holifield Radiation Ion Beam Facility (HRIBF) at Oak Ridge National Laboratory in the USA [3], Heavy-Ion Research Facility of Lanzhou at Lanzhou in China [4], RIKEN in Japan [5], TRI University Meson Facility (TRIUMF) at Vancouver in Canada, and last but not least, Facility for Antiproton and Ion Research (FAIR) at GSI in Germany [6], etc. India is also planning to install Asia's second largest advanced experimental facility, namely Advanced National Facility for Unstable Rare Ion Beams (ANURIB) [7, 8]. This facility will help to collect information regarding the nuclear equation of state (NEOS). With the advancements in various radioactive ion beam facilities, the nuclear Physics community has shifted its interest toward isospin physics. These facilities help in understanding about the isospin content of a nucleus and lead to the discovery of new isotopes. These facilities also help to understand nuclear forces at

A. Sharma (✉)
Department of Physics, GDC Billawar, Jammu 185204, India
e-mail: arungaur.ju@gmail.com

R. Kumar · R. K. Puri
Department of Physics, Panjab University, Chandigarh 160014, India

R. K. Puri et al. (eds.), *Advances in Nuclear Physics*, Springer Proceedings in Physics 257, https://doi.org/10.1007/978-981-15-9062-7_4

the nucleonic level and shed light on the collective behavior of a nucleus in terms of n/p ratio in a nucleus. The term isospin means a pair of identical particles (nucleons), i.e., either neutrons or protons in nuclear matter when their electric charges aren't taken into consideration. In nuclear physics, isospin degree of freedom makes its presence in various phenomenon (like Multifragmentation, Collective Flow, Nuclear Stopping, etc.) and thus attracts the attention of the nuclear physics community. Experimental studies involving neutron-rich nuclei have set an excellent platform to investigate the isospin dependence in the dynamics of heavy-ion collisions at intermediate energies. It has also opened up a new field that helps in extracting the information regarding the equation of state of asymmetric nuclear matter, a topic of current debate. A large number of studies have been reported in the literature where authors have investigated the role of the isospin degree of freedom in heavy-ion collisions at intermediate energies [7, 9–12]. The prime aim of such studies is to probe the properties of nuclear matter in the region between symmetric nuclear matter and pure neutron matter. This information also helps to understand various Astrophysics phenomena such as Supernova, Neutron Stars, etc.

It is worth mentioning that now large information has been collected on fragmentation but the major problem is that all experimental measurements are done when the matter is cold. Therefore, it is obvious that to understand the phenomenon of fragmentation, there is a need for authentic theoretical models/approaches. Broadly, theoretical models/approaches can be divided into statistical and dynamical models/approaches. The dynamical models are of two types 1. Isospin-independent like VUU, QMD, AMD, etc. [8, 13–19] 2. Isospin-dependent like ImQMD, IBUU, IQMD, etc. [20–22]. The isospin-dependent Quantum Molecular Dynamics (IQMD) Model [22] is the improved version of the QMD [8] model with isospin degree of freedom incorporated in it. With the inclusion of isospin degree of freedom, the IQMD model has symmetry potential, isospin dependence nucleon–nucleon cross section and improved Pauli blocking. Also, it is worth mentioning that dynamics of symmetric and asymmetric reactions is quite different [7, 12, 23]. Therefore, it would be interesting to probe isospin effects using IQMD model on the dynamics of asymmetric heavy-ion collisions at intermediate energies. This chapter is organized as follows: in Sect. 4.2, we will give a brief introduction to the formalism used to carry out calculations. Then, in Sect. 4.3, we will review our recently published results for multifragmention. Finally, Sect. 4.4 is devoted to a summary of the chapter.

4.2 Isospin-Dependent Quantum Molecular Dynamics (IQMD) Model

The IQMD model [22] treats different charge states of nucleons, deltas, and pions explicitly, as inherited from the Vlasov–Uehling–Uhlenbeck (VUU) model [24]. The IQMD model has been used successfully for the analysis of a large number of observables from low to relativistic energies. The isospin degree of freedom enters into the calculations via symmetry potential, cross sections, and Coulomb interaction.

In this model, baryons are represented by Gaussian-shaped density distributions

$$f_i(\mathbf{r},\mathbf{p},t) = \frac{1}{\pi^2\hbar^2}\exp\left(-[\mathbf{r}-\mathbf{r_i}(t)]^2\frac{1}{2L}\right)\times\exp\left(-[\mathbf{p}-\mathbf{p_i}(t)]^2\frac{2L}{\hbar^2}\right). \quad (4.1)$$

Nucleons are initialized in a sphere with radius R = 1.12 $A^{1/3}$ fm, in accordance with liquid drop model. The Gaussian Width (interaction range of nucleons) in this model is system size-dependent and its value varies from 8.66 fm^2 for $^{197}_{79}$Au to 4.33 fm^2 for $^{40}_{20}$Ca. This procedure of choosing system size-dependent value of Gaussian width "L" leads to maximum stability of the density profile of the given nucleus. Each nucleon occupies a volume of h^3, so that phase space is uniformly filled. The initial momenta are randomly chosen between 0 and Fermi momentum ($\mathbf{p}_F$). The nucleons of the target and projectile interact by two- and three-body Skyrme forces, Yukawa potential, Coulomb interactions, and momentum-dependent interactions (MDI). In addition to the use of explicit charge states of all baryons and mesons, a symmetry potential between protons and neutrons corresponding to the Bethe–Weizsacker mass formula has been included. The hadrons propagate using Hamilton equations of motion:

$$\frac{d\mathbf{r_i}}{dt} = \frac{d\langle H\rangle}{d\mathbf{p_i}};\ \frac{d\mathbf{p_i}}{dt} = -\frac{d\langle H\rangle}{d\mathbf{r_i}} \quad (4.2)$$

with

$$\begin{aligned}\langle H\rangle &= \langle T\rangle + \langle V\rangle \\ &= \sum_i \frac{p_i^2}{2m_i} + \sum_i\sum_{j>i}\int f_i(\mathbf{r},\mathbf{p},t)V^{ij}(\mathbf{r}',\mathbf{r}) \\ &\quad \times f_j(\mathbf{r}',\mathbf{p}',t)d\mathbf{r}\,d\mathbf{r}'\,d\mathbf{p}\,d\mathbf{p}'. \end{aligned} \quad (4.3)$$

The baryon potentialV^{ij}, in the above relation, reads as

$$\begin{aligned}V^{ij}(\mathbf{r}'-\mathbf{r}) &= V^{ij}_{Skyrme} + V^{ij}_{Yukawa} + V^{ij}_{Coul} + V^{ij}_{mdi} + V^{ij}_{sym} \\ &= \left[t_1\delta(\mathbf{r}'-\mathbf{r}) + t_2\delta(\mathbf{r}'-\mathbf{r})\rho^{\gamma-1}\left(\frac{\mathbf{r}'+\mathbf{r}}{2}\right)\right] \\ &\quad + t_3\frac{\exp(|(\mathbf{r}'-\mathbf{r})|/\mu)}{(|(\mathbf{r}'-\mathbf{r})|/\mu)} + \frac{Z_iZ_je^2}{|(\mathbf{r}'-\mathbf{r})|} \\ &\quad + t_4\ln^2[t_5(\mathbf{p}'-\mathbf{p})^2+1]\delta(\mathbf{r}'-\mathbf{r}) \\ &\quad + t_6\frac{1}{\varrho_0}T_{3i}T_{3j}\delta(\mathbf{r_i}'-\mathbf{r_j}). \end{aligned} \quad (4.4)$$

Here, Z_i and Z_j denote the charges of ith and jth baryon, and T_{3i} and T_{3j} are their respective T_3 components (i.e., 1/2 for protons and −1/2 for neutrons). The parameters μ and t_1,....,t_6 are adjusted to the real part of the nucleonic optical potential. For the density dependence of the nucleon optical potential, standard Skyrme-type parametrization is employed. We also use the isospin and energy-dependent cross section $\sigma = 0.8\ \sigma_{nn}^{free}$. The details about the elastic and inelastic cross sections for Proton–proton and proton–neutron collisions can be found in [22]. The cross sections for neutron–neutron collisions are assumed to be equal to the proton–proton cross sections. Explicit Pauli blocking is also included, i.e., Pauli blocking of the neutrons and protons is treated separately. We assume that each nucleon occupies a sphere in coordinate and momentum space. This trick yields the same Pauli blocking ratio as an exact calculation of the overlap of the Gaussians will yield. We calculate the fractions P_1 and P_2 of the final phase space for each of the two scattering partners that are already occupied by other nucleons with the same isospin as that of scattered ones. The collision is blocked with the probability

$$P_{block} = 1 - [1 - min(P_1, 1)][1 - min(P_2, 1)], \tag{4.5}$$

and, correspondingly is allowed with the probability 1 - P_{block}. For a nucleus in its ground state, we obtain an averaged blocking probability $\langle P_{block} \rangle = 0.96$. Whenever an attempted collision is blocked, the scattering partners maintain the original momenta prior to scattering. The IQMD model briefly discussed above is a primary code and it only generates phase space of nucleons. Therefore, after generating phase space, there is a need for secondary models to clusterize the nucleons into fragments. In this article, we use a minimum spanning tree (MST) method [8, 23, 25] for this purpose. The basic procedure of this method to clusterize the nucleons into fragments is very simple as it considers two nucleons share the same fragment if their centroids are closer than a distance R_{clus}, i.e.,

$$|\ \mathbf{r}_i - \mathbf{r}_j\ | \leq R_{clus}, \tag{4.6}$$

where r_i and r_j are the spatial positions of both nucleons.

4.3 Results and Discussion

In the present chapter, we study some of the isospin effects on the dynamics of fragmentation in asymmetric reactions. To check the effects of isospin degree of freedom (as it leads to refine ingredients), in IQMD model, first of all, we will robust IQMD model against the available experimental data of emulsion experiments for asymmetric reactions [26, 27]. The data of emulsion experiments is of particular interest because it validates any theoretical model designed for studying multifragmentation phenomenon at intermediate energies as a range of beam energies taken for these experiments cover fusion, multifragmentation, and vaporization phenomenon.

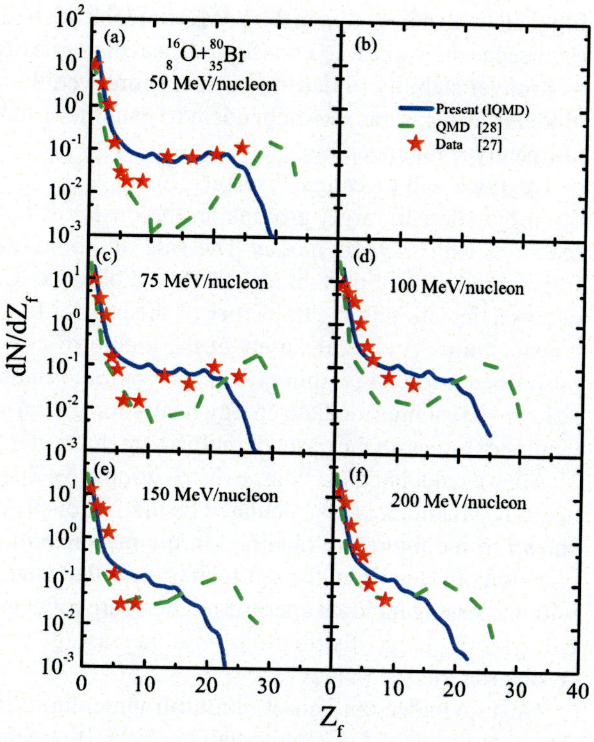

Fig. 4.1 The charge distributions for the central reaction of $^{16}_{8}O + ^{80}_{35}Br$ at incident energies between 50 and 200 MeV/nucleon. Solid lines represent our present calculations using IQMD model. Stars represent experimental data and are extracted from [27]. The results of previous theoretical attempts using QMD model (see dashed lines) [28] are also displayed for comparison (see [7] for original discussion)

We simulated the reaction of $^{16}_{8}O + ^{80}_{35}Br$ at different bombarding energies ranging from 50 to 200 MeV/nucleon and the calculated unnormalized charge distributions (see solid lines) are plotted in Fig. 4.1. The available measurements for same beam energies are shown by solid stars. One can see that present combination of refine ingredients (with the inclusion of isospin degree of freedom) is able to reproduce the measurements for asymmetric reactions. To see the comparison, the calculations of previous attempts [28] using QMD model (as it doesn't have isospsin degree of freedom) are also shown in the figure. One can clearly see that QMD model fails to handle the dynamics of asymmetric reactions. There are many other studies in the literature where authors have reported failure of QMD model to handle the dynamics of asymmetric reactions. For example, one can see the references [29, 30]. To get a clear picture, we have carried out an extensive study as reported in [23] and it has been observed that with the inclusion of refine ingredients due to isospin degree of freedom and high Fermi momentum in IQMD model, one can handle the difficulties reported in reproducing results with QMD approaches for asymmetric reactions.

It is worth mentioning that the Fermi momentum in the IQMD model is not a free parameter at all. Rather, it is determined from the ground-state density, evaluated using Fermi gas model. On the other hand, the Fermi momentum in the QMD model is determined using local binding energies and thus leads to a comparatively lower average value when compared with the Fermi momentum in the IQMD model. The higher Fermi momentum in the IQMD model leads to unbound nucleons at the surface (and hence their spurious emission). But at the same time, higher Fermi pressure leads to stronger stability of density profile. Moreover, the interaction range in the IQMD model is also system size-dependent to gain more stability of the density profile and ultimately stable nuclei.

Next, we will investigate isospin effects via Coulomb forces, i.e., forces between the projectile and target protons on the onset of mutifragmentation in asymmetric reactions using IQMD model. The role of Coulomb forces on the onset of multi-fragmentation is important to understand because it was reported in [31] that there occurs a deviation in the trajectory of the projectile when these forces are taken into consideration. A careful survey of the literature shows that many studies have been performed to probe isospin effects via coulomb interactions on various observables [12, 31–33] at intermediate energy heavy-ion collisions. To check isospin effects via Coulomb forces on the onset of multifragmentation in asymmetric reactions of $^{84}_{36}$Kr + $^{197}_{79}$Au, we calculated the charge distributions ($3 \leq Z_f \leq 12$) at different beam energies using IQMD model and calculated results are displayed in Fig. 4.2. Note that results shown in the figure are obtained in the presence of Coulomb forces using the soft equation of state. From the figure, it is clear that charge distributions become steeper with increasing incident energy and therefore reflects the violence of collisions. This behavior of charge distributions with increasing beam energy is consistent with the results reported in [34–36].

Next, to understand onset of multifragmentation, we fit charge distributions with power law $\propto Z_f^{-\tau}$ for intermediate mass fragments ($3 \leq Z_f \leq 12$). Also, we have extracted values of τ and these values are plotted as a function of incident energy in Fig. 4.3. From the figure, we find that there is no minimum in the extracted values of τ when plotted against incident energies. This pattern of τ values with beam energies in the presence of Coulomb forces reflects the onset of multifragmentauon at beam energies less than 35 MeV/nucleon.

To investigate this point, we have calculated the charge yields ($3 \leq Z_f \leq 12$) for the asymmetric and highly charged system of $^{84}_{36}$Kr + $^{197}_{79}$Au in the absence of Coulomb forces and results are shown in Fig. 4.4.

Again, the calculated charge distributions have been fitted with power law fits $\propto Z_f^{-\tau}$ and extracted values of parameter τ are plotted against incident energies in Fig. 4.5. From the figure, one can clearly see a sharp minimum for soft equation of state at an incident energy of $\simeq$ 55 MeV/nucleon. A remarkable effect of Coulomb forces is observed when compared with the results shown in Fig. 4.4. To get more information, we have also performed calculations for the asymmetric reactions of $^{40}_{18}$Ar + $^{64}_{29}$Cu in the absence of Coulomb forces. For this asymmetric system, the extracted values of τ show minimum at an incident energy of 17 MeV/nucleon. Thus, for a highly charged asymmetric system of $^{84}_{36}$Kr + $^{197}_{79}$Au, minimum occurs at higher incident energies when compared with the light charged asymmetric system of $^{40}_{18}$Ar + $^{64}_{29}$Cu. This clearly shows the dependence of the onset of multifragmentation on the range of Coulomb forces and reaction asymmetry.

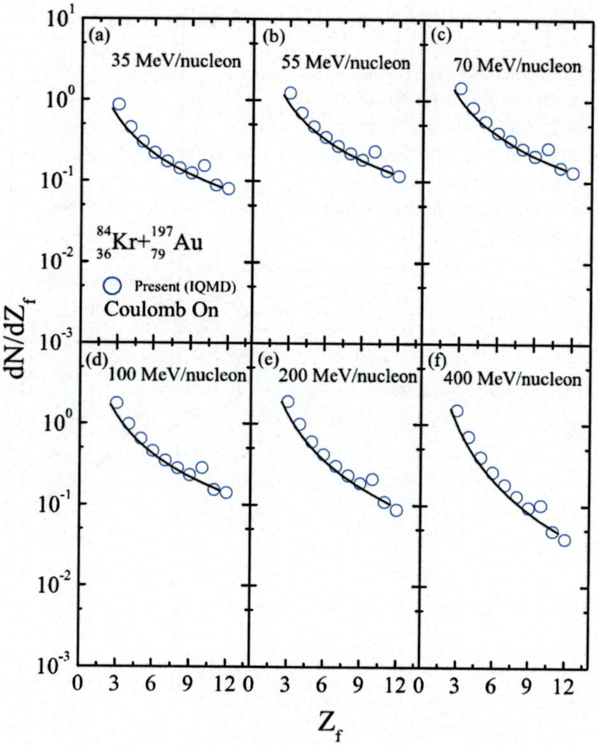

Fig. 4.2 The charge distributions calculated for the central collisions of $^{84}_{36}$Kr + $^{197}_{79}$Au using IQMD model (open circles) at incident energies between 35 and 400 MeV/nucleon in the presence of Coulomb forces. Solid lines represent power law fits over intermediate mass fragments ($3 \leq Z_f \leq 12$). With kind permission (Order Number: 501530607, dated: Nov. 26, 2019) of The European Physical Journal (EPJ), the data shown in the figure has been taken from [12]

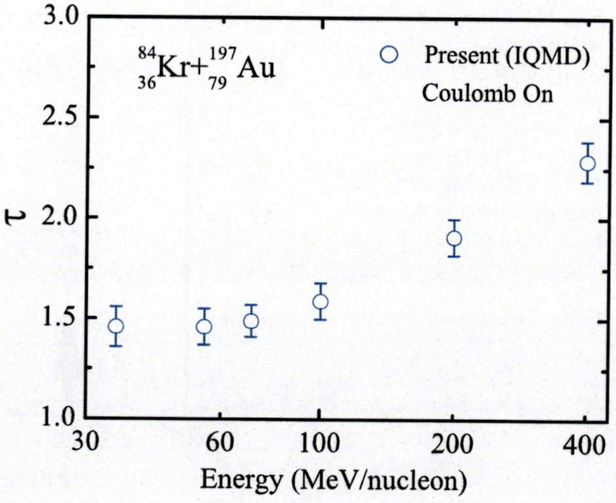

Fig. 4.3 Extracted values of parameter τ (with error bars) obtained from the power law fits of intermediate mass fragments ($3 \leq Z_f \leq 12$) as done in Fig. 4.3 for the central collisions of $^{84}_{36}$Kr + $^{197}_{79}$Au. With kind permission (Order Number: 501530607, dated: Nov. 26, 2019) of The European Physical Journal (EPJ), the data shown in the figure has been taken from [12]

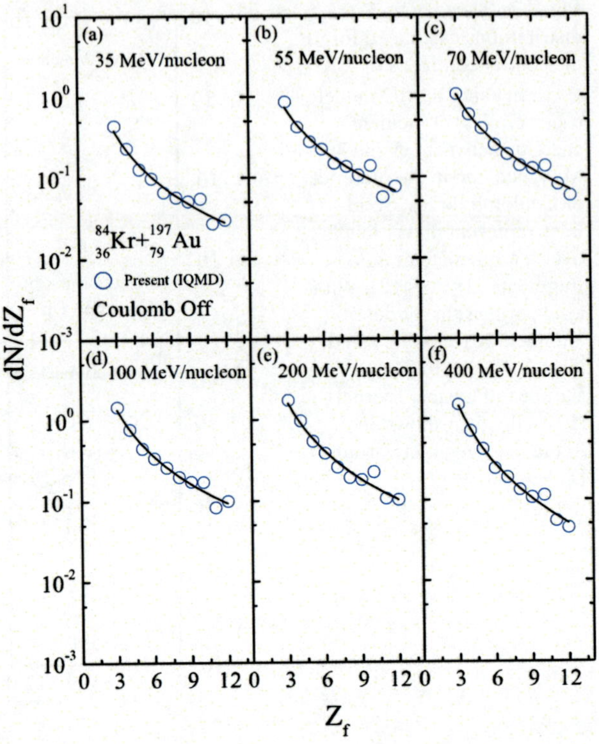

Fig. 4.4 Same as Fig. 4.3, but in the absence of Coulomb forces. With kind permission (Order Number: 501530607, dated: Nov. 26, 2019) of The European Physical Journal (EPJ), the data shown in the figure has been taken from [12]

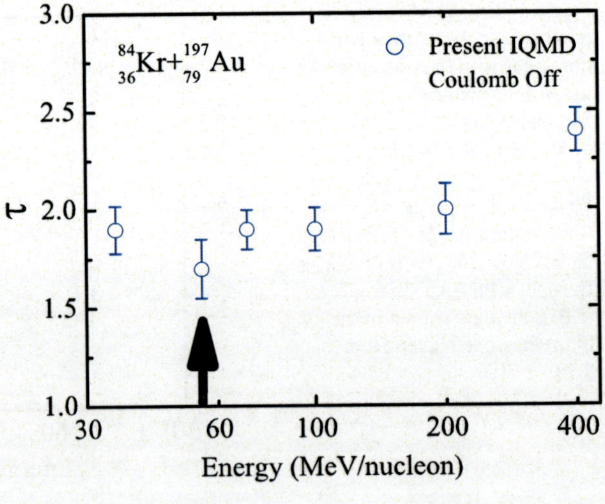

Fig. 4.5 Same as Fig. 4.3, but in the absence of Coulomb forces. With kind permission (Order Number: 501530607, dated: Nov. 26, 2019) of The European Physical Journal (EPJ), the data shown in the figure has been taken from [12]

4.4 Summary

We presented that isospin effects along with high Fermi momentum in IQMD model with basic cluster algorithm can handle the dynamics of asymmetric reactions reasonably well around the Fermi energy domain. Also, it has been shown that isospin effects via Coulomb forces affect the onset of multifragmentation.

Acknowledgements A.Sharma thanks Prof. Rajeev K. Puri for fruitful discussions and support. The author would also like to thank Prof. Arun Bharti and Dr. Sakshi Gautam for valuable discussions during the course of this work.

References

1. See, e.g., http://www.nscl.msu.edu/
2. INFN-LNL Report, pp. 238 (2012)
3. G.D. Alton, J.R. Beene, The holified radioactive ion beam facility at the oak ridge national laboratoty: present status and future plans. J. Phys. G: Nucl. Part. Phys. **24**, 1347 (1998). http://www.phy.ornal.gov
4. J.W. Xia, Y.J. Yuan, Y. Liu, In: *Proceedings of PAC09, Vancouver, BC, Canada*, pp. 3048-3052 (2009); X. Cai et al., J. Phys. Con. Ser. **163**, 012113 (2009)
5. Y. Yano, The RIKEN RI beam factory project: a status report. Nucl. Inst. Meth. B **261**, 1009 (2007)
6. See, e.g., http://www.gsi.de/fair/index-e.html
7. A. Sharma, *Study of Asymmetric and Neutron-Rich Heavy-Ion Collisions Using Isospin-Dependent Quantum Molecular Dyanmics Model* (University of Jammu, Jammu, 2016). Ph. D. Thesis
8. J. Aichelin, Quantum molecular dynamics-a dynamical microscopic n-body approach to investigate fragment formation and the nuclear equation of state in heavy ion collisions. Phys. Rep. **202**, 233–260 (1991); R. Kumar, S. Gautam, R.K. Puri, Multifragmentation within a clusterization algorithm based on thermal binding energies. Phys. Rev. C **89**, 064608 (2014)
9. A. Jain et al., Influence of charge asymmetry and isospin-dependent cross section on nuclear stopping. Phys. Rev. C **84**, 057602 (2011); *ibid.* Influence of charge asymmetry and isospin-dependent cross section on elliptical flow. Phys. Rev. C **85**, 064608 (2012)
10. F. Fu et al., Nuclear stopping and compression in heavy-ion collisions at intermediate energies. Phys. Lett. B **666**, 359–363 (2008); B. Hong et al, Stopping and radial flow in central $^{58}Ni+^{58}Ni$ collisions between 1 A and 2 A GeV. Phys. Rev. C **57**, 244–253 (1998)
11. J.Y. Liu et al., Nuclear Stopping as a Probe for In-Medium Nucleon-Nucleon Cross Sections in Intermediate Energy Heavy Ion Collisions. Phys. Rev. Lett. **86**, 975–978 (2001); ibid. Isospin effect on the process of multifragmentation and dissipation at intermediate energy heavy ion collisions. Phys. Rev. C **63**, 054612 (2001)
12. A. Sharma, A. Bharti, Isospin effects via Coulomb forces on the onset of multifragmentation in light and heavily charged systems. Eur. Phys. J. A **52**, 42 (2016)
13. T. Maruyama, K. Niita, A. Iwamoto, Extension of quantum molecular dynamics and its application to heavy-ion collisions. Phys. Rev. C **53**, 297–304 (1996)
14. N. Wang, Z. Li, X. Wu, Improved quantum molecular dynamics model and its applications to fusion reaction near barrier. Phys. Rev. C **65**, 064608 (2002)
15. N. Wang, Z. Li, X. Wu, Further development of the improved quantum molecular dynamics model and its application to fusion reactions near the barrier. Phys. Rev. C **69**, 034608 (2004)
16. J.P. Bondorf et al., Statistical multifragmentation of nuclei: (I). Formulation of the model. Nucl. Phys. A **443**, 321–347 (1985)

17. H. Feldmeier, Fermionic molecular dynamics. Nucl. Phys. A **515**, 147–172 (1990); H. Feldmeier, J. Schnack, Fermionic molecular dynamics SCV252SCV133 V2. Prog. Part. Nucl. Phys. **39**, 393–442 (1997)
18. R.K. Puri et al., Temperature-dependent mean field and its effect on heavy-ion reactions. Nucl. Phys. A **575**, 733–765 (1994)
19. A. Bohnet et al., Multifragmentation near the threshold. Phys. Rev. C **44**, 2111–2129 (1991)
20. B.A. Li, S.J. Yennello, Isospin nonequilibrium in heavy-ion collisions at intermediate energies. Phys. Rev. C **52**, 1746–1749(R) (1995)
21. B.A. Li et al., Isospin dependence of collective flow in heavy-ion collisions at intermediate energies. Phys. Rev. Lett. **76**, 4492–4495 (1996)
22. C. Hartnack et al., Modelling the many-body dynamics of heavy ion collisions: Present status and future perspective. Eur. Phys. J. A **1**, 151–169 (1998)
23. A. Sharma, A. Bharti, S. Gautam, R.K. Puri, Multifragmentation of nearly symmetric and asymmetric reactions within a dynamical model. Nucl. Phys. A **945**, 95–111 (2016)
24. H. Kruse, B.V. Jacak, H. Stocker, Steady-state chemical kinetics on fractals: segregation of reactants. Phys. Rev. Lett. **54**, 289–291 (1985); J.J. Molitoris, H. Stocker, Time dependence of electron scattering in a beam-plasma system. Phys. Rev. C **32**, 346–348 (R) (1985)
25. R.K. Puri, S. Kumar, Binary breakup: Onset of multifragmentation and vaporization in Ca-Ca collisions. Phys. Rev. C **57**, 2744–2747 (1998); S. Kumar, R.K. Puri, Medium mass fragment production due to momentum dependent interactions. Phys. Rev. C **78**, 064602 (2008)
26. B. Jakobsson et al., On the observation of a transition from fusion to multifragmentation in high multiplicity ^{16}O induced reactions. Nucl. Phys. A **509**, 195–220 (1990)
27. S.R. Souza et al., A dynamical model for multifragmentation of nuclei. Nucl. Phys. A **571**, 159–184 (1994)
28. J. Singh, S. Kumar, R.K. Puri, Momentum dependent interactions and the asymmetry of the reaction: multifragmentation as an example. Phys. Rev. C **63**, 054603 (2001)
29. R. Donangelo, S.R. Souza, Radial flow and multifragmentation within the framework of molecular dynamics approaches. Phys. Rev. C **52**, 326–330 (1995)
30. G.F. Peaslee, et al., Energy dependence of multifragmentation in $^{84}Kr +^{197} Au$ collisions Phys. Rev. C **49**, 2271–2275 (R) (1994)
31. S.K. Chragi, S. Gupta, K. Coulomb-modified Glauber, model description of heavy-ion reaction cross sections, Phys. Rev. C **41**, 1610–1618; Alkahzov, G. et al., (1977) Elastic and inelastic scattering of 1.37 GeV α-particles from $^{40,42,44,48}Ca$. Nucl. Phys. A **280**, 365–376 (1990)
32. J.Y. Liu et al., Isospin effect of Coulomb interaction on the dissipation and fragmentation in intermediate energy heavy ion collisions. Phys. Rev. C **70**, 034610 (2004)
33. H. Zhou, Z. Li, Y. Zhuo, Momentum dependent Vlasov-Uehling-Uhlenbeck calculation of mass dependence of the flow disappearance in heavy-ion collisions. Phys. Rev. C **50**, R2664–2667 (1994)
34. D.R. Bowman et al., Intermediate mass fragment emission as a probe of nuclear dynamics. Phys. Rev. C **46**, 1834 (1992)
35. T. Li et al., Mass dependence of critical behavior in nucleus-nucleus collisions. Phys. Rev. C **49**, 1630–1634 (1994)
36. C. Williams et al., Fragment distributions for highly charged systems. Phys. Rev. C **55**, R2132–2136 (1997)

Chapter 5
Isospin Effects: Nuclear Fragmentation as a Probe

Preeti Bansal, Sakshi Gautam, and Rajeev K. Puri

Abstract With the advancement in radioactive ion beam facilities, the interest of nuclear physics community is shifted toward investigating the role of isopin dependence of equation of state in reaction dynamics of heavy-ion collisions. Here, we probe isospin effects via nuclear multifragmentation. Using the isospin-dependent quantum molecular dynamics (IQMD) model, we present the study of reactions of isotopic and isobaric pairs. Here, we studied the $E_{c.m.}^{max}$ (the energy corresponding to maximal production of fragments) for various fragments produced in multifragmentation. It has been observed that the peak energy is governed by symmetry potential in comparison to isospin dependence of NN cross section. We further continued the study by dividing the final fragmentation pattern into free particles (A = 1) and fragments (A > 1) by labeling them as gas and liquid, respectively. We studied the energy dependence of gas/liquid content for different reactions. Our investigation revealed an existence of *cross-over energy* (the energy corresponding to equal contribution of gas and liquid contents) which is greatly influenced by symmetry energy rather than isospin dependence of nn cross section. Further, our study revealed that *cross-over energy* is sensitive to density dependence of nuclear symmetry energy as well, thus can be used as a good probe for the same at supra-saturation densities.

5.1 Introduction

The heavy-ion collisions bring about various phenomena at different beam energies going from fusion–fission at lower energies to complete disassembly of colliding matter at high energies. At very low energies, compound nucleus is formed with the emission of light particles while quarks and gluons are formed at very high incident energies. At intermediate energies, hot and compressed nuclear matter with

---

P. Bansal (✉) · S. Gautam · R. K. Puri
Department of Physics, Panjab University, Chandigarh 160014, India
e-mail: preetibansal1111@gmail.com

P. Bansal
Department of Physics, Chandigarh University, Mohali 140413, Punjab, India

R. K. Puri et al. (eds.), *Advances in Nuclear Physics*, Springer Proceedings in Physics 257, https://doi.org/10.1007/978-981-15-9062-7_5

densities of 2–3 times the normal nuclear matter density can be formed in a controlled laboratory environment. This compressed phase of nuclear matter is followed by the expansion in which compressed matter cools down and density drops much lower than the normal nuclear matter density. At intermediate energies, the reactions produce fragments of all sizes which include free nucleons (FNs, $A_{frag} = 1$), light clusters (LCs, $2 \leq A_{frag} \leq 4$), medium mass fragments (MMFs, $5 \leq A_{frag} \leq 9$), intermediate mass fragments (IMFs, $5 \leq A_{frag} \leq 30\%A$), and heavy mass fragments (HMFs, $15\%A \leq A_{frag} \leq 30\%A$), where A is the mass of either projectile or target and A_{frag} is the mass of produced fragment. This phenomenon is called as multifragmentation. From the experimental point of view, multifragmentation is viewed as a promising observable to investigate the liquid–gas phase co-existence of nuclear matter [1]. Various studies showed that the structure of fragmentation is influenced by the incident energy [2, 3], impact parameter [2], mass of the colliding nuclei [4–7] as well as by the asymmetry [8, 9] of reaction pairs. In literature, there are many studies present on the multiplicity of IMFs as a function of incident energy [2, 4–6, 10–19] which concluded that the yield of IMFs first increases with energy attains a maxima and then starts decreasing with further increase in energy. Recently, one of us and collaborators [20] studied the behavior of yield of HMFs and MMFs with incident energy and noticed the rise and fall behavior with energy.

At the same time, due to the existing and upcoming radioactive ion beam (RIB) facilities [21], an increasing interest has been seen in the nuclei away from the line of stability. Hence, the present study is also concerned about the reactions including neutron-rich nuclei and in this direction, isospin effects on the peak energy production of IMFs [10, 22–26] will be accounted for. The isospin effects on the yield of IMFs at 40 AMeV incident energy for isotopic reaction pairs of ^{112}Sn + ^{112}Sn and ^{124}Sn + ^{124}Sn are studied by the MSU group [22]. Dempsey et al. [23] suggested isospin effects in the neck fragmentation by studying the semi-peripheral collisions of ^{124,136}Xe + ^{112,124}Sn at 55 AMeV. Miller et al. [24] studied the isospin dependence of fragment production in the reactions of ^{58}Fe + ^{58}Fe and ^{58}Ni + ^{58}Ni at incident energies ranging between 45 and 105 AMeV. It is seen that more IMFs are emitted as a function of charged particle multiplicity in case of neutron-rich system in comparison to isospin symmetric system at energies between 45 and 75 AMeV, while at 105 AMeV, the distinction between the yields obtained from two systems disappears. Kaur et al. [10] also studied the isospin effects on the multiplicity of IMFs for isotopic reaction pairs ^{40}Ca + ^{40}Ca, ^{44}Ca + ^{44}Ca, ^{52}Ca + ^{52}Ca, and ^{60}Ca + ^{60}Ca and isobaric reaction pairs ^{120}Xe + ^{120}Xe, ^{120}Pd + ^{120}Pd, and ^{120}Zr + ^{120}Zr. At the same time, medium and heavy mass fragments are also produced in multifragmentation and their dynamics maybe quite different in comparison to light clusters owing to their different origin. From the literature survey, one finds that unfortunately, there is no study on the effect of neutron content of colliding pairs on the production of MMFs and HMFs. It would be interesting to see the role of isospin degree of freedom on the production of different mass fragments, for example, HMFs, MMFs and IMFs. Here, we will analyze the effect of isospin asymmetry of reacting partners on the peak energy production of HMFs and MMFs and will try to correlate the effect of addition of neutrons on the energy corresponding to maximum production of HMFs and MMFs.

Further, we will proceed with our study on isospin effects via nuclear fragmentation by dividing the final fragmentation pattern into free particles (A = 1) and fragments (A > 1) by labeling them as gas and liquid, respectively, and will analyze the beam energy dependence of such gas/liquid content for semi-central collisions of different asymmetric reacting partners. The present study is carried out by using Isospin-dependent Quantum Molecular Dynamics (IQMD) model [27, 28] which is discussed in the next section. The isospin effects in our model (IQMD) comes via nuclear mean-field (through Coulomb and symmetry potential) and nn scattering cross section (being isospin-dependent).

5.2 Isospin-Dependent Quantum Molecular Dynamics (IQMD) Model

The Isospin-dependent Quantum Molecular Dynamics (IQMD) model is the advanced version of QMD model introduced by Hartnack et al. [29, 30]. This model deals with distinctive charge states of nucleons, deltas, and pions explicitly. The isospin degree of freedom is incorporated through Coulomb potential, nuclear symmetry potential, and nucleon–nucleon scattering cross section. Here, every hadron propagates under the Hamilton equations of motion:

$$\dot{\mathbf{p}}_i = -\frac{\partial \langle H \rangle}{\partial \mathbf{r}_i}; \qquad \dot{\mathbf{r}}_i = \frac{\partial \langle H \rangle}{\partial \mathbf{p}_i}, \tag{5.1}$$

where $\mathbf{r}_i(t)$ and $\mathbf{p}_i(t)$ represent the centroids of the Gaussian wave packet in coordinate and momentum-space, respectively, of ith nucleon at time t. The average value of the total Hamiltonian reads as follows:

$$\begin{aligned} \langle H \rangle &= \langle T \rangle + \langle V \rangle \\ &= \sum_i \frac{\mathbf{p}_i^2}{2m_i} + V^{Sky} + V^{Yuk} + V^{Coul} + V^{MDI} + V^{Sym}. \end{aligned} \tag{5.2}$$

Here, V^{Sky}, V^{Yuk}, V^{Coul}, V^{MDI}, and V^{Sym} represent the local (two- and three-body) Skyrme, Yukawa, Coulomb, momentum-dependent, and symmetry potentials, respectively.

Further, there is likewise a need of secondary model to clusterize the nucleons into fragments. Here, we use Minimum Spanning Tree (MST) method [28, 31–36] to clusterize the nucleons. In this method, a cluster is characterized based on the correlations in the coordinate space, or in other words, two nucleons are allowed to be a part of the same cluster if their centroids are closer than a specific distance R_{clus} called clusterization distance.

$$|\mathbf{r}_i - \mathbf{r}_j| \leq R_{clus}, , \quad (5.3)$$

where r_i and r_j are the spatial positions of the two nucleons. Here, it is worth mentioning that the distinctive values of R_{clus} do not affect the structure of cluster at the freeze-out stage, where all movements and binary NN collisions cease to exist [33]. The value of R_{clus} can vary between 2 and 4 fm. In the present work, the clusterization distance is considered to be 2.8 fm.

5.3 Results and Discussion

To study the role of isospin effects and to see how the maximal production alters with neutron content of the colliding pairs, we simulated several thousand events for the semi-central collisions of isotopic pairs of $^{40}Ca + ^{40}Ca$ (N/Z = 1.0), $^{48}Ca + ^{48}Ca$ (N/Z = 1.4), and $^{60}Ca + ^{60}Ca$ (N/Z = 2.0) and isobaric pairs of $^{48}Cr + ^{48}Cr$ (N/Z = 1.0), $^{48}Ca + ^{48}Ca$ (N/Z = 1.4), and $^{48}S + ^{48}S$ (N/Z = 2.0) at different incident energies between 30 and 150 AMeV using soft equation of state.

In Fig. 5.1, we display the center-of-mass energy dependence of production of heavy mass fragments (HMFs) for the isotopic and isobaric reactions. The circles correspond to our theoretical calculations and lines signify quadratic fits to the theoretical points. We observed a rise and fall behavior in the production of HMFs with incident energy. This rise and fall behavior can be understood in terms of compressional energy. At low energies, the HMF emission is very minute. This may be because the system does not have enough energy to break a colliding pair into a large number of HMFs since the larger fraction of the initial energy is carried away by the emission of pre-equilibrium nucleons. With the increment in the beam energy, more energy is available to break the colliding nuclei into a large number of HMFs. With further increment in the energy, much more energy will be accessible causing breakage of HMFs into light clusters and free nucleons. For isotopic reaction pairs, it is noticed that when we go from $^{40}Ca + ^{40}Ca$ to $^{48}Ca + ^{48}Ca$, the energy corresponding to peak production of HMFs increases. This is because with an increase in system mass, more energy is required to break the colliding matter into HMFs while moving toward $^{60}Ca + ^{60}Ca$, the energy of peak production of HMFs decreases which is contrary to system mass effects. The factor responsible for such behavior can be repulsive symmetry potential which pushes the system to boil at lower energy value which will be discussed in detail in Fig. 5.4. In isobaric reaction pairs, we noticed that the energy of peak production decreases on moving from $^{48}Cr + ^{48}Cr$ to $^{48}S + ^{48}S$ and the peak production is not varying much with an increase in neutron content. As each colliding pair has same mass, therefore, the decrease in peak energy can be because of symmetry potential or Coulomb potential. Going from $^{48}Cr + ^{48}Cr$ to $^{48}S + ^{48}S$, the repulsive symmetry potential increases and Coulomb potential decreases. Since the decrease in peak energy production signifies the dominance of symmetry potential over Coulomb potential which will be discussed in detail later. We additionally see that the maximum yield of HMFs is almost two for all cases which indicates

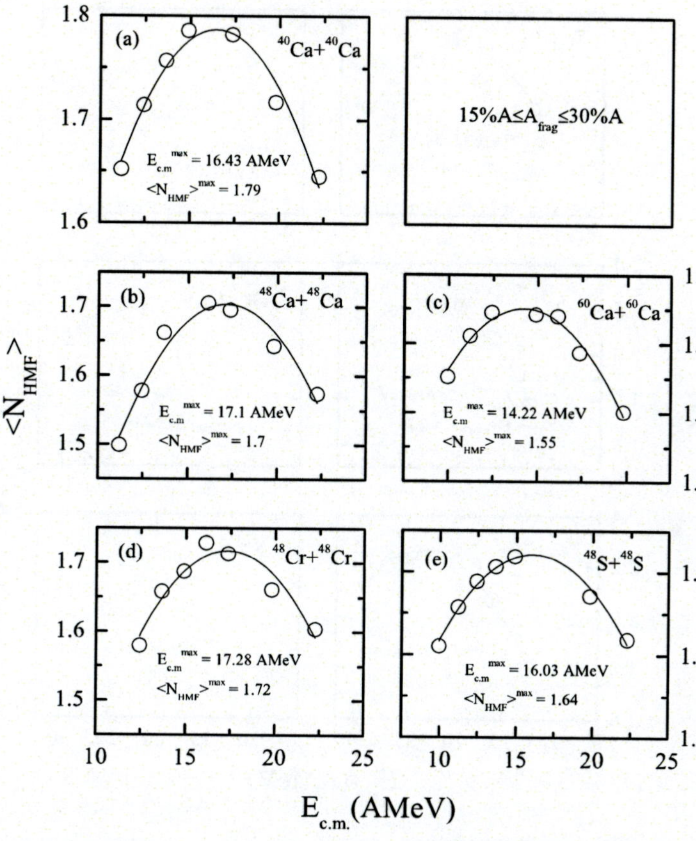

Fig. 5.1 The center-of-mass energy dependence of the multiplicity of heavy mass fragments for isotopic pairs of Ca + Ca reactions and isobaric pairs with mass 96 units

that the HMFs are originated from each remnant of projectile and target, accordingly showing their origin from spectator region.

In Fig. 5.2, we show the production of medium mass fragments (MMFs) as a function of center-of-mass energy. It is clear from the figure that peak energy increases with increment in neutron content. This behavior is in agreement with the system mass effects as with an increase in system mass, more energy is required to break the system into fragments. Also, we notice that the maximal yield of MMFs increases with an increase in system mass, thus showing their origin from the mid-rapidity or participant region. In isobaric reactions, the MMFs show similar behavior as in case of HMFs. Here also, the system starts boiling at lower energies with an increase in neutron content due to the repulsive nature of symmetry potential. However, the yield is nearly constant because of the constant system mass.

In Fig. 5.3, we display the dependence of yield of IMFs on center-of-mass energy. In case of isotopic reaction pairs, the energy corresponding to the maximal generation of IMFs first increments and afterward starts diminishing with increment in neutron

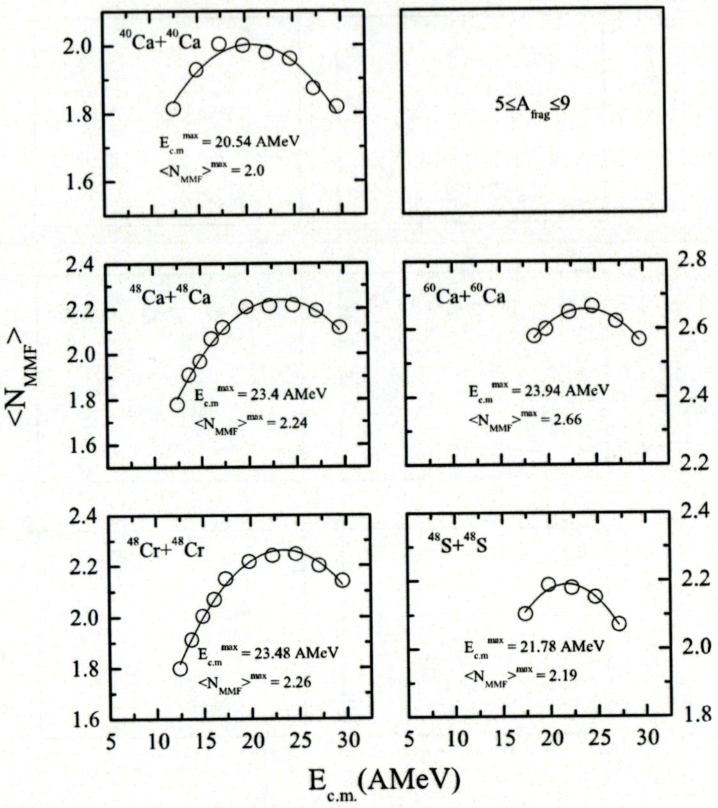

Fig. 5.2 Same as Fig. 5.1, but for MMFs

content. This behavior is similar to the case of HMFs and furthermore detailed in earlier studies [10]. The yield of MMFs at peak production increments with the increment in mass of the system. While in isobaric reactions, we found that the peak energy corresponding to maximum production of IMFs diminishes with increase in N/Z ratio, while maximal yield of IMFs does not change much with the increment in neutron content indicating same behavior as revealed in earlier studies [10].

The factor which could be responsible for such behavior of peak energy production ($E^{max}_{c.m.}$) and peak production ($< N >^{max}$) are Coulomb potential, symmetry potential or isospin-dependent nucleon–nucleon cross section. As discussed before, had Coulomb potential been administering the dynamics, then $E^{max}_{c.m.}$ would have enhanced when we go to higher neutron content in isobaric reaction pairs as a result of decreased Coulomb potential, for example, $E^{max}_{c.m.}$ for ^{48}S + ^{48}S ought to be higher than ^{48}Cr + ^{48}Cr as weaker Coulomb potential in the former would heat up the system at higher energies which is contrary to what we have seen in isobaric pairs. On the other hand in isotopic series, because of the same number of protons, the Coulomb potential does not influence the dynamics. In this way, we will examine the role

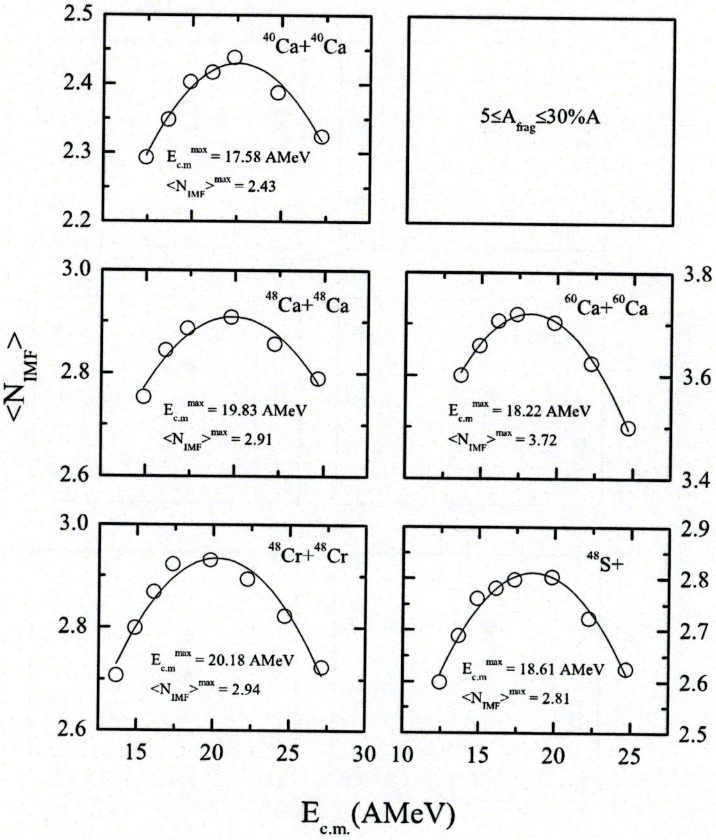

Fig. 5.3 Same as Fig. 5.1, but for IMFs

of isospin impacts by means of symmetry potential and isospin-dependent nn cross section only.

To examine their relative role, we performed the calculations by considering isospin-independent cross section (marked as $\sigma_{non-iso}$), i.e., cross section for neutron–proton collisions is same as proton–proton or neutron–neutron collisions ($\sigma_{np} = \sigma_{nn} = \sigma_{pp}$) and by switching off the symmetry potential (marked as without E_{sym}) and the results are shown by half-filled circles and solid circles, respectively, in Figs. 5.4 and 5.5. In Fig. 5.4, we display the peak center-of-mass energy of HMFs (upper panel), MMFs (middle panel), and IMFs (lower panel) as a function of N/Z ratio for isotopic (left panels) and isobaric (right panels) colliding pairs. From the left upper panel, we observe that the peak energy production of HMFs increases from ^{40}Ca + ^{40}Ca to ^{48}Ca + ^{48}Ca and afterward decreases when we go from ^{48}Ca + ^{48}Ca to ^{60}Ca + ^{60}Ca, while one clearly observes continuous increment in the peak energy production of MMFs (left middle panel) because of the increase in system mass.

Further, we proceed with our study on isospin effects via nuclear fragmentation by dividing the final fragmentation pattern into free particles and fragments. In lit-

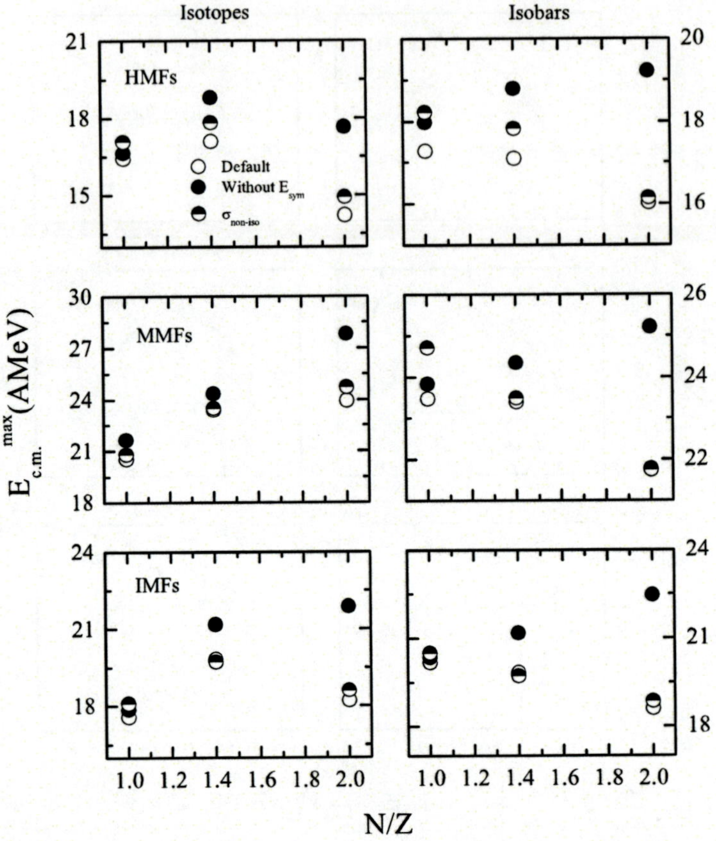

Fig. 5.4 The peak center-of-mass energy of HMFs (upper panel), MMFs (middle panel), and IMFs (lower panel) as a function of N/Z ratio for isotopic and isobaric colliding pairs

erature, there are numerous studies [37–40] available on the isospin fractionation (where unequal partitioning of neutrons and protons of isospin asymmetric system between low- and high-density phases is considered) which can be utilized as an observable to probe the symmetry energy. The studies on isospin fractionation are concerned with isospin content of gas and liquid phases and these studies are limited in incident energy range. Here, we will divide the system into gas (low-density) and liquid (high-density) phases over broad energy range for asymmetric reactions. We will also notice a transition between liquid and gas phases when gas/liquid yield is plotted as a function of incident beam energy. The energy at which both phases have equal contribution is called as *cross-over energy* and is seen as highly sensitive to nuclear symmetry energy and therefore, one can utilize this as an observable to probe the nuclear symmetry energy. In literature, numerous ways are accounted for to characterize the liquid and gas phases. For example, Liu et al. [39, 40] have characterized the gas phase comprising of free particles produced during simulation and all the fragments with charge number within 2 to $(Z_P + Z_T)/2$ at freeze-out (Z_P and

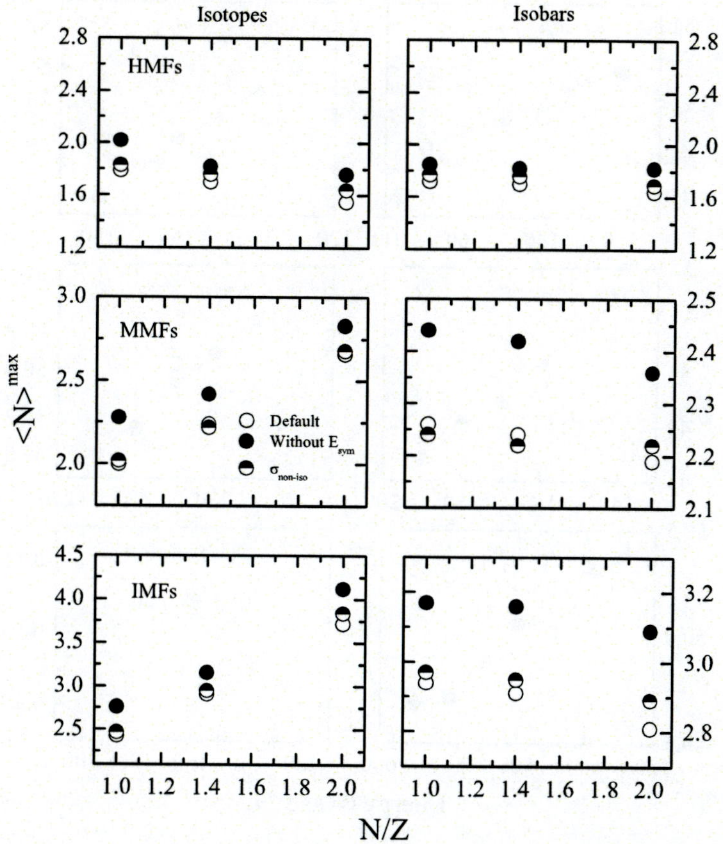

Fig. 5.5 The peak multiplicity of HMFs (upper panel), MMFs (lower panel),and IMFs (middle panel) as a function of N/Z ratio for isotopic and isobaric colliding pairs

Z_T are the charges of projectile and target, respectively), are covered in liquid phase. B. A. Li [41] evaluated the definition of bound (liquid phase) and free (gas phase) as nucleons with local densities greater and less than $\rho_0/8$, respectively, where ρ_0 is the density of nuclear matter at ground state. Investigation of Guo et al. utilizing IBUU04 transport model [42] characterized nucleons with local density less (greater) than $\rho_0/10$ as gas (liquid). The more reasonable definition of gas and liquid phases is forwarded by them for IBUU04 and UrQMD model, where A = 1 was viewed as gas and A > 1 bound nucleons (or pieces) were considered as liquid.

We [43] studied the isospin effects on the *cross-over energy* for isotopic, isobaric, and isotonic reaction pairs and observed the dominance of symmetry potential over isospin dependence of NN scattering cross section. Next, we intend to see the impact of mass asymmetry on the *cross-over energy* and further explore the role of isospin degree of freedom on *cross-over energy* for asymmetric reaction pairs. For this, we simulated thousands of events for the reactions of ^{50}Cr + ^{50}Cr ($\eta = 0$), ^{40}Ca + ^{60}Co ($\eta = 0.2$), ^{31}P + ^{69}Ga ($\eta = 0.4$), ^{20}Ne + ^{80}Kr ($\eta = 0.6$), and ^{11}B + ^{89}Y ($\eta = 0.8$).

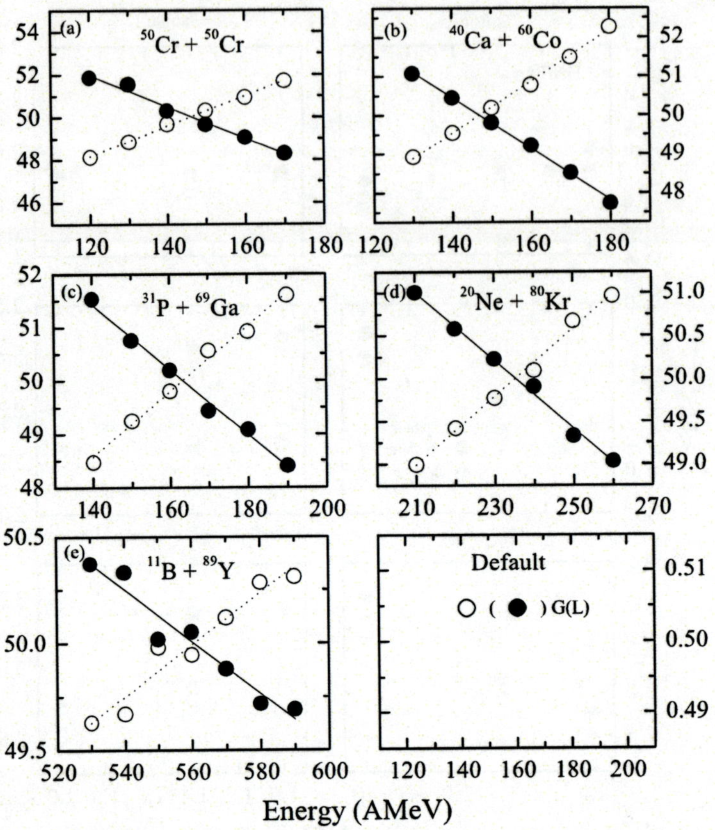

Fig. 5.6 The energy dependence of the gas and liquid content for the reaction pairs having mass asymmetry $\eta = 0, 0.2, 0.4, 0.6$, and 0.8. Various symbols are explained in the text

These reactions are simulated at incident energies between 120 and 600 AMeV. Here, η is varied from 0 to 0.8 keeping total mass of the system fixed ($A_{TOT} = 100$).

In Fig. 5.6, we display the incident energy dependence of gas and liquid contents for $\eta = 0 to 0.8$ covering symmetric and asymmetric reactions. Open (solid) circles correspond to the gas (liquid) content. From the figure, we see that the *cross-over energy* is enhanced with increase in mass asymmetry. Also, we observe that the sharp increment in *cross-over energy* is obtained when we go from $\eta = 0.6$ to 0.8. The reason for this behavior is that for symmetric and nearly symmetric reactions, there is more compression which drives the nuclear matter into the participant zone and thus more free nucleons are emitted even at lower energies. However, for mass asymmetric reaction partners, the number of nucleon–nucleon collisions are reduced because of decreased compressional energy causing a suppressed emission of free nucleons. Therefore, more energy is required to break the system into free nucleons and thus, cross-over is obtained at higher energies.

Next, we will check the role of isospin degree of freedom via symmetry potential and NN scattering cross section on the *cross-over energy*. Here, we simulated the reactions without considering symmetry potential and results are shown in Fig. 7. From the figure, we see that the individual gas content is diminished without symmetry potential. One can clearly see from the figure that the cross-over is acquired at higher energies when we neglect symmetry potential, as expected.

Next, we will explore the role of NN scattering cross section by considering isospin-independent cross section in Fig. 8. Again, the yield of individual gas content is suppressed by considering $\sigma_{non-iso}$ and in this way increment in *cross-over energy* is observed for every asymmetric reaction pair.

Next, we will summarize our results of isospin effects on the *cross-over energy* for mass asymmetric reactions in Fig. 9. Here, solid circles correspond to default calculations. Open circles and half filled circles represent calculations without symmetry potential and isospin independent cross-section, respectively. We found that the *cross-over energy* is increased in absence of symmetry potential as well as by utilizing isospin independent cross-section. Here, we notice that the rise in *cross-over energy* is almost comparable when calculations are performed in absence of symmetry potential or for isospin-independent cross-section, for each colliding pair, independent of its mass asymmetry.

Next, we check the role of density dependence of nuclear symmetry energy on *cross-over energy* for the reactions of ^{31}P + ^{69}Ga ($\eta = 0.4$) and ^{20}Ne + ^{80}Kr ($\eta = 0.6$) by simulating these reactions with soft ($\gamma = 0.5$) and stiff ($\gamma = 1.5$) forms of nuclear symmetry energy. In Fig. 10, we show the energy dependence of gas and liquid yield for the reaction of 31P + 69Ga and 20Ne + 80Kr with soft form of symmetry energy (upper panel) and stiff form of symmetry energy (lower panel). Here, diamonds correspond to the reaction of ^{31}P + ^{69}Ga and triangles represent the reaction of ^{20}Ne + ^{80}Kr. From the figure, we see that the cross-over occurs at lower energy for stiff symmetry energy when contrasted with soft symmetry energy. The occurrence of cross-over at higher energy for soft symmetry energy is because of the fact that the effective strength of symmetry energy is weak in case of soft density dependence at supra-saturation densities (which come into the picture in present collisions) and in this way, weak repulsive forces in case of soft symmetry energy will make the system to break at higher energy. In case of stiff symmetry energy, the system requires a lesser amount of energy to break down because of stronger (repulsive) symmetry potential. It is also clear from the figure that the *cross-over energy* value is higher for the reaction of ^{20}Ne + ^{80}Kr ($\eta = 0.6$) in contrast with ^{31}P + ^{69}Ga ($\eta = 0.4$), as expected. Further, we observed that with increment in mass asymmetry, the sensitivity of density dependence of nuclear symmetry energy on *cross-over energy* is increased. Thus, we conclude that *cross-over energy* is a sensitive probe to study nuclear symmetry energy as well as its density dependence at supra-saturation densities.

5.4 Summary

In summary, we have studied the effect of isospin degree of freedom on the peak energy production and multiplicity of heavy mass fragments (HMFs), medium mass fragments (MMFs), and intermediate mass fragments (IMFs). We observed that the peak energy of HMFs, MMFs, and IMFs decreases with an increase in the neutron content for isobaric pairs, while the different behavior is observed for the isotopic partners. The multiplicity of HMFs remains insensitive to neutron content of colliding pair, while production of MMFs and IMFs increases with neutron content for isotopic pairs. Further, we found that symmetry potential is responsible for the decrease in the peak energy of HMFs, MMFs, and IMFs for isobaric neutron-rich colliding nuclei.

Also, we have studied the yield of gas and liquid content as a function of incident energy in the fragmentation of asymmetric reactions by varying the mass asymmetry from 0 to 0.8. Here, the total system mass was fixed (100 units). Our results indicate that the mass asymmetry influences the *cross-over energy* significantly. The *cross-over energy* is enhanced with mass asymmetry. This is because of the decrease in the participant zone for asymmetric reactions. We again noticed that the symmetry potential has more influence on the *cross-over energy* in comparison to NN scattering cross section. Thus, our study concludes that the *cross-over energy* is a sensitive probe to study the density dependence of nuclear symmetry energy.

Acknowledgements The author (PB) is thankful to Prof. Rajeev K. Puri for giving access to various computer codes developed by him for the present work.

References

1. J. Pochodzalla et al., Probing the nuclear liquid-gas phase transition. Phys. Rev. Lett. **75**, 1040–1044 (1995)
2. M.B. Tsang et al., Onset of nuclear vaporization in ^{197}Au+ ^{197}Au collisions. Phys. Rev. Lett. **71**, 1502–1505 (1993)
3. K.S. Vinayak, S. Kumar, Multifragmentation around the transition energy in intermediate-energy heavy-ion collisions. Phys. Rev. C **83**, 034614 (2011)
4. Y.K. Vermani, R.K. Puri, Mass dependence of the onset of multifragmentation in low energy heavy-ion collisions. J. Phys. G: Nucl. Part. Phys. **36**, 105103 (2009)
5. D. Sisan et al., Intermediate mass fragment emission in heavy-ion collisions: energy and system mass dependence. Phys. Rev. C **63**, 027602 (2001)
6. S. Kaur, A.D. Sood, Model ingredients and peak mass production in heavy-ion collisions. Phys. Rev. C **82**, 054611 (2010)
7. J.K. Dhawan, R.K. Puri, Multifragmentation at the energy of vanishing flow in central heavy-ion collisions. Phys. Rev. C **74**, 054610 (2006)
8. S. Goyal, Multifragmentation at the balance energy of mass-asymmetric colliding nuclei. Phys. Rev. C **84**, 044614 (2011)
9. V. Kaur, S. Kumar, Systematic study of multifragmentation in asymmetric colliding nuclei. Phys. Rev. C **81**, 064610 (2010)
10. S. Kaur, R.K. Puri, Isospin effects on the energy of peak mass production. Phys. Rev. C **87**, 014620 (2013)

11. R.T. de Souza et al., Multifragment emission in the reaction ^{36}Ar + ^{197}Au at E/A = 35, 50, 80, and 110 MeV. Phys. Lett. B **268**, 6–11 (1991)
12. C.A. Ogilvie et al., Rise and Fall of Multifragment Emission. Phys. Rev. Lett. **67**, 1214–1217 (1991)
13. N.T.B. Stone et al., Evidence for the decay of nuclear matter toroidal geometries in nucleus-nucleus collisions. Phys. Rev. Lett. **78**, 2084–2087 (1997)
14. Y.G. Ma, W.Q. Shen, Onset of multifragmentation in intermediate energy light asymmetrical collisions. Phys. Rev. C **51**, 710–715 (1995)
15. G.F. Peaslee et al., Energy dependence of multifragmentation in ^{84}Kr + ^{197}Au collisions. Phys. Rev. C **49**, R2271–R2275 (1994)
16. C. Williams et al., Fragment distributions for highly charged systems. Phys. Rev. C **55**, R2132–R2136 (1997)
17. R. Sun et al., Isotropic emission components in splintering central collisions: (17–115)A MeV ^{40}Ar + Cu, Ag. Au. Phys. Rev. C **61**, 061601(R) (2000)
18. S.R. Souza et al., A dynamical model for multifragmentation of nuclei. Nucl. Phys. A **571**, 159–184 (1994)
19. J.Y. Liu et al., Isospin effect on the process of multifragmentation and dissipation at intermediate energy heavy ion collisions. Phys. Rev. C **63**, 054612 (2001)
20. P. Bansal, S. Gautam, R.K. Puri, On the peak mass production of different fragments in intermediate-energy heavy-ion collisions. Eur. Phys. J. A **51**, 139 (2015)
21. Y. Yano, The RIKEN RI beam factory project: a status report. Nucl. Inst. Methods Phys. Res. B **261**, 1009–1013 (2007)
22. G.J. Kunde et al., Multifragment production in reactions of ^{112}Sn + ^{112}Sn and ^{124}Sn + ^{124}Sn at E/A = 40 MeV. Phys. Rev. Lett. **77**, 2897–2900 (1996)
23. J.F. Dempsey et al., Isospin dependence of intermediate mass fragment production in heavy-ion collisions at E/A = 55 MeV. Phys. Rev. C **54**, 1710–1719 (1996)
24. M.L. Miller et al., Disappearance of the isospin dependence of multifragment production. Phys. Rev. Lett. **82**, 1399–1401 (1999)
25. Z. Chen et al., Isocaling and the symmetry energy in the multifragmentation regime of heavy-ion collisions. Phys. Rev. C **81**, 064613 (2010)
26. C.A. Dorso et al., Isoscaling and the nuclear EoS. J. Phys. G: Nucl. Part. Phys. **38**, 115101 (2011)
27. Ch. Hartnack, R.K. Puri, J. Aichelin, J. Konopka, S.A. Bass, H. Stöcker, W. Greiner, Modelling the many-body dynamics of heavy ion collisions: present status and future perspective. Eur. Phys. J. A **1**, 151–169 (1998)
28. J. Aichelin, Quantum molecular dynamics a dynamical microscopic n-body approach to investigate fragment formation and the nuclear equation of state in heavy ion collisions. Phys. Rep. **202**, 233–360 (1991)
29. C. Hartnack, L. Zhuxia, L. Neise, G. Peilert, A. Rosenhauer, H. Sorge, J. Aichelin, H. Stöcker, W. Greiner, Relativistic heavy-ion colllslons. Nucl. Phys. A **495**, 303c–319c (1989)
30. C. Hartnack, J. Aichelin, H. Stöcker, W. Greiner, Out of plane squeeze of clusters in relativistic heavy ion collisions. Phys. Lett. B **336**, 131–135 (1994)
31. A. Sharma, A. Bharti, S. Gautam, R.K. Puri, Multifragmentation of nearly symmetric and asymmetric reactions within a dynamical model. Nucl. Phys. A **945**, 95–111 (2016)
32. G. Peilert et al., Multifragmentation, fragment flow, and the nuclear equation of state. Phys. Rev. C **39**, 1402–1419 (1989)
33. J. Singh, S. Kumar, R.K. Puri, Model ingredients and multifragmentation in symmetric and asymmetric heavy ion collisions. Phys. Rev. C **62**, 044617 (2000); J. Singh, R.K. Puri, Mass dependence in the production of light fragments in heavy-ion collisions. Phys. Rev. C **65**, 024602
34. R.K. Puri, S. Kumar, Binary breakup: Onset of multifragmentation and vaporization in Ca-Ca collisions. Phys. Rev. C **57**, 2744–2747 (1998)
35. S. Kumar, S. Kumar, R.K. Puri, Medium mass fragment production due to momentum dependent interactions. Phys. Rev. C **78**, 064602 (2008)

36. Y. Zhang, Z. Li, S. Zhou, M.B. Tsang, Effect of isospin-dependent cluster recognition on the observables in heavy ion collisions. Phys. Rev. C **85**, 051602(R) (2012)
37. E. Martin, R. Laforest, E. Ramakrishnan, D.J. Rowland, A. Ruangma, E.M. Winchester, S.J. Yennello, Transition in isospin behavior between light and heavy fragments emitted from excited nuclear systems. Phys. Rev. C **62**, 027601 (2000)
38. D.V. Shetty, S.J. Yennello, E. Martin, A. Keksis, G.A. Souliotis, Energy dependence of the isotopic composition in nuclear multifragmentation. Phys. Rev. C **68**, 021602(R) (2003)
39. J.Y. Liu et al., Isospin effect of Coulomb interaction on the dissipation and fragmentation in intermediate energy heavy ion collisions. Phys. Rev. C **70**, 034610 (2004)
40. J.Y. Liu et al., Isospin fractionation in the nucleon emissions and fragment emissions in the intermediate energy heavy ion collisions. Nucl. Phys. A **726**, 123–133 (2003)
41. B.A. Li, Neutron-proton differential flow as a probe of isospin-dependence of the nuclear equation of state. Phys. Rev. Lett. **85**, 4221–4224 (2000)
42. W.M. Guo et al., Normal or abnormal isospin-fractionation as a qualitative probe of nuclear symmetry energy at supradensities. Phys. Lett. B **738**, 397–400 (2014)
43. P. Bansal, S. Gautam, R.K. Puri, Isospin effects in nuclear fragmentation of isotopic, isobaric, and isotonic reactions. Phys. Rev. C **98**, 024604 (2018)

Chapter 6
On the Fragment Production and Phase Transition Using QMD + SACA Model

S. Sood, Rohit Kumar, Arun Sharma, and Rajeev K. Puri

Abstract In the present study, we have shown the role of different clusterization algorithms on the signals of liquid–gas phase transition in the multifragmentation for the central reactions of ^{40}Ar + ^{45}Sc. We have used the quantum molecular dynamics (QMD) model to generate the phase space of the nucleons and clusterization algorithms based on spatial constraints and its variants, and the energy-based clusterization algorithm. We also present the correlations among fragments within the events via constructing correlation function. We find that the energy-based clusterization algorithm, i.e., simulated annealing clusterization algorithm (SACA) is the most successful among all the available clusterization algorithms. We also find that the event-by-event analysis unfolds and helps to understand reaction picture much better than the quantities constructed by averaging over events.

6.1 Introduction

When the nucleus is excited to more than its binding energy, it decays into number of small chunks known as fragments and the phenomenon is known as multifragmentation. Over the last few decades, an extensive experimental and theoretical efforts have helped us to understand that the fragment number, fragment size (charge), and other fragment properties depend crucially on various entrance channels such as incident energy of projectile, mass of colliding nuclei, impact parameter of a reaction, and isospin content of the colliding nuclei. [1–3].

The nucleon–nucleon interaction is constituted by two parts: long range attractive part and short-range repulsive part. Differing by five orders of magnitude, the

S. Sood · R. Kumar (✉) · R. K. Puri
Department of Physics, Panjab University, Chandigarh 160014, India
e-mail: rohitksharma.pu@gmail.com

R. K. Puri
e-mail: drrkpuri@gmai.com

A. Sharma
Department of Physics, GDC Billawar, Jammu 185204, India
e-mail: arungaur.ju@gmail.com

R. K. Puri et al. (eds.), *Advances in Nuclear Physics*, Springer Proceedings in Physics 257, https://doi.org/10.1007/978-981-15-9062-7_6

nuclear interactions resemble the Van der Waals forces in molecules. This leads to the prediction of liquid–gas phase transition in nuclear matter similar to the one we observe in classical fluids [4]. The phase transition in nuclear matter can be investigated using two different approaches: (i) kinetic methods that predict the phase transition based on breaking of nuclei at subnormal density, and (ii) studying the decay mechanism of the nuclei as a function of excitation energy. The former one is purely theoretical concept, whereas later has an advantage and can be studied both in theory and experiments. We will be sticking to the later method to understand the liquid–gas phase transition in nuclear matter [5–8].

In heavy-ion reactions, it was observed that the fragment charge distribution exhibits power-law fit ($\propto Z^{-\tau}$) near critical point. This behavior was in accordance with earlier predictions by Fisher's droplet model. The interest in this prediction further increased when EoS collaboration observed the power-law behavior in the nuclear fragmentation experiments of Au projectile on C target at 1 GeV/nucleon [9]. Later, ALADIN collaboration studied the caloric curve of Au+Au reactions at 600 MeV/nucleon and found a plateau of the temperature over a wide excitation range—the behavior in accordance with a liquid–gas system [10]. Later on, other than power-law behavior, quantities such as normalized second moment of fragment charges (S_2), variance of fragment charges (γ_2), charge of the second largest cluster ($< Z_{max2} >$, normalized variance of the charge of the largest cluster, derivative of the largest fragment size, and multiplicity derivative of the fragments were also introduced to search critical behavior in multifragmentation [6–8, 11–14]. These quantities show enhanced fluctuations near the critical point.

Various studies have been conducted using the above-mentioned critical parameters to understand the liquid–gas phase transition in nuclear matter, e.g., Li et al. using Michigan State University 4π Array have performed a study of ^{40}Ar+^{45}Sc reactions in the incident energy of 15–115 MeV/nucleon [5]. They have fitted the charge spectra of IMFs [$3 \leq Z \leq 12$] and plotted the τ values as a function of incident energy of the projectile. They have predicted the critical point to occur at 23.9$\pm$0.7 MeV/nucleon. They have also presented the Percolation model calculations for theoretical explanation. In the other study, Belkacem et al., studied the Au+Au reactions at 35 MeV/nucleon for complete impact parameter range using MULTICS-MINIBALL apparatus to investigate critical behavior [15]. They predicted the critical behavior at peripheral geometry. In other study, Ma et al., investigated the reactions of ^{40}Ar+^{27}Al, ^{48}Ti, and ^{58}Ni at an incident energy of 47 MeV/nucleon using Neutron Ion Multi-detector for Reaction Oriented Dynamics (NIMROD) experimental setup [12]. They have analyzed all the critical exponents simultaneously to predict the critical behavior. Very recently, Lin et al., have investigated the critical behavior within the statistical multifragmentation model (SMM) [16]. In another study, Liu et al., studied the total multiplicity derivative, second normalized moment, intermediate mass fragments (IMFs) multiplicity, power-law exponents, Zipf's law, etc. [17]. They found a strong correlation of critical signals with the source size. In our previous attempt, we have studied the central reactions of ^{40}Ar+^{45}Sc using the isospin-dependent quantum molecular dynamics (IQMD) model and its isospin independent version (QMD) coupled with spatial clusterization algorithm to obtain fragments [7]. We obtain signal of phase transition within QMD model and obtained no critical signal in IQMD calculations. The Coulomb forces were found to be the main reason for eliminating the critical signal [8]. Earlier Ma et al., have obtained critical behavior within the QMD model for the reaction of ^{40}Ar+^{27}Al [14]. So we see that the studies with both the statistical models and dynamical do show signatures of phase-transition. Our present study is using the dynamical model QMD.

Further, in some experiments, the higher order correlations among fragments were also studied to better understand the physics near the critical point. For example, Borderie et al., studied the reactions of ^{129}Xe+^{119}Sn at an incident energy of 32 MeV/nucleon using the 4π-multidetector at INDRA [18]. They observed an increase in the multiplicity of equal-sized fragments as a signal of first-order phase transition. These results evident the spinodal decomposition scenario for multifragmentation. The studies were also extended toward N/Z dependence [19].

From the above discussion, it is clear that many facets of the liquid–gas phase transition are already explored but still a clear picture is missing. In the present work, we will present calculations using QMD model coupled with various definitions of fragments for the reactions of ^{40}Ar+^{45}Sc in the incident energy range of 10–115 MeV/nucleon. In particular, we will use energy- based clusterization algorithm, i.e., simulated annealing clusterization algorithm (SACA) [20]. In order to better understand this method's utility, we will also present the calculations with the clusterization algorithms that construct the fragments via using the local correlations among nucleons in coordinate and/or momentum space or based on binding energy conditions to find stable fragments. The liquid–gas phase transition will be predicted using various critical parameters. Also, we present the correlations among the fragments within events and on event-by-event basis. This will be done to look for the behavior change for event-by-event correlations near the critical energies.

A brief detail of the n-body model and the various clusterization is given in Sect. 6.2. In Sect. 6.3, we will present our results and discussion. In Sect. 6.4, we will give summary of our study.

6.2 Methodology

6.2.1 Quantum Molecular Dynamics Model

The quantum molecular dynamics (QMD) model is a dynamical model that uses n-body theory to simulate the reactions on an event-by-event basis. In this model, the reactions are studied via following each individual nucleon where each individual nucleon is represented by a Gaussian wave packet of constant width in coordinate and momentum space [21]. To propagate the nucleons, the classical Hamilton's equations of motion are used. These equations read as

$$\dot{\mathbf{r}}_i = \frac{\partial H}{\partial \mathbf{p}_i};\ \dot{\mathbf{p}}_i = -\frac{\partial H}{\partial \mathbf{r}_i}, \tag{6.1}$$

where H represents the average Hamiltonian and consists of kinetic energy and potential terms. During the propagation, the nucleons interact with each other via two- and three-body mean field interactions and collisions, therefore, move at curved trajectories. Within this model, large fluctuations are allowed for nucleons in coordinate and momentum space necessarily for realistic description for cluster formation.

Over the last three decades using this model, various experimental observations over a wide range of entrance channel have been explained. In particular, this model is applicable for incident energy of projectile starting from approximately 10 MeV/nucleon–2 GeV/nucleon. One can find the detailed formulation of this model in [21].

6.2.2 Fragment Recognition

It is well known now that as soon as the nucleons come out of the condensed phase via adiabatic expansion and ultimately reaches a stage of 'freeze-out' the clusterization algorithms are evoked to obtain the fragments. Here, we use two different types of clusterization algorithms, i.e., minimum spanning tree (MST), minimum spanning tree with momentum cut (MSTP), minimum spanning tree with binding energy check (MSTB), minimum spanning tree with thermal binding cut (MSTBT), and simulated annealing clusterization algorithm (SACA). The former ones use the local correlations among the nucleons in spatial and/or momentum space, whereas the latter one uses correlations at global level. Let us understand few details of these clusterization methods.

- Fragments using local correlations: In this category, local correlations among the nucleons are checked to sort fragments. If r_i (p_i) and r_j (p_j) are the centroids of the ith and jth nucleon in coordinate (momentum) space, respectively, then the following conditions are implemented

$$|r_i - r_j| \leq 4fm, \tag{6.2}$$
$$|p_i - p_j| \leq P_{Fermi}. \tag{6.3}$$

If only first condition is evoked, the method is dubbed as minimum spanning tree (MST) [21]. If nucleons fulfill this condition, they are said to be part of the same cluster C_s, whereas if both the conditions are implemented simultaneously, the method is dubbed as minimum spanning tree with momentum cut (MSTP) [22]. We choose P_{Fermi} equal to average momentum of the nucleons within the nucleus, i.e., 150 MeV/c. Due to additional cut, the later method identifies the fragments much faster compared to former method.

In the other method within this category, the clusters are checked for their binding energies. First, pre-clusters are sorted using the (6.2). Then the binding energy of cluster C_s is calculated as

$$E_{int}^{C_s} = \sum_{j \epsilon C_s} K_j^{cm} + \sum_{j,k \epsilon C_s; j \le k} V_{j,k}. \tag{6.4}$$

Here, K_j^{cm} is the kinetic energy of the fragment in its center-of-mass and $V_{j,k}$ is its potential energy. This binding energy is compared with the binding energies calculated for mass equal to fragment size with the bindings obtained using the liquid-drop formula. If cold bindings are used, the method is minimum spanning tree with binding energy cut (MSTB) [23–25] and if the temperature-dependent binding energies are used, then method is minimum spanning tree with thermal binding cut (MSTBT) [26, 27]. These methods eliminate the fragments which are loosely bound or over excited. We have shown in [26] that MSTBT method should be followed to filter stable fragments and with MSTB method, otherwise, we will get spurious fragments.

- Fragments using global correlations: In this method, the correlations among nucleons in coordinate and momentum space are considered on global level and the fragments are constructed using simulated annealing technique coupled with the metropolis procedure. This method is dubbed as simulated annealing clusterization algorithm (SACA) [20]. This method of keeping the energy of the clusters at center point obtains the most stable fragment configuration. Within the SACA method, the total binding energy of the clusters $E_{\{C_s\}}$ for cluster set $\{C_s\}$ is calculated at each step:

$$E_{\{C_s\}} = \sum_i E_{int}^{C_s}. \tag{6.5}$$

Here, $E_{int}^{C_s}$ is calculated using (6.4). The clusters are checked for their stability to fasten the method, i.e., C_s fragment is stable if its binding is ≤ -4 MeV for cluster size $\ge$4 and 0.0 otherwise.
After the first configuration is obtained, the clusters are allowed to emit or absorb nucleons such that the total sum of the binding energy of the fragments increases. After millions of iterations that cluster configuration is accepted which is most stable. The obtained cluster configuration is well correlated in coordinate and momentum space via minimization of both kinetic and potential terms. It is also found that fragments thus obtained are well in their ground state. Implementing additional binding energy check has insignificant effect on final results.

The MST method or its variants have simple structure which leads its wide acceptability and utilization, but, for certain entrance channels these methods fail to explain observations such as for asymmetric and peripheral reactions [21, 26–28]. Whereas the SACA method has been found exceptionally consistent in reproducing experimental data for wide entrance channels [20, 28, 29]. On one hand, this method explains experimental observations such as multiplicity of fragments and size of the largest cluster, on the other hand it is also able to explain physics of event-by-event-based observables such as multiplicity probability and probability distribution of the first three largest clusters. The bottleneck of this algorithm is its capacity to reduce the freeze-out time down to ~60 fm/c. The gain by a factor of six in time compared to MST method and its variants also allow one to understand the fragments when they are well within high-density region. The only reason for lesser utility of the SACA method is its running time. We will be utilizing the MST and its variants, MSTP, MSTBT (local correlations), and SACA (global correlations) methods to understand liquid–gas phase transition and fragment-fragment correlations at and near the critical energy.

6.3 Results and Discussion

For the present study, we have simulated the reactions of ^{40}Ar+^{45}Sc in the incident energy range of 10–115 MeV/nucleon for central geometries using soft equation of state. The energy-dependent nucleon-nucleon cross-section is used in the present work. Throughout the article, we will restrict our discussion at freeze-out time only. For the MST, MSTP, MSTBT, and SACA methods, the freeze-out times are 300 fm/c and 60 fm/c, respectively. Before discussing the results let us understand the mechanism of fragment formation within different clusterization methods.

In a typical heavy-ion collision, the nuclei approach each other and pump the energy into the system with which they are boosted initially. The nuclear matter then compresses and excites. The hot piece of matter then releases its energy via rapid expansion. The matter then shatters into correlated nucleons. The clusterization methods based on local correlations are not applicable untill the nuclear matter is well diluted. The MST method uses spatial information and may form fragments if consecutive nucleons are within defined distances. Therefore, can form fragments even if the nucleons are spread over space, leading to lesser stability. The MSTP method puts additional cuts on nucleons and excludes the fast moving nucleons from fragments. This de-excites the fragments to some extend and saturates its structure early in time. It also helps to separate the overlapping fragments. In MSTBT method, the thermal binding cut is implemented on MST fragments to sort the stable fragments. It removes the fragments which are loosely bound. Whereas, in SACA method, the coordinate and momentum space nucleons are used to obtain stable fragments. As in this method, one minimizes the binding energy of the clusters, therefore, the final cluster configuration is obtained at early times as soon as fragments are formed. All the above-discussed clusterization methods will be used to analyze the signals of phase transition in fragmentation.

6.3.1 Nuclear Liquid–Gas Phase Transition

As discussed in introduction, there are various different methods to predict liquid–gas phase transition in nuclear matter. We use following procedure and signatures to predict the critical point of the liquid–gas phase transition:

- Functions to fit the fragment charges: In this method, the fragment charge spectra of the IMFs [$3 \leq Z \leq 12$] at various incident energies are fitted with power-law function [Y(Z)$\propto Z^{-\tau}$] and exponential function [Y(Z)$\propto e^{-\lambda Z}$]. The obtained values of τ (λ) are plotted as a function of incident energy of projectile and fitted with the fourth-order polynomial. The minima in the values correspond to the critical point of liquid–gas phase transition [5, 7].

- Campi introduced the other powerful methods to characterize the critical behavior in fragmentation [11]. These methods are based on conditional moments of asymptotic cluster charge distribution. In general, the kth moment of charges is calculated on event-by-event basis using the following definition:

$$M_k = \sum_{Z \neq Z_{max}} Z_i^k n_i(Z), \tag{6.6}$$

where $n_i(Z)$ is the multiplicity of the clusters of charge Z in the event except the charge of the largest cluster. Using the moments Campi constructed reduced second moment of charges (S_2):

$$S_2 = \frac{\sum_{Z \neq Z_{max}} Z_i^2 n_i(Z)}{\sum_{Z \neq Z_{max}} n_i(Z)}. \tag{6.7}$$

The exclusion of the largest cluster in determining S_2 is to make its value proportional to the compressibility κ_T that shows singularity at critical point.
- Another quantity proposed by Campi to investigate critical behavior is the relative variance γ_2 defined as

$$\gamma_2 = \frac{M_2 M_0}{M_1^2}. \tag{6.8}$$

- Ma et al., proposed to use mean charge of second largest cluster ($< Z_{max2} >$) to predict the critical point of the liquid–gas phase transition [13].
- In other study, Ma et al., proposed to use the cross-section of different IMF multiplicities and plotted it against incident energy to predict critical point [14].

All the above quantities are expected to show peak around critical point (except 'τ' and 'λ' which show mimima). It has been observed that all the above-discussed quantities do indeed provide a clear signature of co-existence of liquid–gas phase transition (except the last one which is never been investigated in other studies). The

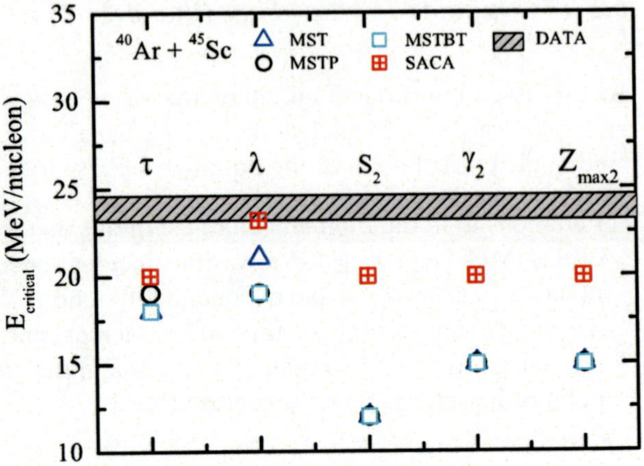

Fig. 6.1 We display the values of various critical exponents, i.e., τ, λ, S_2, γ_2 and $< Z_{max2} >$ for the fragments using clusterization methods MST, MSTP, MSTBT, and SACA

evidence of the criticality is the minima in the value of τ and the maximum in the values of S_2, γ_2 and $< Z_{max2} >$.

In Fig. 6.1, we display the results obtained for the above-mentioned critical parameters using MST, MSTP, MSTB, and SACA methods. We also display the experimental prediction for energy where phase transition is predicted, i.e., 23.9 $\pm$ 0.7 MeV/nucleon using power-law fit (shaded region). This data is taken from [5]. The results of MST, MSTP, MSTBT, and SACA are represented by triangles, circles, squares, and crossed-squares, respectively. From the figure, we see that the MST, MSTP, and MSTBT methods predict the phase transition from power-law fit to be at 18.03, 19.04, and 18.03 MeV/nucleon, respectively [6]. The SACA result from power-law fit is 20.1 MeV/nucleon which is more close to experimental value compared to any other clusterization algorithms. Looking at these results one can say that SACA improved the consistency with experimental data only slightly. But if we compare the experimental value of power-law exponent τ, we get much clearer picture and utility of SACA method. The measured value of τ in experiment is 1.21$\pm$0.01. The MST, MSTP, and MSTBT methods predict the values of τ to be 0.08, 0.16, and 0.09, respectively. These values are far away from the measured value. On the other hand, if we see SACA value, it is 1.39 which is much closer to experimental value. Earlier, Percolation model calculations have predicted the value of $\tau = 1.5 \pm 0.1$ and critical energy at 28$\pm$0.4 MeV/nucleon. So the present calculations with QMD+SACA are the best reproduction of experimental data till now. It is to keep into mind that SACA identifies fragments at 60 fm/c, therefore, one has advantage to study fragments well within/or near compressed/excited nuclear state [30].

Now, if we look at the predictions of critical point using exponential fit parameter λ, S_2, γ_2 and $< Z_{max2} >$ for all the clusterization methods. We get two results:

1. We definitely get signals of phase transition using all critical parameters τ, λ, S_2, γ_2, and Z_{max2} for fragments identified using all clusterization algorithms.
2. The predicted energy do not have significant sensitivity toward any of the clusterization algorithm MST, MSTP, and MSTBT. Whereas, the SACA method

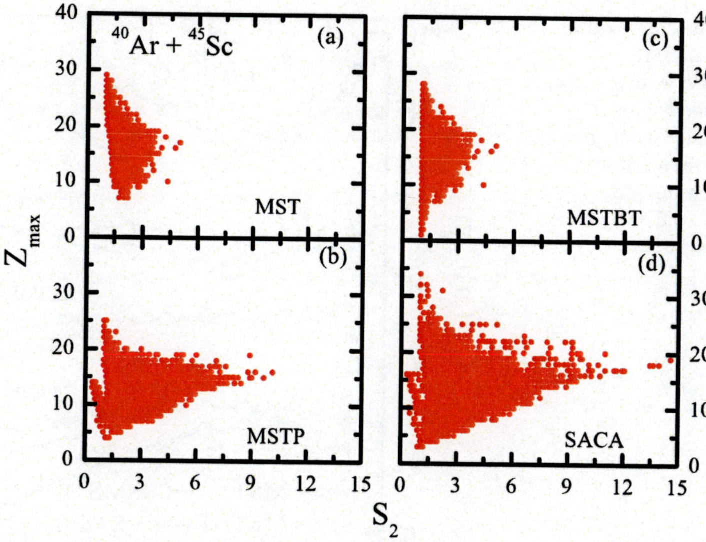

Fig. 6.2 The Campi plots for fragments identified using **a** MST **b** MSTP **c** MSTBT, and **d** SACA methods for the central reactions of ^{40}Ar+^{45}Sc at their respective critical points

improves the predictions if one compares to experimental data. Otherwise, the physics of phase transition, i.e., energy where critical point is expected, is not influenced by clusterization algorithm.

Campi suggested and is now well-known characteristic of the systems that undergoes the continuous phase transition that has the largest fluctuations [11]. These fluctuations can be observed from the event-by-event plots between the size of the largest cluster charge (Z_{max}) and the normalized second moment (S_2). These plots are known as Campi plots and are proved to be very instructive in the previous studies [11].

In Fig. 6.2, we have shown Campi plots for the fragments obtained using the MST, MSTP, MSTBT, and SACA methods. These results are shown at incident energies where critical point is observed within the given clusterization algorithm. We see that the Campi plots differ significantly for the different fragment identification methods. We see least fluctuation in MST method followed by MSTBT method. The MSTP method shows maximum fluctuation among the local correlation category. This shows that although signals of phase transition appear at almost same incident energy for MST, MSTP, and MSTBT methods, the MSTP method shows the strongest signals of phase transition. Again, among all the given algorithms the SACA method best preserves the characteristic signal of liquid–gas phase transition, i.e., the fluctuations are maximum. Thus, SACA is the most suitable for analyzing signals of liquid–gas phase transition [30].

Now, we will use the last-mentioned signal to characterize liquid–gas phase transition. As mentioned earlier in the section, this method was proposed by Ma et al.,

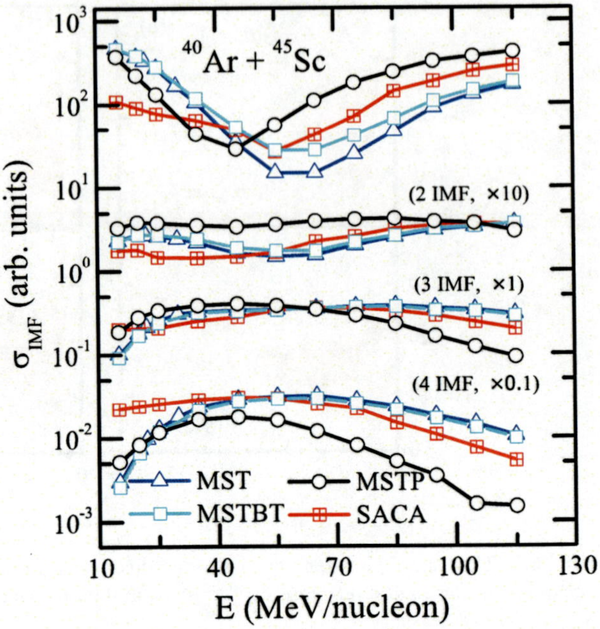

Fig. 6.3 The cross-section of events with IMF multiplicity one, two, three, and four productions at different incident energies for the central reactions of ^{40}Ar+^{45}Sc using MST, MSTP, MSTBT, and SACA fragment identifiers

and is based on the multiplicities of IMFs. They plotted the cross-section of the events having one, two, three, and four multiplicity values as a function of incident energy. They proposed that the cross-section of $\sigma_{IMF} = 1$ ($\sigma_{IMF} = 2$) has minima (maxima) at critical point. This observation was never tested in other studies.

In Fig. 6.3, we displayed the cross-sections of intermediate-mass fragments (IMFs) with one, two, three, and four multiplicity values in the incident energy range between 15 and 115 MeV/nucleon. Symbols are keeping the same meaning as in Fig. 6.1 and lines are drawn just to guide the eyes. We observe the minima for $\sigma_{IMF} = 1$ at 55, 45, 55 MeV/nucleon for MST, MSTP, and MSTBT methods, respectively. The minima are broader 55–65 MeV/nucleon for the case of MST and MSTBT identified fragments, and sharper for the MSTP and SACA identified fragments. But the critical point (as discussed earlier) is far away from these energy values. Therefore, even though we do not exclude the possibility of observation of such minima as a phase transition signal away from critical point, the present study does show that not much can be gained to pin down the exact critical point. On the other hand, the events with $\sigma_{IMF} = 2$ multiplicity do not show sharp maxima or the characteristic signal as was proposed in [14]. If we look at the behavior of higher multiplicity events, we see no peculiar behavior at or near the critical point. We also see differences in the values of cross-sections which is very much obvious and arise due to structure of cluster definitions.

From the above discussion, we find that MSTP method is best among the MST-based clusterization algorithms. Considering this, in the upcoming part of the present study, we will limit to MSTP and SACA methods only. Among these, the MSTP method uses coordinate+moemntum space correlations among nucleons on local level and SACA uses the same information on global level. It is interesting to compare the correlations among the fragments within the events and correlations among fragments on event-by-event basis.

6.3.2 Correlations Among Fragments Within Events

To understand the physics of the MSTP and SACA methods, we look for the radii of the fragments in the coordinate (r) and momentum space (p) from the center-of-mass of the system as a function of their charge, and the relative difference among their

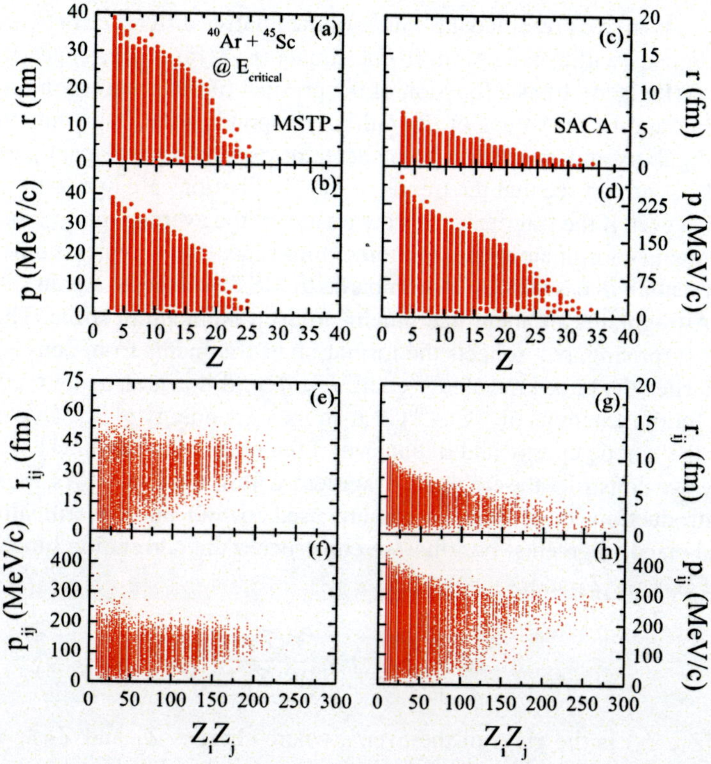

Fig. 6.4 (Top four panels) The radii of the fragments from the center-of-mass of the system in coordinate space (r-fm) and momentum space (p-MeV/c) as a function of the fragment charges and (bottom four panels) relative radii among fragments in coordinate (r_{ij}) and momentum space (p_{ij}) as a function of the product of their charges for the reactions of ^{40}Ca+^{40}Ca (b = 3 fm) at an incident energy of 35 MeV/nucleon. The results of fragments with the MST and SACA methods are displayed in left and right panels, respectively

radii in coordinate (r_{ij}) and momentum space (p_{ij}) as a function of product of the corresponding charges of the fragments. The results are displayed in Fig. 6.4 for the fragments of the reactions of ^{40}Ar+^{45}Sc at the corresponding critical energies with the MSTP and SACA methods. The left and right panels correspond to the results of the MSTP and SACA methods, respectively. One can see from the figure, that the spatial radii (r) of fragments from the center-of-mass of the system is larger for the MSTP method compared to the SACA method, whereas opposite behavior is seen in momentum space. We have maximum value of radius in coordinate (momentum) space near to ~40 fm (~40 MeV/c). On the other hand, the SACA method has radius values in coordinate (momentum) space equal to ~7.5 fm (~250 MeV/c). The difference in the values is due to the structure of the algorithms and their freeze-out times. For both algorithms, we see that with the increase in the size of the fragment, the radii have comparatively lesser values (larger values) in coordinate (momentum) space.

In Fig. 6.4e–h, we presented the values of the relative difference between the radii of the fragments within each event in coordinate (r_{ij}) (Fig. 6.4e, g) and momentum space (p_{ij}) (Fig. 6.4f, h) as a function of the product of their corresponding charges. The smaller and larger values of the radii correspond to the fragments that are the closest neighbors and originated from spectator parts and participant parts, respectively. Also, one can see that the trends of the distributions are horizontal, reflecting the fact that even if the fragment sizes are different, the average relative distances are almost same in coordinate space. In momentum space, the fragments have large relative momentum in case of SACA compared to MSTP methods. Again showing that the SACA fragments are identified much earlier in momentum space. These results may look surprising, but, reflects the formation of fragments from non- equilibrated source. Earlier, the same kind of non-equilibrium condition is also observed by Furuta and Ono for the reactions of ^{40}Ca+^{40}Ca at an incident energy of 35 MeV/nucleon by studying the kinetic energy and radial size of the reaction system [31].

Lastly, we construct the fragment-fragment within the events. As we mentioned in the introduction, these correlations are used to find out the criticality signals, i.e., equal-sized fragments. For this, we constructed the correlation function among fragment charges Z_1 and Z_2 as

$$1 + R(Z_1, Z_2) = \frac{Y(Z_1, Z_2)}{Y'(Z_1, Z_2)}, \tag{6.9}$$

here, $Y(Z_1, Z_2)$ is the yield of the events where charges Z_1 and Z_2 appear within the events and $Y'(Z_1, Z_2)$ is the yield of the correlated and uncorrelated events. The results of correlation function using MSTP and SACA methods are displayed in Fig. 6.5. We see a large difference for the correlation function for the two clusterization algorithms. In the MSTP method, we see lesser correlated events, whereas in the SACA method we see a strong correlations for the lower charge values. We also note that the correlations are more distributed in MSTP fragments compared to SACA fragments.

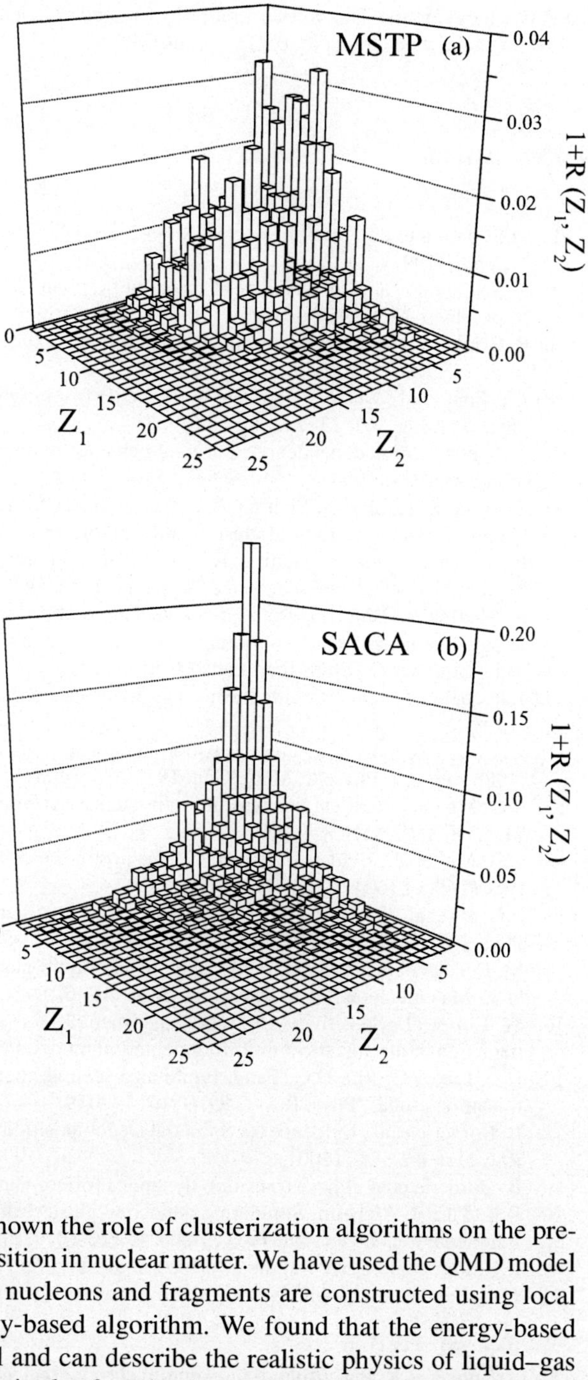

Fig. 6.5 Correlation between the fragment charges within events for the MSTP (top) and SACA (bottom) at the corresponding critical energies

6.4 Summary

In the present study, we have shown the role of clusterization algorithms on the prediction of liquid–gas phase transition in nuclear matter. We have used the QMD model to simulate the phase space of nucleons and fragments are constructed using local correlation methods and energy-based algorithm. We found that the energy-based clusterization is best among all and can describe the realistic physics of liquid–gas phase transition. This algorithm is also found to be consistent with the experimental observations. Later, we showed fragment-fragment correlations within the events and event-by-event correlations.

Acknowledgements This work is funded by Council of Scientific and Industrial Research (CSIR), Govt. of India, vide Grant No. 03(1388)/16/EMR-II.

References

1. M.B. Tsang et al., Onset of nuclear vaporization in ^{197}Au+ ^{197}Au collision. Phys. Rev. Lett. **71**, 1502 (1993)
2. A. Schüttauf et al., Universaility of spectator fragmentation at relativistic bombarding energies. Nucl. Phys. A **607**, 457 (2000)
3. B. Borderie et al., Nuclear multifragmentation and phase transition for hot nuclei. Prog. Part. Nucl. Phys. **51**, 551 (2008)
4. J.E. Finn et al., Nuclear fragment mass yields from high-energy proton-nucleus interactions. Phys. Rev. Lett. **49**, 1321 (1982)
5. T. Li et al., Mass dependence of critical behavior in nucleus-nucleus collisions. Phys. Rev. C **49**, 1630–1634 (1994)
6. S. Sood, R. Kumar, A. Sharma, R.K. Puri, Cluster formation and phase transition in nuclear dissembly using a variety of clusterization algorithms. Phys. Rev. C **99**, 054612 (2019)
7. R. Kumar, S. Sood, A. Sharma, R.K. Puri, On the multifragmentation and phase transition in the perspectives of different n-body dynamical models. Act. Phys. Pol. B **49**, 301 (2018)
8. A. Sharma, A. Bharti, Isospin effects via Coulomb forces on the onset of multifragmentation in light and heavily charged systems. Eur. Phys. J. A **52**, 42 (2016)
9. M.L. Gilkes et al., Phys. Rev. Lett. **73**, 1590 (1994)
10. J. Pochodzalla et al., Probing the nu clear liquid-gas phase transition. Phys. Rev. Lett. **75**, 1040 (1995)
11. X. Campi, Signals of a phase transition in nuclear multifragmentation, Phys. Lett. B **208**, 351 (1988); ibid., J. Phys. A: Math. Gen. **19**, L917 (1986)
12. Y.G. Ma et al., Critical behavior in light nuclear systems: experimental aspects. Phys. Rev. C **71**, 054606 (2005)
13. Y.G. Ma, et al., Evidence of critical behavior in the disassembly of nuclei with A=36, Phys. Rev. C **69**, 031604 (R) (2004)
14. Y.G. Ma et al., Onset of multifragmentation in intermediate energy light asymmetyric collisions. Phys. Rev. C **51**, 710 (1995)
15. M. Belkacem et al., Searching for the nuclear liquid-gas phase transition in Au+Au collisions at 35 MeV/nucleon. Phys. Rev. C **54**, 2435 (1996)
16. W. Lin et al., Sensitivity study of experimental measures for the nuclear liquid-gas phase transition in the statistical multifragmentation model. Phys. Rev. C **97**, 054615 (2018)
17. H.L. Liu, Y.G. Ma, D.Q. Fang, Finite-size scaling phenomenon of nuclear liquid-gas phase transition probes. Phys. Rev. C **99**, 054614 (2019)
18. B. Borderie et al., Evidence for Spinodal Decomposition in Nuclear Multifragmentation. Phys. Rev. Lett. **86**, 3252 (2001)
19. B. Borderie et al., Phase transition dynamics for hot nuclei. Phys. Lett. B **782**, 291 (2018)
20. R.K. Puri, J. Aichelin, Simulating annealing clusterization algorithm for studying the multifragmentation. J. Comput. Phys. **162**, 245 (2000); J. Phys. G: Nucl. Part. Phys. **37**, 015105; R.K. Puri et al., Early fragment formation in heavy-ion collisions. Phys. Rev C **54**, R28
21. J. Aichelin et al., Quantum molecular dynamic: A dynamical microscopic n-body approach to investigate fragment formation and the nuclear equation of state in heavy-ion collisions. Phys. Rep. **202**, 233 (1991)
22. S. Kumar, R.K. Puri, Role of momentum correlations in fragment formation. Phys. Rev. C **58**, 320 (1998)
23. S. Kumar, R.K. Puri, Stability of fragments formed in the simulations of central heavy ion collisions. Phys. Rev. C **58**, 2858 (1998)

24. S. Goyal, R.K. Puri, Formation of fragments in heavy-ion collisions using a modified clusterization method. Phys. Rev. C **83**, 047601 (2011)
25. R. Kumar, S. Gautam, Influence of different liquid-drop based bindings on lighter mass fragments and entropy production. Eur. Phys. J. A **52**, 112 (2016)
26. R. Kumar, S. Gautam, R.K. Puri, Multifragmentation within a clusterization algorithm based on thermal binding energies. Phys. Rev. C **89**, 064608 (2014)
27. R. Kumar, S. Gautam, R.K. Puri, Influence of different binding energies in clusterization approach: fragmentation as an example. J. Phys. G: Nucl. Part. Phys. **43**, 025104 (2016)
28. Y.K. Vermani et al., Microscopic approach to the spectator matter fragmentation from 400 to 1000 MeV/nucleon. Eur. Phys. Lett. **85**, 62001 (2009)
29. R. Kumar, R.K. Puri, Using experimental data to test an n-body dynamical model coupled with an energy-based clusterization algorithm at low incident energies. Phys. Rev. C **97**, 034624 (2018)
30. S. Sood, R.K. Kumar, A. Sharma, R.K. Puri, On the critical behaviorin in light and heavily charges systems. J. Phys. G: Nucl. Part. Phys. (submitted) (2020)
31. T. Furuta, A. Ono, Relevance of equilibrium in multifragmentation. Phys. Rev. C **79**, 14608 (2009)

Chapter 7
Role of Mass Asymmetry on the Energy of Peak Intermediate Mass Production and Its Related Dynamics

Sakshi Sharma, Rohit Kumar, and Rajeev K. Puri

Abstract We study various reactions of different projectiles with fixed targets of ^{124}Sn and ^{197}Au using Isospin-dependent Quantum Molecular Dynamics (IQMD) model. Here, the mass dependence of energy of peak production and peak IMFs multiplicity is investigated for ^{124}Sn and ^{197}Au target reactions. It is observed that the energy of peak production increases on moving toward low mass projectiles for both ^{124}Sn and ^{197}Au target reactions with different slopes. Further, we study the mass dependence of various fragments at the peak energy and power law dependence of these fragments is observed. The study is further continued by investigating the behavior of various other quantities (maximum and average density, collision rate, maximum, and average temperature) at the energy of peak production. Our findings indicate the power law dependence of these quantities at peak energy. Also, the behavior of participant and spectator matter at peak energies for ^{124}Sn and ^{197}Au target reactions is explored.

7.1 Introduction

Heavy-ion collisions (HICs) act as a significant probe to explore nuclear equation of state (EOS) and provide information about the microscopic features of nuclear matter. On the basis of energy, HICs are classified into different energy regimes, i.e., low, intermediate, and high energies. In low energy regime, the nuclear dynamics is mainly governed by mean field, and fusion, fission, cluster radioactivity, formation of super heavy nuclei are studied [1–6]. At high energies, nucleon-nucleon collisions dominate the reaction dynamics and phenomena such as particle production, jet

S. Sharma · R. Kumar (✉) · R. K. Puri
Department of Physics, Panjab University, Chandigarh 160014, India
e-mail: rohitksharma.pu@gmail.com

S. Sharma
e-mail: sharmasakshi.pu@gmail.com

R. K. Puri
e-mail: drrkpuri@gmail.com

R. K. Puri et al. (eds.), *Advances in Nuclear Physics*, Springer Proceedings in Physics 257, https://doi.org/10.1007/978-981-15-9062-7_7

formation, quark gluon plasma, etc. are observed [7–9]. In case of intermediate energy, both mean field and nucleon-nucleon collisions have contribution. At these energies, phenomena such as collective flow, sub-threshold particle production and multifragmentation are the important ones [10–12]. In the present work, we are focussing on various aspects of reaction dynamics at intermediate energies.

The intermediate energy reactions give us an excellent opportunity to understand the behavior of nuclear matter at high densities and temperature. This information is vital to describe the physics of nuclear star formation, supernovae explosion, and understanding the conditions of early stages of universe evolution. We expect this information via studying fragmentation, collective flow and particle production. Among these the fragmentation, which is breaking of nuclei into multiple small and medium mass nuclei when excitation energy exceeds their binding energies, is the most promising one. The extensive studies showed that the structure of fragments is affected by different entrance channels such as incident energy, impact parameter, mass of colliding nuclei, isospin asymmetry, and mass asymmetry [13–21]. It is well known that the dynamics of reaction differ significantly for symmetric and asymmetric reactions. The former has greater share of energy as compression whereas later has more energy as excitation/thermal. Therefore, to understand the various aspects of reaction dynamics, mapping of both symmetric and asymmetric reactions is important.

In previous studies, it was found that the multiplicity of intermediate mass fragments (IMFs) shows a rise and fall behavior when plotted against incident energy of projectile. The energy at which maximum multiplicity of IMFs is observed is termed as $< E^{max}_{c.m.} >$ and the corresponding IMF's multiplicity as $< N^{max}_{IMFs} >$. In experiments, such behavior was observed by Peaslee *et al.* for the reactions of ^{84}Kr + ^{197}Au in the incident energy range of 35–400 MeV/nucleon [21]. The maximum IMF production was observed at $\sim$ 100 MeV/nucelon. They also used Quantum Molecular Dynamics (QMD) and QMD + Statistical Multifragmentation Model (SMM) model to explain the results. Later, the Michigan State University (MSU) group performed more systematic study with the reactions of ^{40}Ar + ^{45}Sc, ^{58}Ni + ^{58}Ni, and ^{86}Kr + ^{93}Nb and reported a linear dependence of $< E^{max}_{c.m.} >$ and power law dependence of $< N^{max}_{IMFs} >$ when plotted against system mass [22]. On the theoretical front, the percolation model calculations were used to explain the observations but remain unsuccessful. Later, Puri and co-workers were successful in explaining the experimental observations. They did so, first using Quantum Molecular Dynamics (QMD) model with advanced microscopic clusterization algorithm based on binding energy, and using the improved QMD, i.e., Isospin-dependent Quantum Molecular Dynamics (IQMD) model with basic spatial correlations-based clusterization algorithm [23, 24]. They also studied the role of different definitions of IMFs on the peak center of mass energy and found it to be insignificant [25]. Also, the role of various model ingredients such as equation of state (EOS), nucleon-nucleon cross-section, gaussian width and isospin effects on energy of peak production of IMFs was investigated. They reported insensitive behavior of $< E^{max}_{c.m.} >$ toward these model ingredients except isospin effects [24, 26].

In case of asymmetric reactions various experimental and theoretical attempts have also been done. On the experimental front, Bowmann *et al.* studied fragment distribution of various asymmetric reactions of ^{129}Xe projectile with ^{12}C, ^{27}Al, ^{51}V, ^{nat}Cu, ^{89}Y, and ^{197}Au using 4π detector. They observed that the fragment multiplicity distributions and charge distributions are target independent [27]. In another study, the ALADIN collaborators performed study on asymmetric reactions of ^{129}Xe, ^{197}Au, and ^{238}U projectiles with Be, C, Al, Cu, In, Au, and U targets and reported a linear rise in the energy of maximum fragment production with the projectile mass [20]. Very recently, Puri *et al.* investigated the mass dependence of $< E^{max}_{c.m.} >$ and $< N^{max}_{IMFs} >$ in various mass asymmetric reactions and successfully reproduced the experimental data using the IQMD model. They found that the slope of the linear rise decreases on shifting from mass symmetric to asymmetric reactions. The study also showed the immense capability of $< E^{max}_{c.m.} >$ of the asymmetric reactions to constrain the density dependence of the symmetry energy at subsaturation densities [28]. In earlier studies, the role of mass asymmetry on the energy of vanishing flow (EVF), geometry of vanishing flow (GVF) and transition energy were also reported [29–31].

The rise and fall behavior of IMFs is also observed with impact parameter for symmetric and asymmetric reactions. This behavior is heavily explored in many experimental and theoretical fronts at fixed incident energies. For instance, the SIS facility at GSI reported the rise and fall in multiplicity of IMFs in symmetric reactions of ^{197}Au + ^{197}Au at energy of 100, 250, and 400 MeV/nucleon . They observed maximum multiplicity of IMFs is shifting from central to peripheral geometries with increase in incident energy [16]. In another study by ALADIN group, the rise and fall behavior is observed for various reactions of ^{197}Au on Be, C, Al, and Au at different energies of 400, 600, 800, and 1000 MeV/nucleon. They reported that on decreasing the target mass, maximum fragment multiplicity is shifting toward central collisions and its dependence on the bombarding energy increases [20]. On theoretical front, it was observed that the $< E^{max}_{c.m.} >$ as well as $< N^{max}_{IMFs} >$ decreases with increase in impact parameter for both symmetric and asymmetric reactions but later has greater slope compared to former [32].

The present study is mainly dedicated to investigate the reaction dynamics of asymmetric reactions at their peak energy of IMFs ($5 \leq A \leq A_{total}/6$) production. The study is carried out by exploring the mass dependence of peak energy of IMFs production for fixed targets of ^{124}Sn and ^{197}Au and by varying the projectile mass from ^{44}Ca to ^{197}Au. Further, in our study we will explore the behavior of different fragments such as free nucleons (FNs) [A=1], A=2, light charge particle (LCPs) [$2 \leq A \leq 4$], medium mass fragments (MMFs) [$5 \leq A \leq 9$] and heavy mass fragments (HMFs) [$10 \leq A \leq A_{tot}/6$] with total system mass at the energy of peak fragment production. Here, we also plan to investigate the behavior of various observables related to nuclear dynamics such as average and maximum central density, temperature, collision number and participant and spectator matter at the energy of peak fragment production which is never been investigated at peak energy of fragmentation. The present study is done within the framework of Isospin-dependent Quantum Molecular Dynamics (IQMD) model which is described in detail in [33].

7.2 Results and Discussion

For the present work, we have simulated various reactions with ^{124}Sn and ^{197}Au targets at semi-central geometries. Several thousand events are simulated for the reactions of ^{197}Au + ^{197}Au (E_{lab} = 50–120 MeV/nucleon), ^{112}Sn + ^{197}Au (E_{lab} = 45–145 MeV/nucleon), ^{72}Ge + ^{197}Au (E_{lab} = 45–195 MeV/nucleon), ^{44}Ca + ^{197}Au (E_{lab} = 55–330 MeV/nucleon), ^{124}Sn + ^{124}Sn (E_{lab} = 40–140 MeV/nucleon), ^{112}Sn + ^{124}Sn (E_{lab} = 40–160 MeV/nucleon), ^{72}Ge + ^{124}Sn (E_{lab} = 45–195 MeV/nucleon) and ^{44}Ca + ^{124}Sn (E_{lab} = 60–225 MeV/nucleon). Here, we have used soft equation of state and isospin-dependent nucleon-nucleon (nn) cross-section.

We observe proper rise and fall in the multiplicity of IMFs with incident energy for various projectiles on ^{124}Sn and ^{197}Au target reactions. In Fig. 7.1a, we display energy of peak IMFs production as a function of projectile mass (A_P).

From the figure, one finds that on decreasing the projectile mass, i.e., on increasing mass asymmetry ($\eta = | \frac{A_T - A_P}{A_T + A_P} |$), the energy of peak production is shifting toward the high energies. This increase in peak energy on increasing the mass asymmetry parameter (η) can be understood in terms of compressional and thermal energies. In case of symmetric reactions, as most of the energy is used in compression. Thus, the maximum production of IMFs is obtained at low incident energies. Whereas in asymmetric reactions, energy of peak production is observed at high energies. This

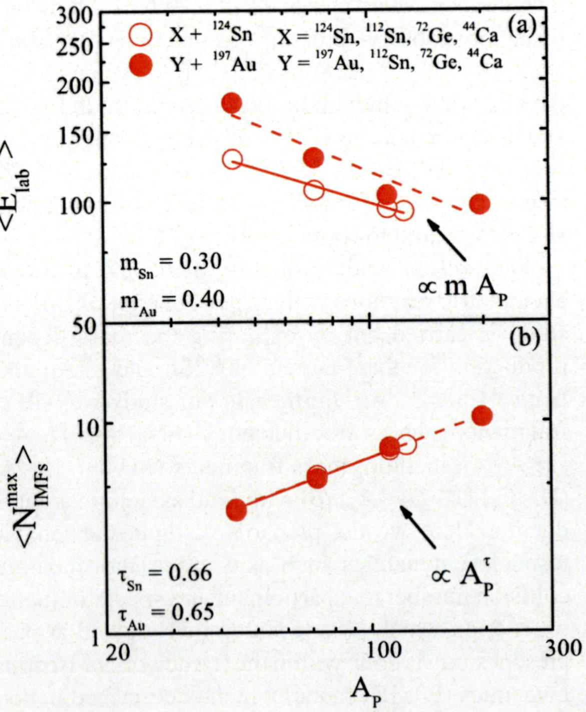

Fig. 7.1 The energy of peak IMFs production (upper panel) and peak IMF multiplicity (lower panel) as a function of projectile mass for ^{124}Sn (open circles) and ^{197}Au (filled circles) target reactions

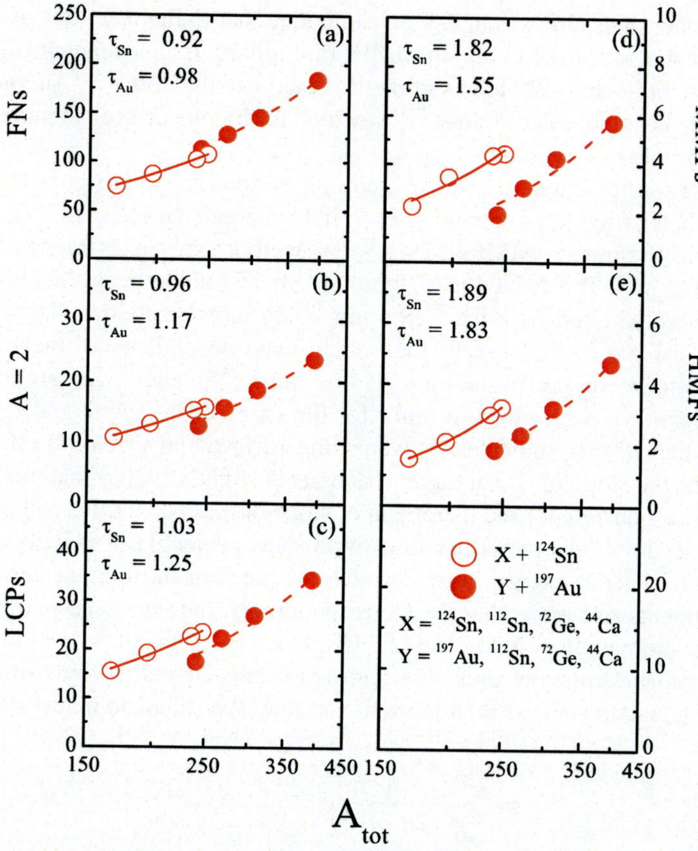

Fig. 7.2 The multiplicity of (**a**) free nucleons, (**b**) fragments with mass A = 2, (**c**) light charged particles (LCPs), (**d**) medium mass fragments (MMFs) and (**e**) heavy mass fragments (HMFs) as a function of the total mass of the system (A_{tot}). Model calculations done for ^{124}Sn (open circles) and ^{197}Au (filled circles) target reactions at their respective peak energies

is because in asymmetric reactions, the thermal part of the energy is more than the compressional part. Therefore, more energy is needed to have maximum production of IMFs.

Now if we compare the results of ^{124}Sn and ^{197}Au targets, the peak energy increases on moving toward the light projectile. In case of ^{197}Au target reactions, the slope is more steeper than ^{124}Sn target reactions. The increment in slope shows that ^{197}Au target reactions need more energy to achieve maximum fragmentation for the same projectiles. This behavior is in accordance to earlier predictions reported in [22–24] that the peak IMFs production increases with rise in system mass.

In Fig. 7.1b, peak IMFs multiplicity ($< N^{max}_{IMFs} >$) as a function of projectile mass (A_P) is displayed for both ^{124}Sn and ^{197}Au target reactions. The value of $< N^{max}_{IMFs} >$ increases as the target mass increases (and η decreases). Peak IMFs multiplicity is found to follow power law ($\propto A_P^{\tau}$) for both ^{124}Sn and ^{197}Au target reactions. In both cases, exponent τ is almost same. One can notice that in the reactions of ^{124}Sn and ^{197}Au the behavior of the maximum IMFs multiplicity is independent of the target for the same projectile. While in symmetric reactions of ^{197}Au + ^{197}Au, the value of $< N^{max}_{IMFs} >$ is higher than ^{124}Sn + ^{124}Sn which is obvious due to the increase in the total system mass.

Now, we further extend the above study for various fragments consisting of free nucleons (FNs), fragments with mass A=2, light charged particles (LCPs) [2≤A≤4], medium mass fragments (MMFs) [5≤A≤9] as well as heavy mass fragments (HMFs) [10≤A≤A_{total}/6]. The behavior of these fragments at the peak energy of IMF's production is explored here for ^{124}Sn and ^{197}Au target reactions. The results are displayed in Fig. 7.2. We observe the system mass dependence of these quantities and follow the power law of the form A^{τ}_{tot} for both ^{124}Sn and ^{197}Au target reactions. The exponent τ is very close to unity for the case of FNs, A=2 and LCPs. Here the linear dependence indicates the vanishing surface and Coulomb effects. In all these cases, the slope of ^{197}Au target reactions is slightly higher than that of ^{124}Sn. The factor responsible for the increment of slope of these quantities is the high peak IMFs energy for ^{197}Au target reactions with same projectile. It is well known that with rise in incident energy more correlations are broken, thus the production of these fragments rise more sharply. Correspondingly, the increased pumped energy causes less production of MMFs and HMFs, as a result their slope rises less sharply.

To extract information about hot and dense nuclear matter, maximum density achieved in reaction acts as an important variable. We calculate matter density as

$$\rho = \sum_{i=1}^{A_{tot}} \frac{1}{(2\pi L)^{3/2}} e^{\{-[\vec{r} - \vec{r}_i(t)]^2/2L\}}. \tag{7.1}$$

Here, A_{tot} is the total system mass. To calculate the density, we take a sphere of 2 fm radius around the center of mass of two colliding nuclei. The average density ($< \rho^{avg} >/\rho_0$) is calculated over the whole sphere and maximum density ($< \rho^{max} >/\rho_0$) reached anywhere in this sphere. In Fig. 7.3, we display the various quantities such as average and maximum density as a function of time for both ^{124}Sn and ^{197}Au target reactions at their respective peak energies. It is evident that the maximal of $< \rho^{avg} >/\rho_0$ and $< \rho^{max} >/\rho_0$ is delayed for symmetric reactions compared to asymmetric reactions. Also, the density zone is wider for symmetric reactions indicating that hot and dense matter exists for longer time. After the compressional phase, the density in central region falls sharply to lower values. In asymmetric reactions the fall is sharp as compared to symmetric reactions. Also, the density profile of ^{197}Au target reactions is wider than ^{124}Sn target reactions reflecting the bulk effects. Another quantity linked with the density is the collision rate (displayed in lower panel). The collision rate (dN_{coll}/dt) is displayed as a function of reaction time. In case of symmetric reactions, the interaction among nucleons continue for longer time thus, the collision rate is high. Whereas in case of asymmetric reactions, due to lesser interaction region the collision rate is less. Again, the difference we see for the ^{124}Sn and ^{197}Au target is due to the difference in their nucleon number and corresponding

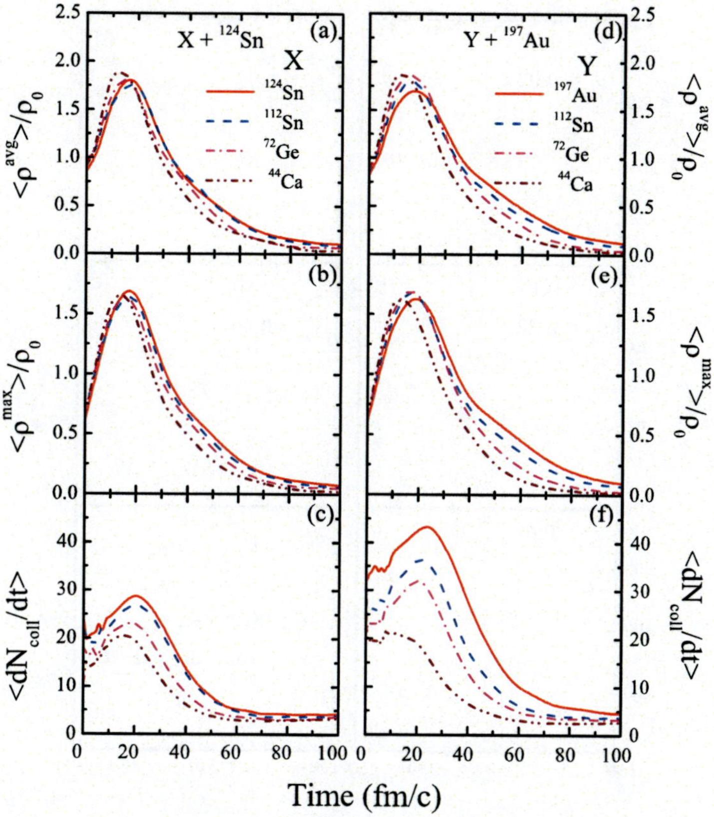

Fig. 7.3 The evolution of $< \rho^{avg} > / \rho_0$, $< \rho^{max} > / \rho_0$ and dN_{coll}/dt as a function of time. Various ^{124}Sn and ^{197}Au target reactions are simulated at their corresponding peak energy of IMFs production

peak energy of IMFs (greater for ^{197}Au than ^{124}Sn). These results clearly show that not only the IMFs production and the peak energies but also their corresponding dynamics change significantly for symmetric to asymmetric reactions.

From Fig. 7.3, we see that the different system masses approach different values for average density ($< \rho^{avg} > / \rho_0$) and maximum density ($< \rho^{max} > / \rho_0$), collision rate (dN_{coll}/dt). To find that how the values change, we note the maximum values corresponding to each system at their corresponding peak energies. Interestingly, the maximal value of average density ($< \rho^{avg} > / \rho_0$) and maximum density ($< \rho^{max} > / \rho_0$) increases with decreasing mass asymmetry of the system and follows power law ($\propto A^{\tau}_{tot}$). It is because in symmetric reactions the peak energy is less, corresponding to that less density is achieved. Whereas in mass asymmetric reactions, the value of density is high as the peak energy is also high for this case. The maximal value of $< \rho^{avg} > / \rho_0$ and $< \rho^{max} > / \rho_0$ follows power law. The mass dependence of nucleon-nucleon collisions ($< N^{coll} >_{max}$) is also displayed in Fig. 7.4c. The linear enhancement is observed with total mass of the system. It is because

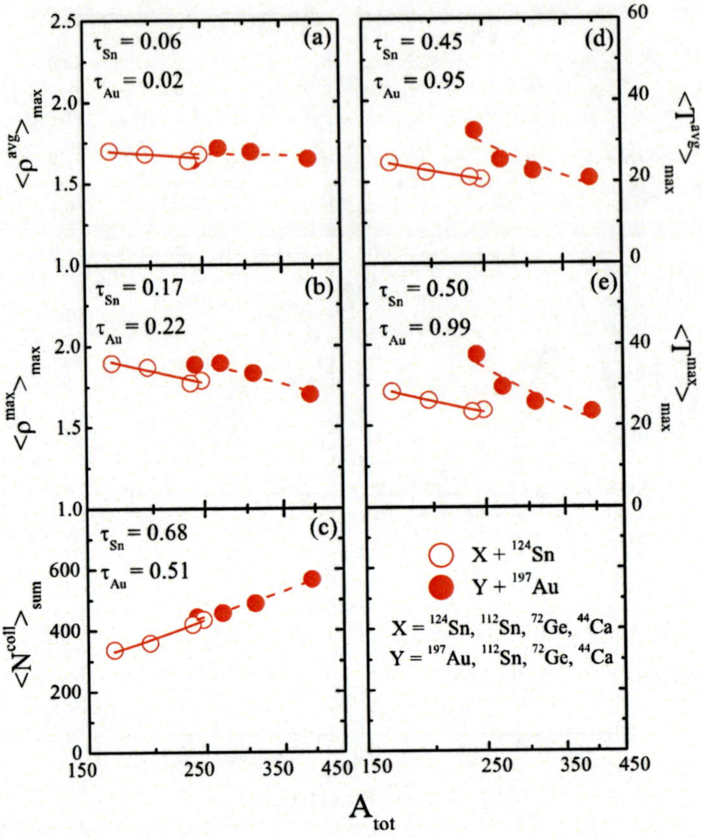

Fig. 7.4 The maximal value of average density ($< \rho^{avg} > / \rho_0$), maximum density ($< \rho^{max} > / \rho_0$), nucleon-nucleon collisions ($< N^{coll} >_{sum}$), average temperature ($< T^{avg} >$) and maximum temperature ($< T^{max} >$) as a function of A_{tot} at their respective peak energies for ^{124}Sn and ^{197}Au target reactions

the nucleon-nucleon (nn) collisions are more in symmetric reactions as compared to the asymmetric ones (due to overlapping region). This enhancement can be parameterized with power law proportional to A^{τ}_{tot}. However, in case of ^{197}Au target reactions power dependence is more strong compared to ^{124}Sn reactions.

The another quantity associated with hot and dense nuclear matter is the temperature. As the nuclear matter in heavy-ion collisions is non-equilibrated, therefore, the temperature is calculated using the local density approximation. The procedure we followed to calculate temperature is described in detail in [34–36]. The maximal value of average temperature ($< T^{avg} >$) and maximum temperature ($< T^{max} >$) as a function of A_{tot} is also plotted in Fig. 7.4. Since the temperature is related to the excitation energy of the system. Thus, in case of symmetric reactions, the peak IMFs production is achieved at low incident energy as a result temperature achieved is less. On the contrary, in mass asymmetric reactions the peak IMFs production is achieved

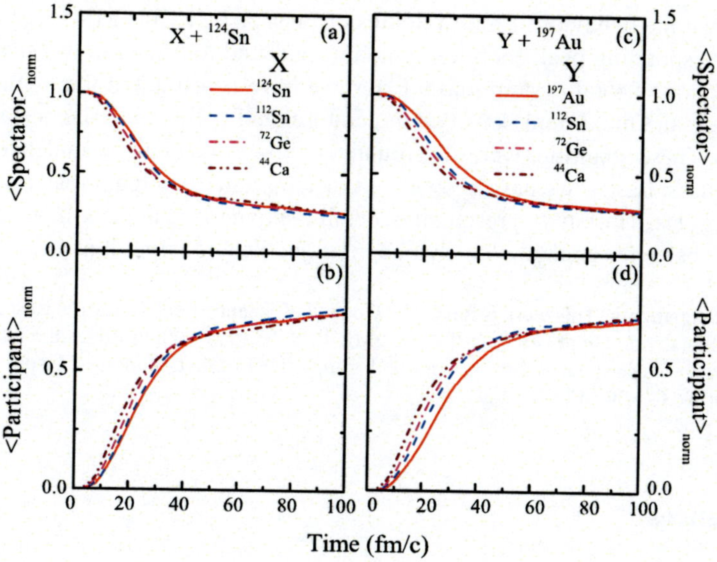

Fig. 7.5 The time evolution of normalized spectator matter (upper panel) and participant matter (lower panel) for ^{124}Sn and ^{197}Au target reactions

at high energies. Thus the maximal value of temperature increases with increase in mass asymmetry. The exponent τ in case of ^{197}Au reactions is more than ^{124}Sn reactions. It is because in case of ^{197}Au target reactions the peak energy required is more, thus temperature raises sharply on moving toward low mass projectiles.

Lastly, in Fig. 7.5, the normalized spectator and participant matter as a function of the reaction time is displayed. These results are displayed at the corresponding peak IMFs production energies. Here, the participant matter is defined using nucleonic concept. Those nucleons which experienced at least one collision are counted as participant matter and the remaining matter is considered as spectator matter. At the starting of the reaction, there is only spectator matter and no participant matter. From the figure, one notices that the transition from spectator to participant matter is slow in case of symmetric reactions, because of their low energy of peak production. On the contrary, in asymmetric reactions this transition is swift and sudden. The obvious reason is the high energy of peak production for the asymmetric reactions.

7.3 Summary

In the present study, we have investigated the behavior of peak IMF production for the reaction of ^{44}Ca, ^{72}Ge, ^{112}Sn, ^{124}Sn, ^{197}Au on the targets of ^{124}Sn and ^{197}Au. We found a linear rise in peak energy of IMFs production and their corresponding multiplicities. The behavior change for the targets of ^{124}Sn and ^{197}Au have been presented. We have also shown the behavior of multiplicities of FNs, A=2, LCPs, MMFs and HMFs at their corresponding peak energies, which were found to follow power law behavior as a function of total system mass. Later, we have investigated the quantities such as average and maximum density, temperature, and collision dynamics at the peak energies. These quantities were also found to follow power law dependence with total system mass. Lastly, we have shown the spectator and participant matter behavior at peak energies. This study presented in detail the dynamics at peak IMFs production and their behavior change with change in projectile and target masses.

Acknowledgements This work is funded by Council of Scientific and Industrial Research (CSIR), Govt. of India, vide Grant No. 03(1388)/16/EMR-II. S. Sharma acknowledges the financial assistance from the Department of Science and Technology (DST), Govt. of India as Inspire Fellowship, vide grant No.C/1486/IFD/2019-20.

References

1. J.R. Birkelund et al., Heavy-Ion Fusion Based on the Proximity Potential and One-Body Friction. Phys. Rev. Lett. **40**, 1123 (1978)
2. J.N. De, W. Stocker, Temperature dependence of fusion barriers. Phys. Rev. C **42**, R819 (1990)
3. M. Sinha et al., Sub-barrier fusion excitation for the system ^{7}Li + ^{28}Si. Phys. Rev. C **78**, 027601 (2008)
4. V.V. Desai et al., Prompt fission neutron spectra in fast-neutron-induced fission of ^{238}U. Phys. Rev. C **92**, 014609 (2015)
5. Z. Ren et al., New perspective on complex cluster radioactivity of heavy nuclei. Phys. Rev. C **70**, 034304 (2004)
6. V.T. Zagrebaev et al., Synthesis of superheavy nuclei: Nucleon collectivization as a mechanism for compound nucleus formation. Phys. Rev. C **64**, 034606 (2001)
7. A. Bonasera, Vlasov description of dense quark matter. Phys. Rev. C **60**, 065212 (1999)
8. S. Terranova, A. Bonasera, Constrained molecular dynamics simulation of the quark-gluon plasma. Phys. Rev. C **70**, 024906 (2004)
9. P. Wang et al., Strange quark matter in a chiral SU(3) quark mean field model. Phys. Rev. C **67**, 015210 (2003)
10. N. Herrmann et al., Collective flow in heavy-ion collisions. Ann. Rev. Nucl. Part. Sci. **49**, 581 (1999)
11. B. Borderie, M.F. Rivet, Nuclear multifragmentation and phase transition for hot nuclei. Prog. Part. Nucl. Phys. **61**, 551 (2008)
12. J.P. Bondrof, Statistical multifragmentation of nuclei. Phys. Rep. **257**, 133 (1995)
13. L. Phair et al., Impact-parameter filters for ^{36}Ar + ^{197}Au collisions at E/A = 50, 80 and 110 MeV. Nucl. Phys. A **548**, 489 (1992)
14. W.J. Llope et al., Autocorrelations and intermediate-mass-fragment multiplicities in central heavy-ion collisions. Phys. Rev. C **51**, 1325 (1995)
15. N. Marie et al., Experimental determination of fragment excitation energies in multifragmentation events. Phys. Rev. C **58**, 256 (1998)
16. M.B. Tsang et al., Onset of Nuclear Vaporization in ^{197}Au + ^{197}Au Collisions. Phys. Rev. Lett. **71**, 1502–1505 (1993)
17. R. Sun et al., Isotropic emission components in splintering central collisions: (17–115) A MeV ^{40}Ar + Cu, Ag. Au. Phys. Rev. C **61**, 061601(R) (2000)
18. C.A. Ogilvie et al., Rise and Fall of Multifragment Emission. Phys. Rev. Lett. **67**, 1214–1217 (1991)

19. M. Begemann-Blaich et al., Quantum molecular dynamics simulation of multifragment production in heavy ion collisions at E/A =600 MeV. Phys. Rev. C **48**, 610–617 (1993)
20. A. Schüttauf et al., Universality of spectator fragmentation at relativistic bombarding energies. Nucl. Phys. A **607**, 457–486 (1996)
21. G.F. Peaslee et al., Energy dependence of multifragmentation in ^{84}Kr + ^{197}Au collisions. Phys. Rev. C **49**, R2271–R2275 (1994)
22. D. Sisan et al., Intermediate mass fragment emission in heavy-ion collisions: Energy and system mass dependence. Phys. Rev. C **63**, 027602 (2001)
23. Y.K. Vermani, R.K. Puri, Mass dependence of the onset of multifragmentation in low energy heavy-ion collisions. J. Phys. G: Nucl. Part. Phys. **36**, 105103 (2009)
24. S. Kaur, R.K. Puri, Isospin effects on the energy of peak mass production. Phys. Rev. C **82**, 054611 (2013)
25. S. Sharma, R. Kumar, R.K. Puri, Does the range of IMF affect rise and fall trend in fragmentation. AIP Conference Proceedings **1953**, 140022 (2018)
26. S. Kaur, A.D. Sood, Model ingredients and peak mass production in heavy-ion collisions. Phys. Rev. C **82**, 054611 (2010)
27. D.R. Bowman et al., Intermediate mass fragment emission as a probe of nuclear dynamics. Phys. Rev. C **46**, 1834–1848 (1992)
28. S. Sharma, R. Kumar, R.K. Puri, *Role of mass asymmetry on the peak energy of intermediate mass fragments production and its influence towards isospin effects* (Phys. Rev, C (Under review), 2020)
29. S. Goyal, R.K. Puri, On the sensitivity of the energy of vanishing flow towards mass asymmetry of colliding nuclei. Nucl. Phys. A **853**, 164–172 (2011)
30. S. Goyal, Role of the mass asymmetry of reaction on the geometry of vanishing flow. Nucl. Phys. A **856**, 154–161 (2011)
31. V. Kaur, S. Kumar, R.K. Puri, On the elliptical flow and mass asymmetry of the colliding nuclei. Phys. Lett. B. **697**, 512–516 (2011)
32. Sharma, S., Kumar, R., Puri, R. K. (2020). On the energy of maximum production of intermediate mass fragments in heavy-ion collisions using microscopic model. Proceedings of the Jangjeon Mathematical Society (Submitted)
33. Ch. Hartnack et al., Modelling the many-body dynamics of heavy ion collisions: Present status and future perspective. Eur. Phys. J. A **1**, 151–169 (1998)
34. D.T. Khoa et al., Microscopic study of thermal properties of the nuclear matter formed in heavy-ion collisions. Nucl. Phys. A **542**, 671–698 (1992)
35. D.T. Khoa et al., In-medium effects in description of heavy-ion collisions with realistic NN interactions. Nucl. Phys. A **548**, 102–130 (1992)
36. R.K. Puri et al., Temperature-dependent mean field and its effect on heavy-ion reactions. Nucl. Phys. A **575**, 733–765 (1994)

Chapter 8
Reaction Dynamics for Stable and Halo Nuclei Reactions at Intermediate Energies

Sucheta, Rohit Kumar, and Rajeev K. Puri

Abstract We have investigated the influence of halo structure of nuclei on the fragment production for different equations of states. For this, we simulate the central reactions of ^{37}Mg + ^{37}Mg halo and ^{36}Mg + ^{36}Mg stable nuclei at different incident energies. Our detailed study indicates that the halo nuclei reactions can be vital in constraining the exact equation of state of nuclear matter.

8.1 Introduction

Nuclear matter is defined as the substance having huge number of protons and neutrons interacting only via nuclear forces. The physics trailing it is to understand the evolution of universe, formation of proto-neutron stars, properties of nuclear matter inside the neutron stars and physics of supernova explosions. To understand the above mentioned phenomena one needs information about the nuclear matter at different densities and temperatures [1, 2]. This information helps one to construct the equation of state (EOS) of nuclear matter. Now, to obtain this information, one studies the nuclear reactions at low, intermediate, and high incident energies. At lower energies, one studies phenomena such as fusion-fission [3], β- decay [4] and behavior of exotic nuclei such as halo nuclei [5]. Whereas at ultrahigh energies, one obtains the information at the quark level [6]. The intermediate energy reactions help to obtain information of nucleons at extreme conditions of temperature and pressure.

Further, based on the incompressibility, the equation of state of nuclear matter has two possibilities, i.e., soft and hard. To predict the exact form of equation of state, at intermediate energies, one uses fragment formation, their multiplicity, their flow, and particle production. These observables depends on various entrance channels

Sucheta (✉) · R. Kumar · R. K. Puri
Department of Physics, Panjab University, Chandigarh 160014, India
e-mail: dsucheta28@gmail.com

R. Kumar
e-mail: rohitksharma.pu@gmail.com

R. K. Puri
e-mail: drrkpuri@gmail.com

R. K. Puri et al. (eds.), *Advances in Nuclear Physics*, Springer Proceedings in Physics 257, https://doi.org/10.1007/978-981-15-9062-7_8

such as incident energy, choice of projectile and target, their isospin-content, mass asymmetry and the structure of projectiles and targets. In this regard, the quantum molecular dynamics (QMD) model was used to study the fragmentation of ^{197}Au + ^{197}Au reactions using soft and hard equations of state [7]. Sood and Puri have also investigated the role of soft and hard equation of state on the energy of vanishing flow for symmetric reactions for ^{12}C to ^{238}U reactions [8]. Only marginal differences were observed. Earlier studies, do show that the momentum-dependent interactions are also required to explain some features of nuclear reactions at intermediate energies. In [7], significant effect of MDI was observed in the peripheral collisions of various nearly symmetric reactions. Vermani et al. [9], also investigated the role of MDI on ^{197}Au + ^{197}Au reactions results at an incident energy of 400 MeV/nucleon.

The past few decades have seen a rapid increase in the interest of structure and reaction dynamics involving exotic nuclei which are far away from the line of stability especially halo nuclei. Generally, halo nuclei is regarded as a threshold phenomena having an abrupt increase in the interaction cross-section and narrow momentum distribution of the nucleons. This exotic phenomena was first studied in 1985 in Berkeley experiments by Tanihata et al. [10, 11], using radioactive nuclear beams through the measurements of the interaction cross-sections. The root behind the halo structure is the weakly bound nucleons which decouple from a tightly bound nuclear core. The weakly bound nucleons makes the interaction cross-sections of these nuclei appreciably larger. These nuclei are also investigated for different nuclear reactions. To mention a few, at low incident energy, Raj Kumari has studied the fusion probabilities of halo nuclei using proximity-based potentials. They had investigated the effect of the halo nuclei such as ^{6}He, ^{11}Be and ^{8}B on the fusion cross-section and the barrier heights [12]. Their findings reveals that the extended sizes of the halo nuclei contributes toward the enhancement of the fusion probabilities and reduction in the barrier heights. They also conclude that the enhancement in the fusion cross-section is much significant at low incident energies and as the energy increases the effect diminishes. To study the role of halo structure in nuclear dynamics at intermediate energies, a number of studies are also done, e.g., within the framework of Isospin-dependent quantum molecular dynamics (IQMD) model, Liu et al. [13] studied the role of loose neutron-halo structure of ^{19}B on the fragment multiplicity and nuclear stopping in the incident energy range of 20 to 150 MeV/nucleon. To perceive the role of halo structure on the reaction dynamics, they had compared the fragment multiplicity and nuclear stopping observables of the ^{19}B halo structured nuclei with the same mass stable ^{19}F nuclei. The effect of the halo structured nuclei on the fragment multiplicity and nuclear stopping is found to be more at lower incident energies but starts decreasing gradually with further increase in the energy. They also concluded that the halo structured nuclei increases the fragment multiplicity but the opposite is observed for the nuclear stopping. In other study, using relativistic mean field (RMF) densities, in the framework of Glauber model formalism, the reaction cross-sections for all the $^{24-40}$Mg isotopes is studied with stable target of ^{12}C at an incident energy of 240 MeV/nucleon [14]. Their results well explained the experimental observables for the reaction cross-sections for all the Mg isotopes at RIBF and RIKEN except for the ^{37}Mg which supports its halo structure [15]. Further, the various properties of ^{37}Mg were examined from extended density distribution. The above-cited studies presented that the halo nuclei reactions can be very rich source of information for the understanding of nuclear matter properties.

Using isospin-dependent quantum molecular dynamics (IQMD) model, we carry out the calculations by comparing the reactions of halo nuclei ^{37}Mg + ^{37}Mg and stable nuclei ^{36}Mg + ^{36}Mg. It is indicated in the experimental measurement that there exists a neutron halo structure in ^{37}Mg [15]. In the present study, we plan to look for enhancement or reduction of fragmentation toward different equation of states. The obtained behavior will also be compared with the nearly mass stable nuclei reactions for better understanding. Note that, though the previous studies have investigated the different equation of states for stable nuclei reactions but none study reported for halo nuclei induced reactions. The chapter is structured as following: In Sect. 8.2, we will present the details of the IQMD model. Section 8.3 is dedicated to results and discussion followed by the conclusions in Sect. 8.4.

8.2 Isospin-Dependent Quantum Molecular Dynamics (IQMD) Model

The isospin-dependent quantum molecular dynamics (IQMD) model [16] is an n-body model which follows heavy-ion reactions on nucleonic level. This model generates the information of the reaction in the form of phase-space information of individual nucleon. Here, each nucleon is represented by a Gaussian wave packet of the form:

$$\phi_i(\mathbf{r}, \mathbf{r}_i(t), \mathbf{p}_i(t)) = \frac{1}{(2\pi L)^{3/4}} e^{\left[\frac{i}{\hbar}\mathbf{p}_i(t)\cdot\mathbf{r} - \frac{(\mathbf{r}-\mathbf{r}_i(t))^2}{4L}\right]}. \tag{8.1}$$

The centroids of each nucleon is propagated using Hamilton's classical equations of motion. These equations reads as

$$\dot{\mathbf{r}}_i = \frac{\partial\langle H\rangle}{\partial\mathbf{p}_i}; \qquad \dot{\mathbf{p}}_i = -\frac{\partial\langle H\rangle}{\partial\mathbf{r}_i}. \tag{8.2}$$

Here, $< H >$ is the average Hamiltonian that includes the kinetic and potential energy terms. The potential part incorporates the Skyrme (V^{Sky}), Yukawa (V^{Yuk}), Coulomb (V^{Coul}), momentum dependent potentials (V^{MDI}), and symmetry potential (V^{sym}). Mathematically, the potential is represented as

$$V = V^{Sky} + V^{Yuk} + V^{Coul} + V^{MDI} + V^{sym}.$$

The baryon-baryon potential V_{ij} is given as

$$V_{ij} = t_1\,\delta(\mathbf{r}_i - \mathbf{r}_j) + t_2\,\delta(\mathbf{r}_i - \mathbf{r}_j)\rho^{\gamma-1}((\mathbf{r}_i + \mathbf{r}_j)/2) + t_3\,\frac{e^{-|\mathbf{r}_i - \mathbf{r}_j|/\mu}}{|\mathbf{r}_i - \mathbf{r}_j|/\mu}$$
$$+\frac{Z_i Z_j e^2}{|\mathbf{r}_i - \mathbf{r}_j|} + t_4\,\ell n^2[t_5\,(\mathbf{p}_i - \mathbf{p}_j)^2 + 1]\,\delta(\mathbf{r}_i - \mathbf{r}_j)$$
$$+t_6\frac{1}{\rho_o}T_{3i}T_{3j}\delta(\mathbf{r}_i - \mathbf{r}_j). \tag{8.3}$$

The parameters t_1, t_2, t_3, ..., t_6 are fitted to obtain static soft (S), static hard (H), soft with momentum-dependent interactions (SMD), and hard with momentum-dependent interactions (HMD) equation of states. The details of the potentials and the parameters can be obtained from [16]. At the end of a event run we get information of phase-space of nucleons which is further subjected to spatial constraints for constructing fragments [17, 18].

8.3 Results and Discussions

For the present work, we have simulated thousands of events of the reactions of ^{37}Mg + ^{37}Mg and ^{36}Mg + ^{36}Mg for central geometries in the incident energy range of 20 to 150 MeV/nucleon. Here, note that we are incorporating the halo structure of ^{37}Mg as the extended radius following the previous studies [12–14, 19]. All reactions are followed till 300 fm/c and the stored phase space of the nucleons is then subjected to clusterization algorithm. Therefore, this study will give us the upper limit to what will be observed by incorporating actual halo structure. In the present study, we are showing the observables related to fragments at freeze-out time only.

In Fig. 8.1, we have displayed the values of mean size of the largest fragment ($< A_{max} >$), and multiplicities of free nucleons ($< N_{FNs} >$) ($A_f = 1$), light charged particles ($< N_{LCPs} >$) ($2 \leq A_f \leq 4$) and intermediate mass fragments ($< N_{IMFs} >$) ($5 \leq A_f \leq A_{tot}/6$) for the reactions of halo nuclei ^{37}Mg + ^{37}Mg and stable nuclei ^{36}Mg + ^{36}Mg. From the figure, we see that the size of the largest fragment ($< A_{max} >$) decreases as we increase the excitation energy of the system. The behavior is similar in stable and halo nuclei reactions. First we discuss, the results of the soft equation of state for the reactions of stable and halo nuclei. We see that at all incident energies the size of $< A_{max} >$ is less for halo nuclei reactions as compared to stable ones. Now, this can be explained as following: the radius of ^{37}Mg is quite large compared to ^{36}Mg, that causes the correlations among nucleons as very weak due to which lesser energy is required to break them. Therefore, if the same energy is supplied, the size of $< A_{max} >$ is smaller for ^{37}Mg + ^{37}Mg reactions compared to ^{36}Mg + ^{36}Mg reactions. Opposite to the behavior of $< A_{max} >$, the multiplicities of free nucleons ($< N_{FNs} >$) and light charged particles ($< N_{LCPs} >$) increases gradually with the incident energy. We see larger values of $< N_{FNs} >$ for halo nuclei reactions as compared to stable nuclei reactions. This behavior can be explained by looking at the size of $< A_{max} >$. Lesser the size of $< A_{max} >$ greater the number of

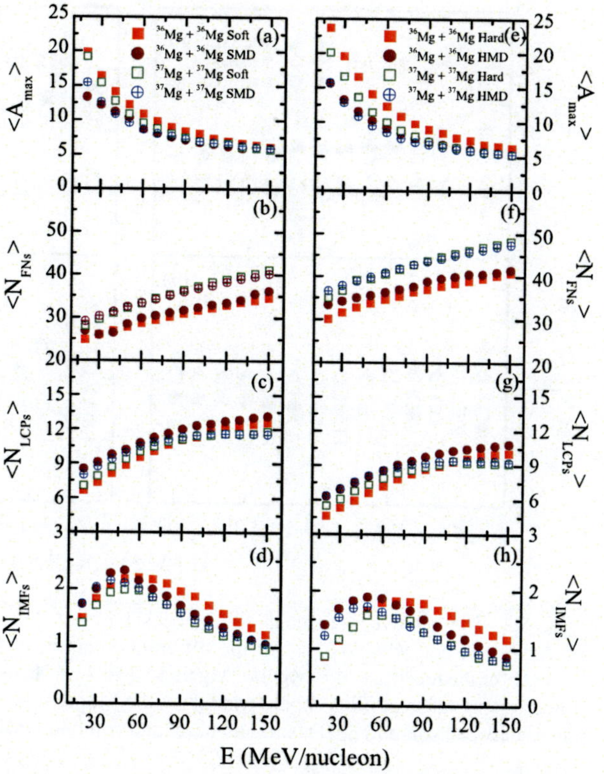

Fig. 8.1 The mean size of the largest fragment ($< A_{max} >$), and multiplicities of free nucleons ($< N_{FNs} >$), light charged particles ($< N_{LCPs} >$) and intermediate mass fragments ($< N_{IMFs} >$), respectively, for the central collisions of ^{36}Mg + ^{36}Mg and ^{37}Mg + ^{37}Mg as a function of incident energy. In the left panels, the filled (open) squares, filled (open) circles represent results with the soft and soft with momentum-dependent equation of state for stable (halo) induced reactions. In right panels, the filled (open) squares, filled (open) circles represent results with the hard and hard with momentum-dependent equation of state for stable (halo) induced reactions

free nucleons. There is difference in the multiplicities of LCPs also. The multiplicity of IMFs shows well-known rise and fall behavior, which is in accordance with Sisan et al. [20] and extensive studies of Puri and collaborators [21, 23, 24]. If we look at the results of ($< N_{IMFs} >$), we see that the multiplicities reach its maximum values at lower incident energy, i.e., $\sim$ 50 MeV/nucleon for halo nuclei reactions compared to $\sim$ 60 MeV/nucleon for stable nuclei reactions. Again, these results points toward the lesser energy requirement for the breaking of correlations among the nucleons in halo nuclei reactions compared to stable nuclei reactions.

Further, if we see the mean sizes of $< A_{max} >$ and values of fragment multiplicities for hard equation of state (right panels), we see that the behavior is similar to what we have observed for soft equation of state. Though, we see slight increase in the values of $< A_{max} >$ at lower incident energies, but the values are almost same at higher incident energies. These results are consistent with the earlier predictions of Sharma et al. [22] and Vermani et al. [9]. Now, if we see the fragment multiplicities, the values of free nucleons are larger with corresponding lesser values of LCPs and IMFs. We also observed that the maximum value of $< N_{IMFs} >$ occurs at same incident energy for stable nuclei reactions, as well as, for halo nuclei reactions. The similar results were reported by Kaur et al. in [23, 24].

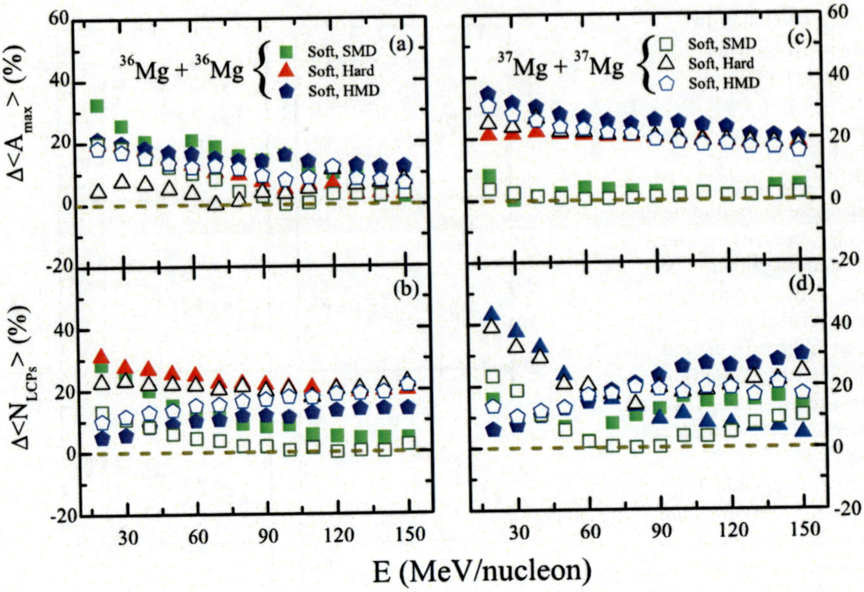

Fig. 8.2 The relative percentage differences between the results of $< A_{max} >$, $< N_{FNs} >$, $< N_{LCPs} >$, and $< N_{IMFs} >$ for the different equation of states with respect to soft equation of state for the reactions of ^{36}Mg + ^{36}Mg and ^{37}Mg + ^{37}Mg as a function of incident energy. Here, filled (open) squares, filled (open) triangles, filled (open) pentagons represents the relative differences between soft and SMD, soft and hard, and soft and HMD for stable (halo) nuclei reactions

For the better understanding of the role played by the different equation of states and how MDI influences the reaction output for stable and halo nuclei reactions, we have calculated relative percentage difference between the results. We calculate the relative percentage difference using the formula:

$$\%difference = \left|\frac{X - Y}{X}\right| \times 100\%, \tag{8.4}$$

where X and Y stands for observed values with one and other equation of state, respectively. The results are displayed in Fig. 8.2. In this figure, we displayed the relative percentage difference among the calculations of soft and SMD (squares), soft and hard (triangles) and soft and HMD (pentagons) for stable (filled) as well as halo (open) nuclei reactions. From the figure, we see that for all the fragments maximum percentage differences are between 0–40% range. We see differences between the different equations of states for the case of halo nuclei reactions and stable nuclei reactions. The differences is due to the change in the nuclear density and Fermi-momentum for halo nuclei compared to stable nuclei, lesser density, and Fermi-momentum values. For example, we note that the differences among the values for soft and SMD equation of states for $< A_{max} >$ is ~32% at 20 MeV/nucleon and decreases gradually to ~12% at 150 MeV/nucleon for stable nuclei reactions. This

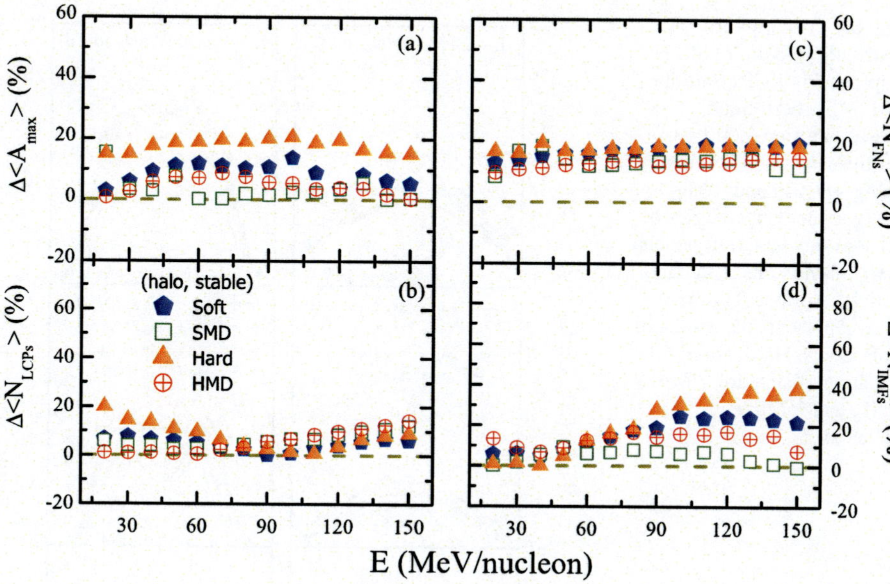

Fig. 8.3 The relative percentage differences among the results of $< A_{max} >$, $< N_{FNs} >$, $< N_{LCPs} >$, and $< N_{IMFs} >$ between the stable and halo nuclei reactions as a function of incident energy. Pentagons, squares, triangles, and circles represents the difference between the results of soft, SMD, hard, and HMD equation of state, respectively

decrease is also seen for halo nuclei reactions from ∼20% to ∼8%. Similarly for $< N_{FNs} >$, $< N_{LCPs} >$ and $< N_{IMFs} >$ the maximum difference is ∼40%. From this we found that the role of different equations of state is reduced when one uses halo nuclei as projectile and target compared to stable ones for $< A_{max}$, FNs and LCPs except for IMFs for both hard and soft EOS, where we see enhanced difference in the results for $< N_{IMFs} >$ at higher incident energies. Therefore, multiplicity of IMFs ($< N_{IMFs} >$) of halo nuclei reactions can be vital to constraint the equation of state.

Next, in Fig. 8.3, we have displayed the difference in the results among different equation of states for stable and halo induced reactions, e.g., soft equation of state for halo and stable induced reactions. We found that the halo structure can change the results of $< A_{max} >$, $< N_{FNs} >$ and $< N_{LCPs} >$ upto ∼20% only. Whereas, this difference is up to ∼40% in case of $<_{IMFs}>$ at higher incident energies. If we compare the results with calculations of Liu et al. [13], which points toward the significant difference in the results at lower incident energies; our results signifies that the halo nuclei can be lead to change the results significantly at higher incident energies also for different equation of states. Also note that in last two decades, both experimental and theoretical studies have shown that the multiplicity of IMF's shows a rise and fall behavior with increase in incident energy of projectile and have a connection with the observation of liquid-gas like behavior of nuclear matter [25, 26]. The use of halo nuclei can severally influence the signature of liquid-gas

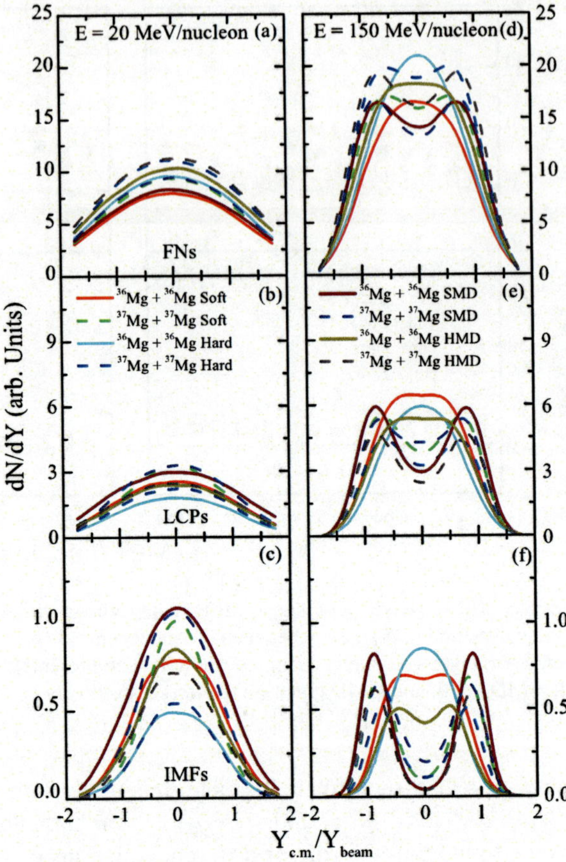

Fig. 8.4 The rapidity distribution (dN/dY) of FNs (top panels), LCPs (middle), and IMFs (bottom) at freeze-out time (300 fm/c) for the central collisions of $^{37}Mg + ^{37}Mg$ and $^{36}Mg + ^{36}Mg$ at incident energies of 20 MeV/nucleon (left panels) and 150 MeV/nucleon (right panels). Solid and dashed lines represents the results of Soft, SMD, Hard, and HMD equation of state for $^{36}Mg + ^{36}Mg$ and $^{37}Mg + ^{37}Mg$, respectively

phase transition. Therefore, will help to indepthen the physics of liquid-gas phase-transition, which will be explored in future studies.

In Fig. 8.4, we have shown the rapidity of various fragments, i.e., free nucleons (FNs), light charges particles (LCPs), and intermediate mass fragments (IMFs) at 20 MeV/nucleon (left) and 150 MeV/nucleon (right panels) for $^{36}Mg + ^{36}Mg$ and $^{37}Mg + ^{37}Mg$. The rapidity of fragments is calculated using the following formula:

$$Y = \frac{1}{2} In \frac{E + p_z}{E - p_z}. \tag{8.5}$$

Here, E and p_z are the total energy and total longitudinal momentum of the fragment. Various lines representing the reactions induced by stable (solid) and halo (dashed) nuclei for different equation of states, i.e., soft, hard, SMD, and HMD as mentioned in the graph. From the figure, we see that at 20 MeV/nucleon for soft equation of state of energy the free nucleons (FNs) have greater contribution from the participant region for $^{37}Mg + ^{37}Mg$ compare $^{36}Mg + ^{36}Mg$. Whereas, the LCPs and IMFs have greater contribution from participant region and lesser from spectator region for former compared to later (Fig. 8.4). This can be understood as following: the greater value of radius for halo nucleus causes lesser value of momentum of its constituting nucleons (compared to what stable nucleus nucleon have). Therefore, when the two nuclei collide the nucleons have enough time to interact and loose its energy and thermalize. Thus, the fragments do not have large rapidity values for halo nuclei reactions compared to stable nuclei reactions. Contrary to it, at 150 MeV/nucleon the nucleons have lesser time to thermalize and also, the loose structure of halo nuclei makes them more transparent to each other. This leads to cause greater share of fragments from spectator (lesser interacting nucleons). That is also clear from the figure, where for halo nuclei reactions one has two peaks at projectile and target regions. For stable nuclei reactions one has greater contribution for participant region. For hard equation of state at 20 and 150 MeV/nucleon the role of halo structure have similar effects on rapidity of fragments. Similarly for SMD and HMD equation of states, we also see the difference in the origin of fragments for stable and halo nuclei reactions. Here, we see the fragments have different origins for stable and halo

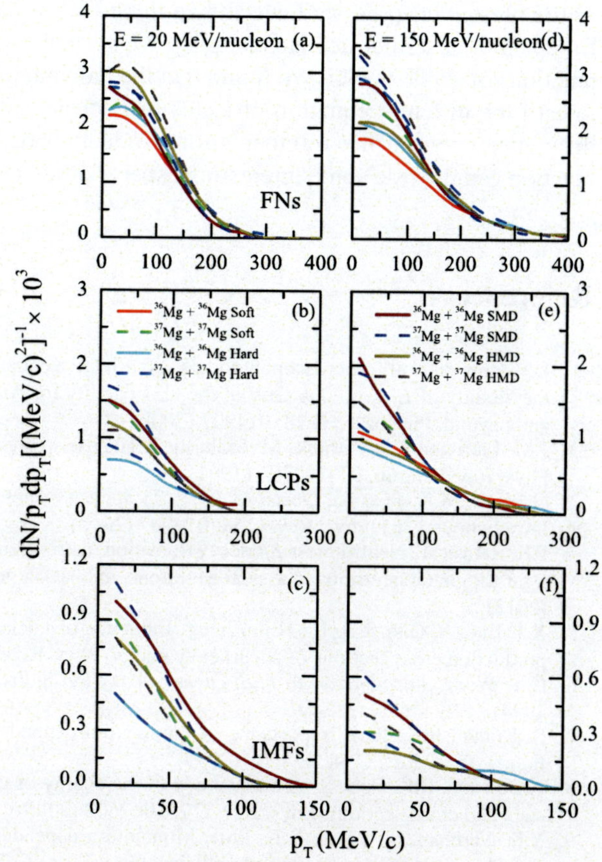

Fig. 8.5 The p_T spectra for central collisions of ^{36}Mg + ^{36}Mg and ^{37}Mg + ^{37}Mg at two different incident energies of 20 MeV/nucleon (left) and 150 MeV/nucleon (right panels). The solid and dashed lines represent the results of Soft, SMD, Hard, and HMD equation of state, respectively

nuclei reaction, therefore, the fragmentation of later can greatly contribute toward the understanding of nuclear matter properties.

Lastly, to further strengthen our point towards the role of halo structured nuclei, in Fig. 8.5 we examine the transverse momentum spectra (p_T-spectra) of free nucleons (FNs), light charged particles (LCPs), and intermediate mass fragments (IMFs) using different equations of state with and without momentum-dependent interactions. The representation of the symbols in the figure is same as in Fig. 8.4. The transverse momentum spectra gives us information about the amount of longitudinal momentum which is converted into the transverse momentum. Here, the results are reflecting the results in earlier paragraph. The fragments produced in the halo nuclei reactions have lesser momentum in the transverse direction. On the other hand, stable nuclei reaction have greater number of fragments in high p_T region compared lower ones, if the results are compared to stable ones.

8.4 Summary

In the present chapter, we have shown the effect of halo structure of nuclei on the fragmentation at intermediate energies. We have also compared our results with the reactions of stable nuclei. We found that the halo nuclei reactions are more sensitive toward the different equation of states of nuclear matter. We have also shown the difference between the fragmentation production between halo and stable nuclei reaction can provide some interesting aspects of nuclear equation of state.

References

1. L.V. Bravina et al., Microscopic models and effective equation of state in nuclear collisions in the vicinity of E_{lab} = 30 A GeV at the GSI Facility for Antiproton and Ion Research (FAIR) and beyond. Phys. Rev. C **78**, 014907 (2008)
2. J.M. Lattimer, C.J. Pethick, M. Prakash, P. Haensel, Direct URCA process in neutron stars. Phys. Rev. Lett. **66**, 2701–2704 (1991)
3. G. Kaur, M.K. Sharma, Decay of ^{150,158}Ti nuclear systems formed in reactions induced by loosely bound ^{6}Li. Phys. Rev. C **78**, 014907 (2013)
4. P.K. Rath et al., Neutrinoless $\beta\beta$ decay transition matrix elements within mechanisms involving light Majorana neutrinos, classical Majorons, and sterile neutrinos. Phys. Rev. C **88**, 064322 (2013)
5. X.J. Bao, Y. Gao, J.Q. Li, H.F. Zhang, Influence of nuclear basic data on the calculation of production cross sections of superheavy nuclei. Phys. Rev. C **92**, 014601 (2015)
6. C.Y. Wong, *Introduction to high-energy heavy-ion collisions* (World Scientific, Singapore, 1994)
7. S. Kumar, R.K. Puri, Importance of momentum dependent interactions in multifragmentationon. Phys. Rev. C **60**, 054607 (1999)
8. A.D. Sood, R.K. Puri, Systematic study of the energy of vanishing flow: role of equations of state and cross-sections. Phys. Rev. C **73**, 067602 (2006)
9. Y.K. Vermani, S. Goyal, R.K. Puri, Momentum dependence of the nuclear mean field and multifragmentation in heavy-ion collisions. Phys. Rev. C **78**, 064613 (2009)

10. I. Tanihata et al., Measurements of Interaction Cross Sections and Nuclear Radii in the Light p -Shell Region. Phys. Rev. Lett. **55**, 2676–2679 (1985)
11. I. Tanihata et al., Measurement of interaction cross sections using isotope beams of Be and B and isospin dependence of the nuclear radii. Phys. Lett. B **206**, 592–596 (1988)
12. R. Kumari, Study of fusion probabilities with halo nuclei using different proximity based potentials. Nucl. Phys. A **917**, 85–91 (2013)
13. J.Y. Liu et al., Special roles of loose neutron-halo nucleus structure on the fragmentation and momentum dissipation in heavy ion collisions. Phys. Lett. B **617**, 24–32 (2005)
14. M.K. Sharma et al., Search for halo structure in ^{37}Mg using the Glauber model and microscopic relativistic mean-field densities. Phys. Rev. C **93**, 014322 (2016)
15. M. Takechi et al., Evidence of halo structure in ^{37}Mg observed via reaction cross sections and intruder orbitals beyond the island of inversion. Phys. Rev. C **90**, 061305 (2014)
16. Ch. Hartnack et al., Modelling the many-body dynamics of heavy ion collisions: Present status and future perspective. Eur. Phys. J. A **1**, 151–169 (1998)
17. J. Aichelin, Quantum molecular dynamics a dynamical microscopic n-body approach to investigate fragment formation and the nuclear equation of state in heavy ion collisions. Phys. Rep. **202**, 233–360 (1991)
18. R. Kumar, S. Gautam, R.K. Puri, Multifragmentation within a clusterization algorithm based on thermal binding energies. Phys. Rev. C **89**, 064608 (2014); R. Kumar, R.K. Puri, Using experimental data to test an -body dynamical model coupled with an energy-based clusterization algorithm at low incident energies. Phys. Rev. C **97**, 034624 (2018)
19. Sucheta, R. Kumar, R.K. Puri, On the study of fragmentation of loosely bound nuclei using dynamical model. Proc. DAE Symp. Nucl. Phys. **63**, 556–557 (2018)
20. D. Sisan et al., Intermediate mass fragment emission in heavy-ion collisions: energy and system mass dependence. Phys. Rev. C **63**, 027602 (2001)
21. P. Bansal, S. Gautam, R.K. Puri, On the peak mass production of different fragments in intermediate-energy heavy-ion collisions. Eur. Phys. J. A **51**, 139 (2015)
22. A. Sharma, A. Bharti, Isospin effects via Coulomb forces on the onset of multifragmentation in light and heavily charged systems. Eur. Phys. J. A **52**, 42 (2016)
23. S. Kaur, A.D. Sood, Model ingredients and peak mass production in heavy-ion collisions. Phys. Rev. C **82**, 054611 (2010)
24. S. Kaur, R.K. Puri, Isospin effects on the energy of peak mass production. Phys. Rev. C **87**, 014620 (2013)
25. S. Sood, R. Kumar, A. Sharma, R.K. Puri, Cluster formation and phase transition in nuclear disassembly using a variety of clusterization algorithms. Phys. Rev. C **99**, 054612 (2019)
26. R. Kumar, S. Sood, A. Sharma, R.K. Puri, On the multifragmentation and phase transition in the perspectives of different n-body dynamical models. Act. Phys. Pol. B **49**, 301–306 (2018)

Chapter 9
PHQMD—A Microscopic Transport Approach for Heavy-Ion Collisions and Cluster Formation

Jörg Aichelin, M. Winn, E. L. Bratkovskaya, Arnaud Le Fèvre, Yvonne Leifels, V. Kireyeu, V. Kolesnikov, and V. Voronyuk

Abstract We present the basic ideas and new results of the recently advanced transport approach called PHQMD (Parton-Hadron-Quantum Molecular Dynamics) to describe heavy-ion collisions and especially the cluster formation in these reactions. This n-body transport approach allows to address physical processes which cannot be addressed by transport approaches which calculate the time evolution of the one-body phase space density or the one-body Greens function. They include the formation of fragments, especially of hyper-fragments, as well as the study of fluctuations as a signal of a possible first-order phase transition which some models predict if the strongly interacting matter is compressed to a high baryon density. This approach is at the same time an extension of the QMD model, which has successfully been used to describe fragments and other observables in heavy-ion collisions at beam energies below 2 AGeV and the PHSD transport approach which is a microscopic covariant dynamical transport approach for the one-body Greens function which successfully described a multitude of single-particle observables in p+p, p+A and A+A collisions from lower SIS ($E_{beam} = 1$ AGeV) to LHC energies ($\sqrt{s} =$ 5.2 TeV). We present the model and discuss how the parametrization of nuclear potential interactions influences the prediction for different observables.

J. Aichelin · M. Winn
SUBATECH, Université de Nantes, IMT Atlantique, IN2P3/CNRS 4 rue Alfred Kastler, 44307 Nantes cedex 3, France

J. Aichelin (✉)
Frankfurt Institute for Advanced Studies, Ruth Moufang Str. 1, 60438 Frankfurt, Germany
e-mail: aichelin@in2p3.fr

E. L. Bratkovskaya · A. Le Fèvre · Y. Leifels
GSI Helmholtzzentrum für Schwerionenforschung GmbH, Darmstadt, Germany

E. L. Bratkovskaya
Institut für Theoretische Physik, Goethe-Universität Frankfurt am Main, Frankfurt, Germany

V. Kireyeu · V. Kolesnikov · V. Voronyuk
Joint Institute for Nuclear Research, Joliot-Curie 6, 141980 Dubna, Moscow region, Russia

R. K. Puri et al. (eds.), *Advances in Nuclear Physics*, Springer Proceedings in Physics 257, https://doi.org/10.1007/978-981-15-9062-7_9

9.1 Introduction

The heavy-ion experiments at the Relativistic Heavy-Ion Collider (RHIC) at Brookhaven and at the Large Hadron Collider (LHC) at CERN have produced ample evidence that a new state of matter, a plasma of quarks and gluons (QGP), is produced [1]. Such a new state of matter has been predicted by lattice gauge calculations in which the thermal properties of strongly interacting matter are calculated by solving the partition function based on the Lagrangian of Quantum Chromo Dynamics (lQCD) [2, 3]. The analysis of these experiments shows that the QGP, produced at these energies, is characterized by a vanishing chemical potential and it is exactly this limit in which lQCD calculations are possible.

At very low beam energies, of the order of 1 AGeV, heavy-ion reactions are well described by hadronic degrees of freedom, baryons, and mesons.

The challenge is now to understand the transition between low energy, hadron dominated, reactions and that at high energy where quarks and gluons, the constituents of hadrons, are the right degrees of freedom to describe the experimental findings. This transition regime is characterized by a finite chemical potential or, in other words, by a finite baryon density. Phenomenological models predicted that for finite chemical potentials the transition between hadronic matter and a plasma of quarks and gluons is a first-order phase transition [4, 5] what increases the interest to study this region.

In order to study nuclear matter at finite baryon densities presently two accelerators are under construction, the Facility for Antiproton and Ion Research (FAIR) in Darmstadt and the Nuclotron-based Ion Collider fAcility (NICA) in Dubna. They will become operational in the next years. Moreover, the presently running BES-II (Beam Energy Scan) at RHIC, which includes a fixed target program, will also provide experimental data in this energy regime.

In order to understand the result of these experiments and to interpret the physical message of the different observables, transport approaches have to be employed which simulate the heavy-ion reaction on a computer. In these approaches, one can understand the origin of the different phenomena which are encoded in the experimental results. The PHQMD model [6] is such a transport approach which has been especially developed to address the seminal challenges of the physics at finite baryon density. These include the origin of the formation of clusters at midrapidity, observed at all energies, despite a temperature of the expanding fireball at midrapidity of the order of 100 MeV, an environment in which clusters, which have a typical binding energy of 8 MeV/nucleon cannot survive. These include as well the origin of the production of hyper-cluster, clusters with at least one strange baryon, the creation of multi-strange baryons and last but not least the identification of observables which signal an eventual first-order phase transition.

The presently available transport approaches are either nonrelativistic, like the QMD approach [7–9], or propagate single-particle Greens functions, like PHSD, are not well suited to address cluster formation. The novel Parton-Hadron-Quantum-Molecular Dynamics (PHQMD) approach overcomes these limitations. It is based on the collision integrals of the PHSD approach [10–14] and density dependent 2-body potential interactions of QMD type models [7, 15, 16]. The model uses as well the dynamical quasi particle description of the QGP of PHSD. The original PHSD mean-field propagation (realized within the parallel ensemble method) is kept as an option, too, which will allow to investigate the differences between both the approaches.

In contradistinction to statistical models or coalescence models, in PHQMD the clusters are formed dynamically. This means that at the end of the heavy-ion reaction the same nucleon-nucleon potential interaction, which is present during the whole time evolution, forms bound clusters of nucleons which are well distinct in phase space from other clusters and free nucleons.

9.2 The PHQMD Approach

The propagation of the Wigner density is determined by a generalized Ritz variational principle [17], which has been developed for the Time Dependent Hartree–Fock approach,

$$\delta \int_{t_1}^{t_2} dt < \psi(t)| i\frac{d}{dt} - H|\psi(t) >= 0. \tag{9.1}$$

In our approach [6] we assume that the n-body Wigner density is the direct product of the single-particle Wigner densities. There are also QMD versions which use a Slater determinant, FMD [17] and AMD [18], but due to the difficulty to formulate collision terms these approaches have only been applied to low energy heavy-ion collisions. Assuming that the wave functions have a Gaussian form and that the width of the wave function is time independent one obtains for the time evolution of the centroids of the Gaussian single particle wave functions two equations which resemble the equation of motion of a classical particle with the phase space coordinates $\mathbf{r}_{i0}$, $\mathbf{p}_{i0}$ [7].

The difference is that here the expectation value of the quantal Hamiltonian is used and not a classical Hamiltonian:

$$\dot{r}_{i0} = \frac{\partial \langle H \rangle}{\partial p_{i0}} \qquad \dot{p}_{i0} = -\frac{\partial \langle H \rangle}{\partial r_{i0}} \ . \tag{9.2}$$

These time evolution equations are specific for Gaussian wave functions. For other choices of wave functions the time evolution equations would be different. The Hamiltonian of the nucleus is the sum of the Hamiltonians of the nucleons, composed of kinetic and two-body potential energy,

$$H = \sum_i H_i = \sum_i (T_i + V_i) = \sum_i \left(T_i + \sum_{j \neq i} V_{i,j} \right) . \tag{9.3}$$

The interaction between the nucleons has two parts, a local Skyrme type interaction and a Coulomb interaction

$$\begin{aligned} V_{i,j} &= V(\mathbf{r_i}, \mathbf{r_j}, \mathbf{r_{i0}}, \mathbf{r_{j0}}, t) = V_{\text{Skyrme}} + V_{\text{Coul}}(+V_{mom}) \\ &= \frac{1}{2} t_1 \delta(\mathbf{r_i} - \mathbf{r_j}) + \frac{1}{\gamma+1} t_2 \delta(\mathbf{r_i} - \mathbf{r_j})\, \rho^{\gamma-1}(\mathbf{r_i}, \mathbf{r_j}, \mathbf{r_{i0}}, \mathbf{r_{j0}}, t) \\ &+ \frac{1}{2} \frac{Z_i Z_j e^2}{|\mathbf{r_i} - \mathbf{r_j}|}, \end{aligned} \tag{9.4}$$

with the density $\rho(\mathbf{r_i}, \mathbf{r_j}, \mathbf{r_{i0}}, \mathbf{r_{j0}}, t)$ defined as

$$\begin{aligned} &\rho(\mathbf{r_i}, \mathbf{r_j}, \mathbf{r_{i0}}, \mathbf{r_{j0}}, t) = \\ &= C \frac{1}{2} \Big[\sum_{j, i \neq j} \left(\frac{1}{\pi L} \right)^{3/2} e^{-\frac{1}{L}(\mathbf{r_i} - \mathbf{r_j} - \mathbf{r_{i0}(t)} + \mathbf{r_{j0}(t)})^2} \\ &+ \sum_{i, i \neq j} \left(\frac{1}{\pi L} \right)^{3/2} e^{-\frac{1}{L}(\mathbf{r_i} - \mathbf{r_j} - \mathbf{r_{i0}(t)} + \mathbf{r_{j0}(t)})^2} \Big], \end{aligned} \tag{9.5}$$

where C is a correction factor explained below.

We define the *interaction* density $\rho_{int}(\mathbf{r_{i0}}, t)$, which for non-relativistic case can be written as

$$\rho_{int}(\mathbf{r_{i0}}, t) = C \sum_{j, j \neq i} \left(\frac{1}{\pi L} \right)^{3/2} \mathrm{e}^{-\frac{1}{L}(\mathbf{r_{i0}}(t) - \mathbf{r_{j0}}(t))^2}. \tag{9.6}$$

The interaction density has twice the width of the particle density, and is obtained by calculating the expectation value of the local Skyrme potential which is $\propto \delta(\mathbf{r_i} - \mathbf{r_j})$. The correction factor C in (9.5) depends on L. It is introduced because nuclear densities are calculated differently in mean-field approaches—for which the Skyrme parametrization has been developed—and QMD approaches. In mean-field transport or hydrodynamical approaches the density, which enters the density-dependent two-body interaction, is obtained by summing over all particles in the system $\rho_{int}^{MF}(\mathbf{r_{i0}}, t) = \sum_j \ldots$. In QMD type approaches we have to exclude self-interactions and, therefore, the density which enters the density dependent interaction is the sum over all nucleons with the exception of that nucleon on which this density dependent potential acts, $\rho_{int}(\mathbf{r_{i0}}, t) = \sum_{j \neq i} \ldots$. Both differ by $(\frac{1}{\pi L})^{3/2}$. To compensate for the lower density in the QMD type approaches compared to the mean-field approaches we introduce the correction factor C which is adjusted numerically to achieve equality of both densities. With this correction factor we can use also for the QMD approach the Skyrme potentials.

The expectation value of the potential energy V_i, $\langle V_i \rangle = \langle V(\mathbf{r_{i0}}, t) \rangle$, of the nucleon i is given by

$$\langle V(\mathbf{r_{i0}}, t)\rangle = \sum_{j, j\neq i} \int d^3r_i d^3r_j d^3p_i d^3p_j V(\mathbf{r_i}, \mathbf{r_j}, \mathbf{r_{i0}}, \mathbf{r_{j0}}) \\ \times f(\mathbf{r_i}, \mathbf{p_i}, \mathbf{r_{i0}}, \mathbf{p_{i0}}, t) f(\mathbf{r_j}, \mathbf{p_j}, \mathbf{r_{j0}}, \mathbf{p_{j0}}, t), \tag{9.7}$$

with f being the Wigner density of the particles

$$f(\mathbf{r_i}, \mathbf{p_i}, \mathbf{r_{i0}}, \mathbf{p_{i0}}, t) = \frac{1}{\pi^3\hbar^3} e^{-\frac{2}{L}(\mathbf{r_i}-\mathbf{r_{i0}}(t))^2} e^{-\frac{L}{2\hbar^2}(\mathbf{p_i}-\mathbf{p_{i0}}(t))^2}. \tag{9.8}$$

Numerical test have shown that the time evolution of the system does not change if we replace $1/2(\rho_{int}(\mathbf{r_{i0}}, t) + \rho_{int}(\mathbf{r_{j0}}, t))$ by $\rho_{int}(\mathbf{r_{i0}}, t)$ or by $\rho_{int}(\mathbf{r_{j0}}, t)$. For the Skyrme potential we can therefore use the analytical form

$$\langle V_{Skyrme}(\mathbf{r_{i0}}, t)\rangle = \alpha\left(\frac{\rho_{int}(\mathbf{r_{i0}}, t)}{\rho_0}\right) + \beta\left(\frac{\rho_{int}(\mathbf{r_{i0}}, t)}{\rho_0}\right)^{\gamma}. \tag{9.9}$$

The expectation value of the Coulomb interaction can also be calculated analytically.

The expectation value of the Hamiltonian which enters in (9.2) is finally given by

$$\langle H\rangle = \langle T\rangle + \langle V\rangle \\ = \sum_i \left(\sqrt{p_{i0}^2 + m^2} - m\right) + \sum_i \langle V_{Skyrme}(\mathbf{r_{i0}}, t)\rangle. \tag{9.10}$$

p-A experiments have shown that the potential energy of nuclear matter does not only depend on the density but also on the relative momentum between proton and heavy ion. This momentum dependence is quite often neglected because it is difficult to implement in transport approaches. It is, however, important if one wants to describe the data quantitatively because momentum dependent interactions influence observables in a different way than static interactions. For example, the maximal density which is reached in heavy-ion reactions varies considerably as a function of the centrality and so does the static interaction. The momentum dependent interaction, on the contrary, remains constant. Here we introduce a potential which presents a fit to the most recent data [19] which are presented in Fig. 9.1. We see that they changed quite a bit as compared to the old data which are represented by a fit. This old fit has been used up to now in all QMD and IQMD calculations [20]. The momentum dependent part of the potential we parametrize by

$$V_{mom}(\mathbf{p_{i0}}, \mathbf{p_{j0}}) = \exp\left(-c\sqrt{(\mathbf{p_{i0}} - \mathbf{p_{j0}})^2}\right)\left(a(\mathbf{p_{i0}} - \mathbf{p_{j0}})^2 + b(\mathbf{p_{i0}} - \mathbf{p_{j0}})^4\right), \frac{\rho}{\rho_0} \tag{9.11}$$

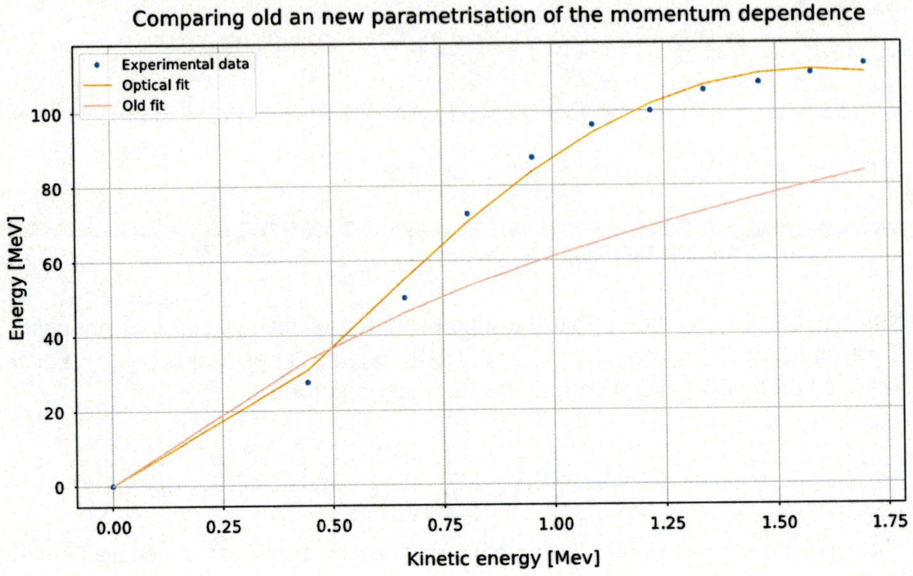

Fig. 9.1 The optical potential as a function of the beam momentum in p-A collisions. We display the old parametrization, used in IQMD calculations, with the new parametrization based on more recent data

with the parameters

- $a = 236.326\,(\text{MeV})^{-1}$
- $b = -20.7304\,(\text{MeV})^{-2}$
- $c = 0.901519\,\text{MeV}^{-1}$

where $\mathbf{p_{i0}}, \mathbf{p_{j0}}$ are the momenta of the nucleons in the center of mass system. We assume that the potential depends linearly on the density. This parametrization is shown as well in Fig. 9.1

The nuclear equation of state (EOS) describes the variation of the energy $E(T = 0, \rho/\rho_0)$ when changing the nuclear density in infinite matter to values different from the saturation density ρ_0 for zero temperature. In infinite matter the density is position independent and we can use (9.11) to connect our Hamiltionian with nuclear matter properties because for a given value of γ the parameters t_1, t_2 in (9.4) are uniquely related to the coefficients α, β of the EoS, (9.11).

Two of the 3 parameters of the static Skyrme potential can be fixed by the condition that the energy per nucleon has a minimum of $\frac{E}{A}(\rho_0) = -16$ MeV at ρ_0.

The third equation is historically provided by fixing the compression modulus K of nuclear matter, the inverse of the compressibility $\chi = \frac{1}{V}\frac{dV}{dP}$, which corresponds to the curvature of the energy at $\rho = \rho_0$ (for $T = 0$ K) is given in Table 9.1.

$$K = -V\frac{dP}{dV} = 9\rho^2 \frac{\partial^2(E/A(\rho))}{(\partial\rho)^2}|_{\rho=\rho_0}. \tag{9.12}$$

Table 9.1 Parameter sets for the nuclear equation of state used in the PHQMD model

EoS	Soft (S)	Hard (H)	Soft Momentum (SM)
α (MeV)	−383.5	−125.3	−478.87
β (MeV)	329.5	71.0	413.76
γ	1.15	2.0	1.10
K (MeV)	200	380	200

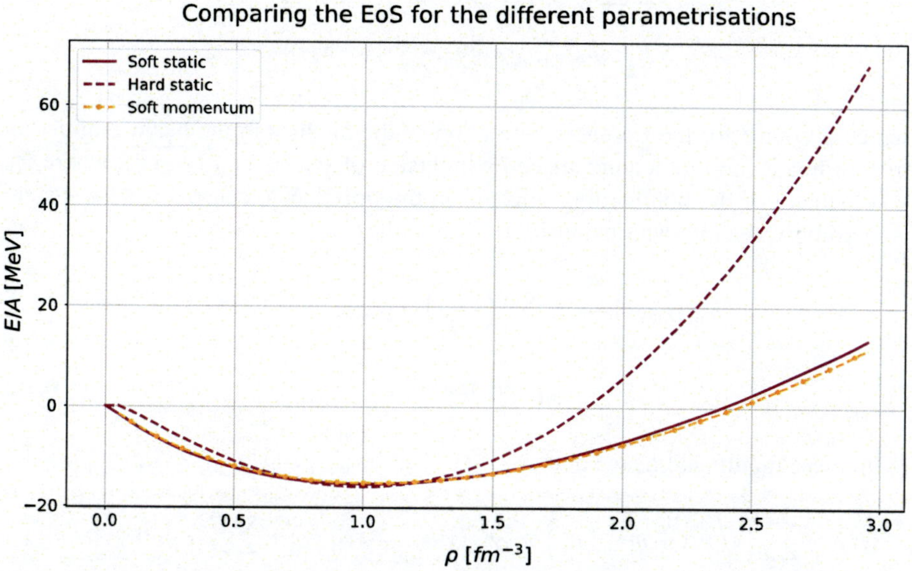

Fig. 9.2 The energy per nucleon for the three EoS: hard (dashed blue line), soft (solid blue line), and soft momentum dependent (dashed orange line)

Here P is the pressure in the system of volume V. An EOS with a rather low value of the compression modulus K yields a weak repulsion against the compression of nuclear matter and thus describes "soft" matter (denoted by "S" or "SM"). A high value of K causes a strong repulsion of nuclear matter under compression (called a hard EoS, "H"). The hard and soft EOS used in this study are illustrated in Fig. 9.2. Values of the parameters for the different model choices can be found in Table 9.1.

The influence of the nucleon potential and hence of the EoS on hadronic observables as well as on the cluster formation in heavy-ion collisions is well established at low energies (cf. e.g., [21]) where the nonrelativistic Hamiltonian formulation of QMD (presented in this section) is applicable. With increasing bombarding energies a relativistic dynamics becomes more important. The relativistic formulation of molecular dynamics has been developed in [15], however, the numerical realization of this method for realistic heavy-ion calculations is still not achievable with present computer power since it takes an about two orders of magnitude longer time to simulate the reaction due to the inversion of high dimensional matrices. Therefore, we are facing the problem of how to extend the nonrelativistic QMD approach to the high energy collisions, considered in this study, within a framework which can be numerically realized.

In order to extend our approach for relativistic energies, we introduce the modified single-particle Wigner density $\tilde{f}$ of the the nucleon i

$$\tilde{f}(\mathbf{r_i}, \mathbf{p_i}, \mathbf{r_{i0}}, \mathbf{p_{i0}}, t) = \tag{9.13}$$
$$= \frac{1}{\pi^3} \mathrm{e}^{-\frac{2}{L}(\mathbf{r_i^T}(t)-\mathbf{r_{i0}^T}(t))^2} \mathrm{e}^{-\frac{2\gamma_{cm}^2}{L}(\mathbf{r_i^L}(t)-\mathbf{r_{i0}^L}(t))^2}$$
$$\times \mathrm{e}^{-\frac{L}{2}(\mathbf{p_i^T}(t)-\mathbf{p_{i0}^T}(t))^2} \mathrm{e}^{-\frac{L}{2\gamma_{cm}^2}(\mathbf{p_i^L}(t)-\mathbf{p_{i0}^L}(t))^2},$$

which accounts for the Lorentz contraction of the nucleus in the beam z-direction, in coordinate and momentum space by inclusion of $\gamma_{cm} = 1/\sqrt{1 - v_{cm}^2}$, where v_{cm} is a velocity of the bombarding nucleon in the initial NN center-of-mass system. Accordingly, the interaction density (9.6) modifies as

$$\tilde{\rho}_{int}(\mathbf{r_{i0}}, t) \to C \sum_j \left(\frac{1}{\pi L}\right)^{3/2} \gamma_{cm}\, \mathrm{e}^{-\frac{1}{L}(\mathbf{r_{i0}^T}(t)-\mathbf{r_{j0}^T}(t))^2} \times \mathrm{e}^{-\frac{\gamma_{cm}^2}{L}(\mathbf{r_{i0}^L}(t)-\mathbf{r_{j0}^L}(t))^2}. \tag{9.14}$$

With these modifications we obtain

$$\langle \tilde{H} \rangle = \sum_i \left(\sqrt{p_{i0}^2 + m^2} - m\right) + \sum_i \langle \tilde{V}_{Skyrme}(\mathbf{r_{i0}}, t) \rangle + \sum_i \langle \tilde{V}_{mom}(\mathbf{r_{i0}}, \mathbf{p_{i0}}, t) \rangle, \tag{9.15}$$

with

$$\langle \tilde{V}_{Skyrme}(\mathbf{r_{i0}}, t) \rangle = \alpha \left(\frac{\tilde{\rho}_{int}(\mathbf{r_{i0}}, t)}{\rho_0}\right) + \beta \left(\frac{\tilde{\rho}_{int}(\mathbf{r_{i0}}, t)}{\rho_0}\right)^{\gamma}, \tag{9.16}$$

and

$$\langle V_{mom}(\mathbf{p_{i0}}, \mathbf{r_{i0}}) \rangle = \sum_{j0 \neq i0} \exp\left(-c\sqrt{(\mathbf{p_{i0}} - \mathbf{p_{j0}})^2}\right) \left(a(\mathbf{p_{i0}} - \mathbf{p_{j0}})^2 + b(\mathbf{p_{i0}} - \mathbf{p_{j0}})^4\right) \times \left(\frac{C\left(\frac{1}{\pi L}\right)^{3/2} \gamma_{cm}\, \mathrm{e}^{-\frac{1}{L}(\mathbf{r_{i0}^T}(t)-\mathbf{r_{j0}^T}(t))^2} \mathrm{e}^{-\frac{\gamma_{cm}^2}{L}(\mathbf{r_{i0}^L}(t)-\mathbf{r_{j0}^L}(t))^2}}{\rho_0}\right) \tag{9.17}$$

with the time evolution equations (9.2).

9.3 Energy Conservation

One of the essential conditions for a successful description of heavy-ion collisions is the energy conservation during the interaction. We use here a 4th order Runge Kutta approach to solve the system of coupled equations with a fixed time step (0.2 fm/c) method. Since all particles move on curved trajectories the time step should be small. Also for the stability of fragments a time step of this order is necessary. Figure 9.3 shows the time evolution of the total energy per nucleon in the nucleus-nucleus center of mass system for a very peripheral Au+Au collisions at a beam energy of 600 AMeV. We display as well separately the kinetic energy and the potential energy. We see that the energy is very well conserved in time. It varies for all equations of state by less than 1% during 160 fm/c. The S and SM equation of state have a very similar initial energy whereas the H equation of state has 3–4 MeV less energy. This is a consequence of the fact that we use here the same initial distribution of the nucleons in coordinate and momentum space for which the different equations of states give different potential energies (only in infinite nuclear matter they agree at saturation density). In order to have the same initial energy, we have to expand radially the initial configuration of the nucleon in coordinate space by a small amount.

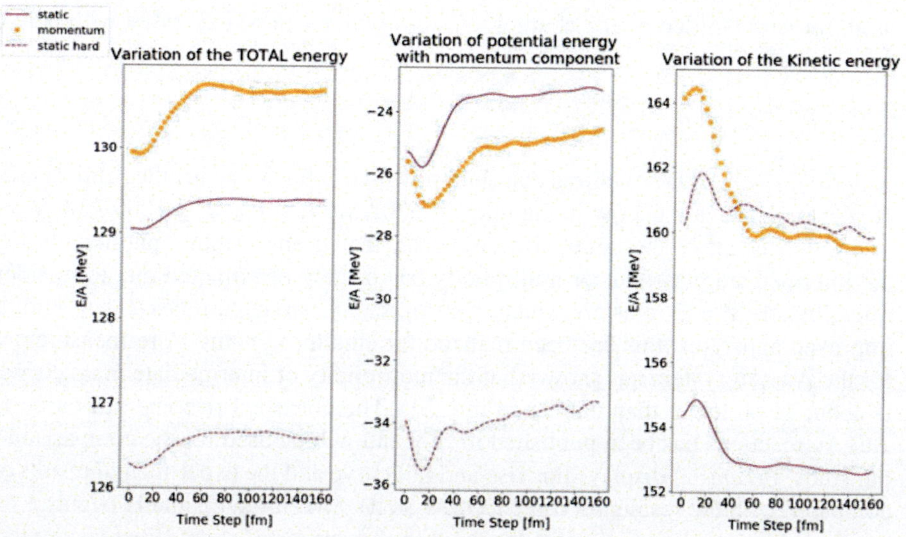

Fig. 9.3 The energy (Total, potential and kinetic) per nucleon for the three EoS: hard (dashed blue line), soft (full blue line), and soft momentum dependent (dotted orange line)

9.4 Results

It is very challenging to develop an approach which propagates clusters and baryons consistently and no satisfying approaches have been developed so far. Therefore, in the PHQMD approach we propagate baryons and mesons only and define a consistent theoretical method which allows to identify clusters consisting of the propagated nucleons. In our approach clusters are formed by the same nucleon–nucleon interaction which is active during the entire heavy-ion reaction. We call this *dynamical* cluster formation in contradistinction to models where fragments are created instantaneously at a given time like in coalescence models. In this article we identify the cluster by either the SACA [22, 23] or the MST algorithm [7]. The former is based on the idea of Dorso and Randrup [24] that the most bound configuration of nucleons and clusters, identified after the violent phase of the reaction, has a large overlap with the final distribution of clusters and free nucleons. The ALADIN collaboration has measured the cluster formation at beam energies between 600 AMeV and 1000 AMeV [25, 26], the largest beam energies for which cluster studies have been performed. Therefore, we use these data to study the dependence of the cluster formation on the nuclear equation of state. One of the key results of the ALADIN collaboration is the "rise and fall" of the multiplicity of intermediate mass clusters $3 \leq Z \leq 30$ emitted in forward direction. This multiplicity is presented as a function of the sum of all forward emitted bound charges, $Z_{\text{bound 2}}$ which can be expressed with help of the Θ function:

$$Z_{\text{bound 2}} = \sum_i Z_i \ \Theta(Z_i - (1+\epsilon)),$$

with $(0 < \epsilon < 1)$. One obtains a distribution which is for Au projectiles almost independent of the beam energy in the interval $600\text{AMeV} \leq E_{beam} \leq 1000$ AMeV and also independent of the target size. We note, that in the original publication [25] the intermediate mass cluster multiplicity has been overestimated due to misidentified, mostly $Z = 3$, clusters which were in reality two α particles. Later, with an improved apparatus, this has been realized for smaller systems. A re-measurement for the Au+Au system has shown that the multiplicity of intermediate mass clusters is about 15 % lower than published in [25] . The corrected rise and fall curve for Au+Au reactions has been published in [27] and will be used for the comparison in our study. Figure 9.4 displays this rise and fall curve and the experimental results are compared with the results for three EOS: S (left), SM (middle), and H (right).

We see a very different result for the three equations of state. Whereas the hard equation of state agrees quite nicely with the data, the soft and soft momentum dependent interaction give a quite different result. It is evident that a soft equation of state makes the nuclei quite unstable, especially in semiperipheral and peripheral interactions where $Z_{\text{bound 2}}$ is large. Although the excitation energy of the projectile and target remnants is small there the perturbation, caused by the interaction between projectile and target, is sufficient to create instabilities.

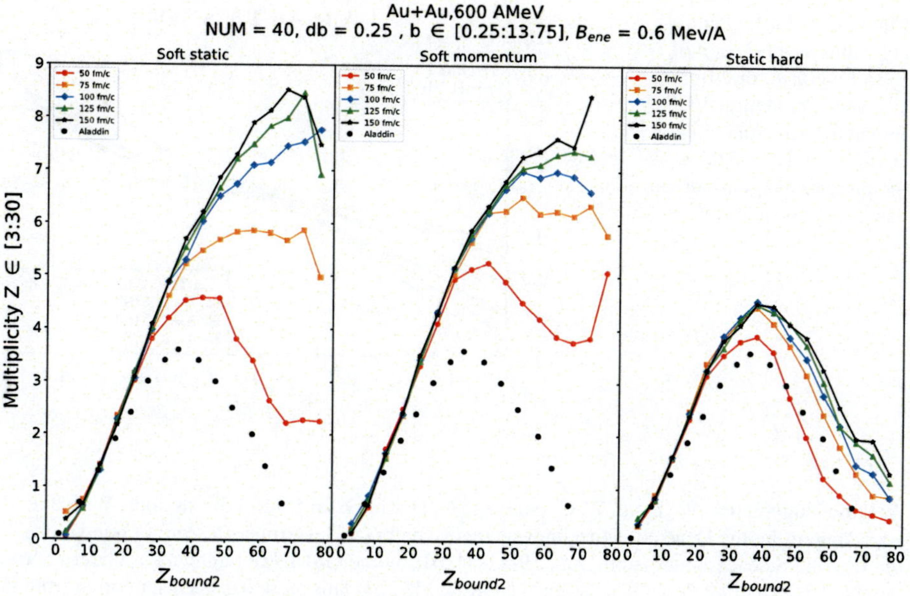

Fig. 9.4 Charged fragment multiplicity as a function of the total bound charge for the soft static interaction (left), the soft momentum interaction (center), and the hard static interaction (right). The fragments are identified with the SACA method

Another observable which is sensitive to the EOS is the in-plane flow, v_1, the first expansion coefficient of the Fourier series of the azimuthal distribution

$$\frac{dN}{d\phi} \propto 1 + 2v_1 cos(\phi) + 2v_2 cos(2\phi)....., \tag{9.18}$$

with $v_1 =< cos(\phi) >$. It has been measured by the FOPI collaboration [28]. If the EOS is harder the repulsive forces are larger but the same is true if the forces become momentum dependent. The results of our calculations for central Au+Au reactions at $E_{beam} = 1.5$ AGeV are displayed in Fig. 9.5. There we plot v_1 as a function of the scaled rapidity y/y_0, where y_0 is the projectile rapidity in the center of mass system. We see that indeed in central collision v_1 for a SM EOS is larger than for a S and comes close to that for a H EOS. We see as well that fragment shows a large v_1 because they are mainly composed of nucleons which passed the transverse surface of the overlap zone where the transverse forces are largest. Nucleons coming from the center of the overlap zone show a smaller v_1.

In conclusion, we presented the new PHQMD approach with special emphasis on the different parametrization of the nuclear equation of state. We see that the nuclear equation of state influences several observables, like the v_1 and the cluster production. We demonstrated as well that clusters have different properties in v_1 as compared to single nucleons. This opens the possibility to use these data to fix experimentally the nuclear EOS from heavy-ion data.

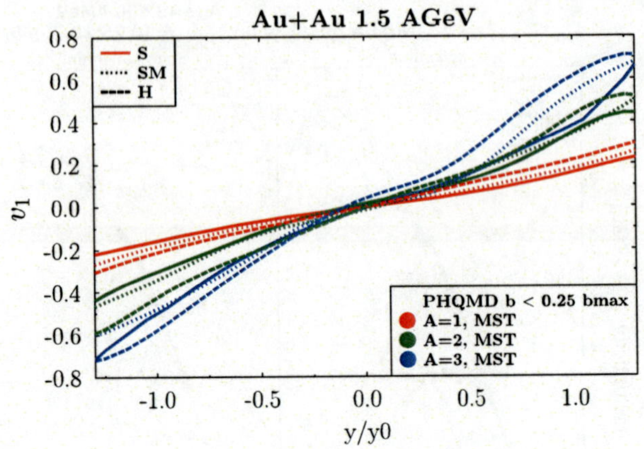

Fig. 9.5 v_1 for for central collisions as a function of y/y_0 for cluster of different size and for the three EOS for central Au+Au collision at $E_{beam} = 1.5$ AGeV. y_0 is the projectile rapidity in the center of mass system

Acknowledgements We thank W. Cassing and C. Hartnack for fruitful discussions. We acknowledge support by the Deutsche Forschungsgemeinschaft (DFG, German Research Foundation), by the Russian Science Foundation grant 19-42-04101, by the GSI-IN2P3 agreement under contract number 13-70 and by the COST Action THOR, CA15213. This work has been supported from the European Unions Horizon 2020 research programe under grant agreement No 824093.

References

1. U. Heinz, O. Evdokimov, P. Jacobs, (eds.), the proceedings of the 26th international conference on ultra-relativistic nucleus-nucleus collisions: quark matter. Nucl. Phys. A **961**, 1 (2017)
2. S. Borsanyi, Z. Fodor, C. Hoelbling, S.D. Katz, S. Krieg, K.K. Szabo, Full result for the QCD equation of state with 2+1 flavors. Phys. Lett. B **730**, 99–104 (2014)
3. A. Bazavov et al., [Hot QCD Collaboration], Equation of state in (2+1)-flavor QCD. Phys. Rev. D **90**, 094503 (2014)
4. M. Asakawa, K. Yazaki, Chiral restoration at finite density and temperature. Nucl. Phys. A **504**, 668–684 (1989)
5. M.A. Stephanov, QCD phase diagram and the critical point. Prog. Theor. Phys. Suppl. **153**, 139–156 (2004)
6. J. Aichelin, E. Bratkovskaya, A. Le Fèvre, V. Kireyeu, V. Kolesnikov, Y. Leifels, V. Voronyuk, Parton-Hadron-Quantum-Molecular Dynamics (PHQMD)—novel microscopic N-body transport approach for heavy-Ion collisions, dynamical cluster formation and hypernuclei production (2019). arXiv:1907.03860 [nucl-th]
7. J. Aichelin, Quantum molecular dynamics: a dynamical microscopic n body approach to investigate fragment formation and the nuclear equation of state in heavy ion collisions. Phys. Rep. **202**, 233–360 (1991)
8. J. Aichelin, A. Bohnet, G. Peilert, H. Stoecker, W. Greiner, A. Rosenhauer, Quantum molecular dynamics approach to heavy ion collisions: description of the model, comparison with fragmentation data, and the mechanism of fragment formation. Phys. Rev. C **37**, 2451 (1988)
9. C. Hartnack, R.K. Puri, J. Aichelin, J. Konopka, S.A. Bass, H. Stoecker, W. Greiner, Modeling the many body dynamics of heavy ion collisions: present status and future perspective. Eur. Phys. J. A **1**, 151–169 (1998)
10. W. Cassing, E.L. Bratkovskaya, Parton transport and hadronization from the dynamical quasiparticle point of view. Phys. Rev. C **78**, 034919 (2008)

11. W. Cassing, From Kadanoff-Baym dynamics to off-shell parton transport. Eur. Phys. J. ST **168**, 3 (2009)
12. W. Cassing, E.L. Bratkovskaya, Parton-Hadron-String Dynamics: an off-shell transport approach for relativistic energies. Nucl. Phys. A **831**, 215–242 (2009)
13. E.L. Bratkovskaya, W. Cassing, V.P. Konchakovski, O. Linnyk, Parton-Hadron-String Dynamics at relativistic collider energies. Nucl. Phys. A **856**, 162–182 (2011)
14. O. Linnyk, E.L. Bratkovskaya, W. Cassing, Effective QCD and transport description of dilepton and photon production in heavy-ion collisions and elementary processes. Prog. Part. Nucl. Phys. **87**, 50–115 (2016)
15. R. Marty, J. Aichelin, Molecular dynamics description of an expanding $q/\bar{q}$ plasma with the Nambu-Jona-Lasinio model and applications to heavy ion collisions at energies available at the BNL relativistic heavy Ion collider and the CERN large hadron collider. Phys. Rev. C **87**, 034912 (2013)
16. R. Marty, E. Bratkovskaya, W. Cassing, J. Aichelin, Observables in ultrarelativistic heavy-ion collisions from two different transport approaches for the same initial conditions. Phys. Rev. C **92**, 015201 (2015)
17. H. Feldmeier, Fermionic molecular dynamics. Nucl. Phys. A **515**, 147–172 (1990)
18. A. Ono, H. Horiuchi, T. Maruyama, A. Ohnishi, Antisymmetrized version of molecular dynamics with two nucleon collisions and its application to heavy ion reactions. Prog. Theor. Phys. **87**, 1185–1206 (1992)
19. S. Hama, B.C. Clark, E.D. Cooper, H.S. Sherif, R.L. Mercer, Global Dirac optical potentials for elastic proton scattering from heavy nuclei. Phys. Rev. C **41**, 2737–2755 (1990). https://doi.org/10.1103/PhysRevC.41.2737
20. J. Aichelin, A. Rosenhauer, G. Peilert, H. Stoecker, W. Greiner, Importance of momentum dependent interactions for the extraction of the nuclear equation of state from high-energy heavy Ion collisions. Phys. Rev. Lett. **58**, 1926–1929 (1987)
21. C. Hartnack, H. Oeschler, Y. Leifels, E.L. Bratkovskaya, J. Aichelin, Strangeness production close to threshold in proton-nucleus and heavy-Ion collisions. Phys. Rep. **510**, 119–200 (2012)
22. R.K. Puri, C. Hartnack, J. Aichelin, Early fragment formation in heavy Ion collisions. Phys. Rev. C **54**, R28–R31 (1996)
23. R.K. Puri, J. Aichelin, Simulated annealing clusterization algorithm for studying the multifragmentation. J. Comput. Phys. **162**, 245–266 (2000)
24. C. Dorso, J. Randrup, Early recognition of clusters in molecular dynamics. Phys. Lett. B **301**, 328–333 (1993)
25. A. Schuttauf et al., Universality of spectator fragmentation at relativistic bombarding energies. Nucl. Phys. A **607**, 457–486 (1996)
26. C. Sfienti et al., [ALADiN2000 Collaboration] Gross properties and isotopic phenomena in spectator fragmentation. Nucl. Phys. A **787**, 627–632 (2007)
27. A. Le Fèvre, Y. Leifels, J. Aichelin, C. Hartnack, V. Kireyev, E. Bratkovskaya, FRIGA, a new approach to identify isotopes and hypernuclei in n-body transport models. Nuovo Cim. C **39**, 399 (2017)
28. W. Reisdorf et al., [FOPI Collaboration] Systematics of azimuthal asymmetries in heavy ion collisions in the 1 A GeV regime. Nucl. Phys. A **876**, 1–60 (2012)

Chapter 10
PHSD—A Microscopic Transport Approach for Strongly Interacting Systems

E. L. Bratkovskaya, W. Cassing, P. Moreau, L. Oliva, O. E. Soloveva, and T. Song

Abstract We present the basic ideas of the Parton–Hadron–String Dynamics (PHSD) transport approach which is a microscopic covariant dynamical model for strongly interacting systems formulated on the basis of Kadanoff–Baym equations for Green's functions in phase-space representation (in 1st order gradient expansion beyond the quasiparticle approximation). The approach consistently describes the full evolution of a relativistic heavy-ion collision from the initial hard scattering and string formation through the dynamical deconfinement phase transition to the strongly interacting quark-gluon plasma (sQGP) as well as hadronization and the subsequent interactions in the expanding hadronic phase. The PHSD approach has been applied to p+p, p+A, and A+A collisions from lower SIS to LHC energies and been successful in describing a large number of experimental data including single-particle spectra, collective flow and electromagnetic probes. Some highlights of recent PHSD results will be presented.

10.1 Introduction

The phase transition from partonic degrees of freedom (quarks and gluons) to interacting hadrons is a central topic of modern high-energy physics. In order to understand the dynamics and relevant scales of this transition laboratory experiments under controlled conditions are performed with relativistic nucleus-nucleus collisions. Hadronic spectra and relative hadron abundances from these experiments reflect important aspects of the dynamics in the hot and dense zone formed in the early phase of the reaction and collective flows provide information on the transport

E. L. Bratkovskaya (✉) · L. Oliva · T. Song
GSI Helmholtzzentrum für Schwerionenforschung GmbH, Darmstadt, Germany
e-mail: elena.bratkovskaya@th.physik.uni-frankfurt.de

E. L. Bratkovskaya · P. Moreau · L. Oliva · O. E. Soloveva
Institut für Theoretische Physik, Goethe-Universität Frankfurt am Main, Frankfurt, Germany

W. Cassing
Institut für Theoretische Physik, Justus-Liebig-Universität Giessen, Giessen, Germany
e-mail: wolfgang.cassing@theo.physik.uni-giessen.de

R. K. Puri et al. (eds.), *Advances in Nuclear Physics*, Springer Proceedings in Physics 257, https://doi.org/10.1007/978-981-15-9062-7_10

properties of the medium generated on short time scales. Since relativistic heavy-ion collisions start with impinging nuclei in their ground states a proper non-equilibrium description of the entire dynamics through possibly different phases up to the final asymptotic hadronic states—eventually showing some degree of equilibration—is mandatory.

About 40 years ago cascade calculations have been employed for the description of nucleus–nucleus collisions in the 1-2 AGeV range [1] which provided already some good idea about the reaction dynamics including essentially nucleons, Δ-resonances, pions, and kaons. These calculations have been based on the Boltzmann equation which, however, is entirely classical and lacks quantum statistics appropriate for fermions and bosons. In particular the Pauli-blocking for nucleons was found to be essential at lower bombarding energies and cascade calculations were extended in line with the Uehling-Ulenbeck equation for fermions [2] incorporating also some mean-field potential calculated in Hartree approximation with various two-body Skyrme forces. These types of transport models are denoted as Boltzmann-Uehling-Uhlenbeck (BUU) or Vlasov-Uehling-Uhlenbeck (VUU) models [3, 4] and are still in use nowadays by some groups. Independently, Quantum Molecular Dynamical (QMD) models [5] have been proposed in which the test particles of the BUU/VUU approaches are replaced by Gaussians allowing for the simulation of single events while keeping the fluctuations. Explicit isospin degrees of freedom have been incorporated in IQMD [6], too. Since these types of models are based on a Hamiltonian with fixed two-body forces one could evaluate the nuclear equation of state (EoS) at zero temperature or in thermal equilibrium and one of the primary issues was to extract the nuclear EoS from heavy-ion data by means of BUU/VUU or QMD calculations. Later on higher baryonic resonances as well as mesons like η, $K^{\pm}$, K^0, $\bar{K}^0$, ρ, ω, ϕ have been incorporated which led to coupled-channel BUU (CBUU) approaches.

Apart from adding more hadronic degrees of freedom in BUU/VUU fully relativistic formulations have been carried out on the basis of some Lagrangian density including a selected set of hadronic degrees of freedom [7–9]. All baryons in such relativistic BUU (RBUU) models were propagated with scalar and vector self energies that were matched to reproduce collective flow data from heavy-ion collisions as well as particle spectra. This was a necessary step to go ahead in bombarding energy to ultra-relativistic p+A and A+A collisions, which were studied experimentally at the CERN SPS in the nineties. However, when increasing the number of degrees of freedom and adding high-mass short-lived resonances a lot of ambiguities entered the RBUU models since the couplings between the different hadronic species were unknown experimentally to a large extent. A way out was to incorporate the particle production by string formation and decay in line with the LUND model [10] which included only a formation time of hadrons ($\tau_F \approx 0.8$ fm/c) and a fragmentation function primarily fitted to hadron spectra from e^+e^- annihilation, where only a single string is formed. Familiar versions are the Hadron-String-Dynamics (HSD) [11, 12] or Ultra-relativistic Quantum Molecular Dynamics (UrQMD) [13] approaches that have been applied to p+A and A+A reactions in a wide range of energies up to the top SPS energy of 158 AGeV. In fact, a direct comparison between these two models for p+p and A+A collisions has provided very similar results for hadron spectra and flows up to $\sqrt{s_{NN}} = 17.3$ GeV [14]. Furthermore, a relativistic extension of the QMD model—based on the NJL Lagrangian—has been proposed in [15] but not followed up further except for a comparative study in [16].

By Legendre transformations the Hamiltonian density could be easily evaluated in the RBUU models and the nuclear EoS in thermal (or chemical) equilibrium, accordingly. However, it was soon noticed that with increasing temperature T and baryon density ρ_B (or baryon chemical potential μ_B) the energy density was likely to exceed some critical energy density ($\sim$1 GeV/fm^3) as indicated by early lattice QCD (lQCD) calculations which also showed that with increasing T a restoration of chiral symmetry should happen as seen from the temperature dependence of the scalar quark condensate $< \bar{q}q > (T)$. Furthermore, the interaction rates of strongly interacting hadrons reached a couple of hundred MeV at high baryon density ρ_B and temperature T such that the on-shell quasiparticle limit—applied in the standard models— became questionable. Furthermore, the spectral evolution especially of vector mesons in a hot and dense environment became of primary interest since the electromagnetic decay of vector mesons into dilepton pairs could be measured experimentally and was considered as a primary probe for the restoration of chiral symmetry in these media. To this end the relativistic transport approach was extended to off-shell dynamics on the basis of the Kadanoff–Baym dynamics in the turn of the Millenium [17–19] and it became possible to calculate the in-medium spectroscopy of vector mesons in heavy-ion collisions [20]. On the other hand, experimental observations at the Relativistic Heavy Ion Collider (RHIC) indicated that a new medium (Quark-Gluon Plasma (QGP)) was created in ultra-relativistic Au+Au collisions that is interacting more strongly than hadronic matter. Moreover, in line with theoretical studies in [21–23] the QCD medium showed phenomena of an almost perfect liquid of partons [24, 25] as extracted from the strong radial expansion and the scaling of elliptic flow $v_2(p_T)$ of mesons and baryons with the number of constituent quarks and antiquarks [24].

The question about the properties of this (non perturbative) QGP liquid became of primary interest as well as dynamical concepts describing the formation of color neutral hadrons from colored partons (hadronization). A fundamental issue for hadronization is the conservation of 4-momentum as well as the entropy problem because by fusion/coalescence of massless (or low constituent mass) partons to color neutral bound states of low invariant mass (e.g., pions) the number of degrees of freedom and thus the total entropy is reduced in the hadronization process [26–28]. This problem—a violation of the second law of thermodynamics as well as the conservation of four-momentum and flavor currents—has been addressed in [29] on the basis of the Dynamical QuasiParticle Model (DQPM) employing covariant transition rates for the fusion of massive quarks and antiquarks to color neutral hadronic resonances or strings. The DQPM is an effective field-theoretical model based on covariant propagators for quarks/antiquarks and gluons that have a finite width in their spectral functions (imaginary parts of the propagators). The determination/extraction of complex self energies for the partonic degrees of freedom has been performed in [30, 31] by fitting lattice QCD (lQCD) data within the DQPM and thus extracting a temperature-dependent effective coupling (squared) $g^2(T/T_c)$, where T_c denotes the critical temperature for the phase transition from hadrons to partons. This transition at low baryon chemical potential was found to be a crossover and the critical temperature T_c could be extracted from the lQCD data. In fact, the DQPM allows for a simple and transparent interpretation of lattice QCD results for thermodynamic quantities as well as correlators and leads to effective strongly interacting partonic quasiparticles with broad spectral functions. For a review on off-shell transport theory and results from the DQPM in comparison to lQCD we refer the reader to [32, 33].

Now a consistent dynamical approach—valid also for strongly interacting systems—could be formulated on the basis of Kadanoff–Baym (KB) equations [17] or off-shell transport equations in phase-space representation, respectively [17–19]. In the KB theory the field quanta are described in terms of dressed propagators with complex selfenergies (as in the DQPM). Whereas the real part of the selfenergies can be related to mean-field potentials (of Lorentz scalar, vector or tensor type), the imaginary parts provide information about the lifetime and/or reaction rates of time-like particles [32]. Once the proper (complex) selfenergies of the degrees of freedom are known the time evolution of the system is fully governed by off-shell transport equations (as described in [17, 32]).

10.2 The PHSD Approach

The Parton–Hadron–String-Dynamics approach is a microscopic covariant transport model that incorporates effective partonic as well as hadronic degrees of freedom and involves a dynamical description of the hadronization process from partonic to hadronic matter. Whereas the hadronic part is essentially equivalent to the conventional HSD approach [12] the partonic dynamics is based on the Dynamical Quasiparticle Model [30, 31] which describes QCD properties in terms of single-particle Green's functions in the form

$$G^R(\omega, \mathbf{p}) = \left(\omega^2 - \mathbf{p}^2 - M^2 + 2i\gamma\omega\right)^{-1}, \tag{10.1}$$

where M denotes the (resummed) mass of the parton and γ its width, while $(\omega, \mathbf{p})$ is the parton four-momentum. With the (essentially three) DQPM parameters for the temperature-dependent effective coupling $g^2(T/T_c)$ fixed by lattice QCD results the approach is fully defined in the partonic phase. We mention in passing that the off-shell transport equations can be solved within an extended test particle Ansatz [17, 32].

One might ask whether the quasiparticle properties—fixed in thermal equilibrium—should be appropriate also for the non-equilibrium configurations. This question is nontrivial and can only be answered by detailed investigations, e.g., on the basis of Kadanoff–Baym equations. We recall that such studies have been summarized in [32] for strongly interacting scalar fields that initially are far off-equilibrium and simulate momentum distributions of colliding systems at high relative momentum. The results for the effective parameters M and γ, which correspond to the time-dependent pole mass and width of the propagator (10.1), indicate that the quasiparticle properties—except for the very early off-equilibrium configuration—are close to the equilibrium mass and width even though the phase-space distribution of the particles is far from equilibrium (cf. Figs. 8 to 10 in [32]). Accordingly, we will adopt the equilibrium quasiparticle properties also for phase-space configurations out of equilibrium as appearing in relativistic heavy-ion collisions. The reader has to keep in mind that this approximation is well motivated, however, not fully equivalent to the exact solution.

On the hadronic side PHSD includes explicitly the baryon and antibaryon octet and decouplet, the 0^-- and 1^--meson nonets as well as selected higher resonances as in HSD [12]. Hadrons of higher masses (> 1.5 GeV in case of baryons and > 1.3 GeV in case of mesons) are treated as "strings" (color-dipoles) that decay to the known (low-mass) hadrons according to the JETSET algorithm [10]. We discard an explicit recapitulation of the string formation and decay and refer the reader to the original work [10].

10.2.1 Hadronization

Whereas the dynamics of partonic as well as hadronic systems is fixed by the DQPM or HSD, respectively, the change in the degrees of freedom has to be specified in line with the lattice QCD equation of state. The hadronization, i.e., the transition from partonic to hadronic degrees of freedom, has been introduced in [29, 34] and is repeated here for completeness. The hadronization is implemented in PHSD by local covariant transition rates, e.g., for $q + \bar{q}$ fusion to a mesonic state m of four-momentum $p = (\omega, \mathbf{p})$ at space-time point $x = (t, \mathbf{x})$:

$$\frac{dN_m(x,p)}{d^4x d^4p} = Tr_q Tr_{\bar{q}}\, \delta^4(p - p_q - p_{\bar{q}})\, \delta^4\left(\frac{x_q + x_{\bar{q}}}{2} - x\right) \omega_q\, \rho_q(p_q)\, \omega_{\bar{q}}\, \rho_{\bar{q}}(p_{\bar{q}})$$
$$\times |v_{q\bar{q}}|^2\, W_m(x_q - x_{\bar{q}}, (p_q - p_{\bar{q}})/2)\, N_q(x_q, p_q)\, N_{\bar{q}}(x_{\bar{q}}, p_{\bar{q}})\, \delta(\text{flavor, color}). \quad (10.2)$$

In (10.2) we have introduced the shorthand notation,

$$Tr_j = \sum_j \int d^4x_j \int \frac{d^4p_j}{(2\pi)^4}\,, \quad (10.3)$$

where $\sum_j$ denotes a summation over discrete quantum numbers (spin, flavor, color); $N_j(x, p)$ is the phase-space density of parton j at space-time position x and four-momentum p. In (10.2) δ(flavor, color) stands symbolically for the conservation of flavor quantum numbers as well as color neutrality of the formed hadronic state m which can be viewed as a color-dipole or "pre-hadron". Furthermore, $v_{q\bar{q}}(\rho_p)$ is the effective quark-antiquark interaction from the DQPM (displayed in Fig. 10 of [31]) as a function of the local parton ($q + \bar{q} + g$) density ρ_p (or energy density). Furthermore, $W_m(x, p)$ is the dimensionless phase-space distribution of the formed "pre-hadron", i.e.,

$$W_m(\xi, p_\xi) = \exp\left(\frac{\xi^2}{2b^2}\right) \exp\left(2b^2(p_\xi^2 - (M_q - M_{\bar{q}})^2/4)\right), \tag{10.4}$$

with $\xi = x_1 - x_2 = x_q - x_{\bar{q}}$ and $p_\xi = (p_1 - p_2)/2 = (p_q - p_{\bar{q}})/2$. The width parameter b is fixed by $\sqrt{\langle r^2 \rangle} = b = 0.66$ fm (in the rest frame) which corresponds to an average rms radius of mesons. We note that the expression (10.4) corresponds to the limit of independent harmonic oscillator states and that the final hadron-formation rates are approximately independent of the parameter b within reasonable variations. By construction the quantity (10.4) is Lorentz invariant; in the limit of instantaneous hadron-formation, i.e., $\xi^0 = 0$, it provides a Gaussian dropping in the relative distance squared $(\mathbf{r}_1 - \mathbf{r}_2)^2$. The four-momentum dependence reads explicitly (except for a factor 1/2)

$$(E_1 - E_2)^2 - (\mathbf{p}_1 - \mathbf{p}_2)^2 - (M_1 - M_2)^2 \leq 0, \tag{10.5}$$

and leads to a negative argument of the second exponential in (10.4) favoring the fusion of partons with low relative momenta $p_q - p_{\bar{q}} = p_1 - p_2$.

Some comments on the hadronization scheme are in order: The probability for a quark to hadronize is essentially proportional to the time step dt in the calculation, the number of possible hadronization partners in the volume $dV \sim 5$ fm^3, and the transition matrix element squared (apart from the gaussian overlap function). For temperatures above T_c the probability is rather small ($\ll 1$) but for temperatures close to T_c and below T_c the matrix element becomes very large since it essentially scales with the effective coupling squared $g^2(T/T_c)$ which is strongly enhanced in the infrared. For a finite timestep dt—as used in the calculations—the probability becomes larger than 1 which implies that the quark has to hadronize with some of the potential antiquarks in the actual timestep if the temperature or energy density becomes too low. Furthermore, the gluons practically freeze out close to T_c since the mass difference between quarks and gluons increase drastically with decreasing temperature and the reaction channel $g \leftrightarrow q + \bar{q}$ is close to equilibrium. This implies that all partons hadronize. Due to numerics some leftover partons may occur at the end of the calculations which are forced to hadronize by increasing the volume dV until they have found a suitable partner. In practice the forced hadronization was only used for LHC energies where the computational time was stopped at ~ 1000 fm/c when partons with rapidities close to projectile or target rapidity did not yet hadronize due to time dilatation ($\gamma_{cm} \approx 1400$).

Related transition rates (10.2) are defined for the fusion of three off-shell quarks ($q_1 + q_2 + q_3 \leftrightarrow B$) to a color neutral baryonic (B or $\bar{B}$) resonances of finite width (or strings) fulfilling energy and momentum conservation as well as flavor current conservation (cf. [34]). In contrast to the familiar coalescence models this hadronization scheme solves the problem of simultaneously fulfilling all conservation laws and the constraint of entropy production. For further details we refer the reader to [29, 34].

10.2.2 Initial Conditions

The initial conditions for the parton/hadron dynamical system have to be specified additionally. In order to describe relativistic heavy-ion reactions we start with two nuclei in their semi-classical groundstate, boosted toward each other with a velocity β (in z-direction), fixed by the bombarding energy. The initial phase-space distributions of the projectile and target nuclei are determined in the local Thomas-Fermi limit as in the HSD transport approach [12] or the UrQMD model [13]. We recall that at relativistic energies the initial interactions of two nucleons are well described by the excitation of two color neutral strings which decay in time to the known hadrons (mesons, baryons, antibaryons) [10]. Initial hard processes—i.e., the short-range high-momentum transfer reactions that can be well described by perturbative QCD—are treated in PHSD (as in HSD) via PYTHIA. The novel element in PHSD (relative to HSD) is the string melting concept as also used in the A Multi-Phase Transport (AMPT) model [28] in a similar context. However, in PHSD the strings (or possibly formed hadrons) are only allowed to melt if the local energy density $\epsilon(x)$ (in the local rest frame) is above the transition energy density ϵ_c which in the DQPM is $\epsilon_c \approx 0.5$ GeV/fm^3. The mesonic strings then decay to quark-antiquark pairs according to an intrinsic quark momentum distribution,

$$F(\mathbf{q}) \sim \exp(-2b^2\mathbf{q}^2) \,, \tag{10.6}$$

in the meson rest frame (cf. (10.2) for the inverse process). The parton final four-momenta are selected randomly according to the momentum distribution (10.6) (with b= 0.66 fm), and the parton-energy distribution is fixed by the DQPM at given energy density $\epsilon(\rho_s)$ in the local cell with scalar parton density ρ_s. The flavor content of the $q\bar{q}$ pair is fully determined by the flavor content of the initial string. By construction the "string melting" to massive partons conserves energy and momentum as well as the flavor content. In contrast to [28] the partons are of finite mass— in line with their local spectral function—and obtain a random color $c = (1, 2, 3)$ or (r, b, g) in addition. Of course, the color appointment is color neutral, i.e., when selecting a color c for the quark randomly the color for the antiquark is fixed by $-c$. The baryonic strings melt analogously into a quark and a diquark while the diquark, furthermore, decays to two quarks. Dressed gluons are generated by the fusion of nearest neighbor $q + \bar{q}$ pairs ($q + \bar{q} \to g$) that are flavor neutral until the ratio of gluons to quarks reaches the value $N_g/(N_q + N_{\bar{q}})$ given by the DQPM for the energy density of the local cell. This recombination is performed for all cells in space during the passage time of the target and projectile (before the calculation continues with the next timestep) and conserves the four-momentum as well as the flavor currents. We note, however, that the initial phase in PHSD is dominated by quark and antiquark degrees of freedom [35].

Apart from proton-proton, proton-nucleus or nucleus-nucleus collisions the PHSD approach can also be employed to study the properties of the interacting hadron/parton system in a finite box with periodic boundary conditions [36]. To this aim the system is initialized by a homogeneous distribution of test particles in a finite box with a momentum distribution close to a thermal one. Note that in PHSD the system cannot directly be initialized by a temperature and chemical potential since these Lagrange parameters can only be determined when the system has reached a thermal and chemical equilibrium, i.e., when all forward and backward reaction rates have become equal; this is easy to check in the transport simulations.

10.2.3 Partonic Cross Sections

On the partonic side the following elastic and inelastic interactions are included in PHSD $qq \leftrightarrow qq$, $\bar{q}\bar{q} \leftrightarrow \bar{q}\bar{q}$, $gg \leftrightarrow gg$, $gg \leftrightarrow g$, $q\bar{q} \leftrightarrow g$, $qg \leftrightarrow qg$, $g\bar{q} \leftrightarrow g\bar{q}$ exploiting detailed-balance with cross sections calculated from the leading Feynman diagrams employing the effective propagators and couplings $g^2(T/T_c)$ from the DQPM [37]. As an example we show in Fig. 10.1 the leading order Feynman diagrams for the $qq' \to qq'$ and $q\bar{q} \to q'\bar{q}'$ processes.

Partonic reactions such as $g + q \leftrightarrow q$ or $g + g \leftrightarrow q + \bar{q}$ have been discarded in the present calculations due to their low rates since the large mass of the gluon leads to a strong mismatch in the energy thresholds between the initial and final channels. In this case q stands for the 4 lightest quarks (u, d, s, c). Furthermore, the evaluation of photon and dilepton production is calculated perturbatively and channels like $g + q \to q + \gamma$ are included. In this case the probability for photon (dilepton) production from each channel is added up and integrated over space and

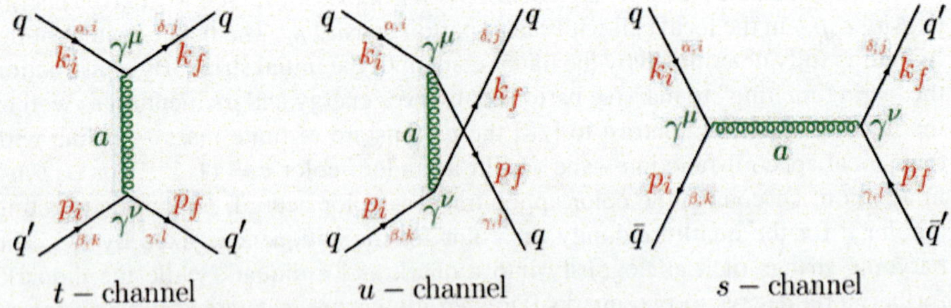

Fig. 10.1 Leading order Feynman diagrams for the $qq' \to qq'$ and $q\bar{q} \to q'\bar{q}'$ processes. The initial and final 4-momenta are k_i and p_i, and k_f and p_f, respectively. The indices $i, j, k, l = 1 - 3$ denote the quark colors, $a = 1 - 8$ the gluon colors while the quark flavor is indicated by the indices $\alpha, \beta, \delta, \gamma = u, d, s, ...$

time [33] without introducing any new parameter in the PHSD approach since the electromagnetic coupling is well known.

Numerical tests of the parton dynamics with respect to conservation laws, interaction rates in and out-off equilibrium in a finite box with periodic boundary conditions have been presented in [39]. In fact, in [39] it was shown that the PHSD calculations in the box give practically the same results in equilibrium as the DQPM. We note in passing that the total energy is conserved in the box calculations up to about 3 digits while in the heavy-ion collisions addressed here in the following the violation of energy conservation is typically less than 1% [34].

10.3 Transport Properties of the Partonic System

The starting point to evaluate viscosity coefficients of partonic matter is the Kubo formalism [38] which was also used to calculate the viscosities within the PHSD in a box with periodic boundary conditions (cf. [39]). We focus here on the calculation of the shear viscosity based on [40] which reads

$$\eta^{\rm Kubo}(T,\mu_q) = -\int \frac{d^4p}{(2\pi)^4}\, p_x^2 p_y^2 \sum_{i=q,\bar{q},g} d_i \frac{\partial f_i(\omega)}{\partial\omega}\, \rho_i(\omega,\mathbf{p})^2 \tag{10.7}$$

$$= \frac{1}{15T}\int \frac{d^4p}{(2\pi)^4}\, \mathbf{p}^4 \sum_{i=q,\bar{q},g} d_i\, ((1\pm f_i(\omega)) f_i(\omega))\, \rho_i(\omega,\mathbf{p})^2,$$

where the notation $f_i(\omega) = f_i(\omega, T, \mu_q)$ is used for the distribution functions, and ρ_i denotes the spectral function of the partons, while d_i stand for the degeneracy factors. We note that the derivative of the distribution function accounts for the Pauli-blocking (–) and Bose-enhancement (+) factors. Following [41], we can evaluate the integral over $\omega = p_0$ in (10.7) by using the residue theorem. When keeping only the leading order contribution in the width $\gamma(T, \mu_B)$ from the residue—evaluated at the poles of the spectral function $\omega_i = \pm\tilde{E}(\mathbf{p}) \pm i\gamma$—we finally obtain

$$\eta^{\rm RTA}(T,\mu_q) = \frac{1}{15T}\int \frac{d^3p}{(2\pi)^3} \sum_{i=q,\bar{q},g} \frac{\mathbf{p}^4}{E_i^2\, \Gamma_i(\mathbf{p},T,\mu_q)}\, d_i\, ((1\pm f_i(E_i)) f_i(E_i))\,, \tag{10.8}$$

which corresponds to the expression derived in the relaxation-time approximation (RTA) [42] by identifying the interaction rate Γ with 2γ as expected from transport theory in the quasiparticle limit [43]. We recall that γ is the width parameter in the parton propagator (1). The interaction rate $\Gamma_i(\mathbf{p}, T, \mu_q)$ (inverse relaxation time) is calculated microscopically from the collision integral using the differential cross sections for parton scattering as described in Section 2.3. We, furthermore, recall that the pole energy is $E_i^2 = p^2 + M_i^2$, where M_i is the pole mass given in the DQPM.

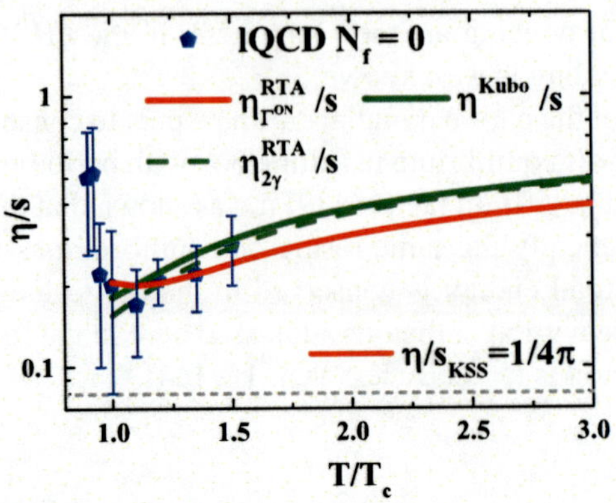

Fig. 10.2 The ratio of shear viscosity to entropy density as a function of the scaled temperature T/T_c for $\mu_B = 0$ from (10.7)–(10.8). The solid green line (η^{Kubo}/s) shows the results from the original DQPM in the Kubo formalism while the dashed green line ($\eta^{\text{RTA}}_{2\gamma}/s$) shows the same result in the quasiparticle approximation (10.8). The solid red line ($\eta^{\text{RTA}}_{\Gamma^{\text{on}}}/s$) results from (10.8) using the interaction rate Γ^{on} calculated by the microscopic differential cross sections in the on-shell limit. The dashed gray line demonstrates the Kovtun-Son-Starinets bound [44] $(\eta/s)_{\text{KSS}} = 1/(4\pi)$, and the symbols show lQCD data for pure SU(3) gauge theory obtained within the Backus-Gilbert method taken from [45] (pentagons)

We use here the notation $\sum_{j=q,\bar{q},g}$ which includes the contribution from all possible partons which in our case are the gluons and the (anti-)quarks of three different flavors (u, d, s).

The actual results are displayed in Fig. 10.2 for the ratio of shear viscosity to entropy density η/s as a function of the scaled temperature T/T_c for $\mu_B = 0$ in comparison to those from lattice QCD [45]. The solid green line (η^{Kubo}/s) shows the result from the original DQPM in the Kubo formalism while the dashed green line ($\eta^{\text{RTA}}_{2\gamma}/s$) shows the same result in the relaxation-time approximation (10.8) by replacing Γ_i by $2\gamma_i$. The solid red line ($\eta^{\text{RTA}}_{\Gamma^{\text{on}}}/s$) results from (10.8) using the interaction rate Γ^{on} calculated by the microscopic differential cross sections in the on-shell limit. We find that—apart from temperatures close to T_c—the ratios η/s do not differ very much and have a similar behavior as a function of temperature. The approximation (10.8) of the shear viscosity is found to be very close to the one from the Kubo formalism (10.7) indicating that the quasiparticle limit ($\gamma \ll M$) holds in the DQPM.

An overview for the ratio of shear viscosity to entropy density η/s as a function of the scaled temperature $T/T_c(\mu_B)$ and μ_B is given Fig. 10.3 in case of the Kubo formalism (a) (10.7) and the relaxation-time approximation (10.8) (b). There is no strong variation with μ_B for fixed $T/T_c(\mu_B)$, however, the ratio increases slightly with μ_B in the on-shell limit while it slightly drops with μ_B in the Kubo formalism for the DQPM. Accordingly, there is some model uncertainty when extracting the shear viscosity in the different approximations.

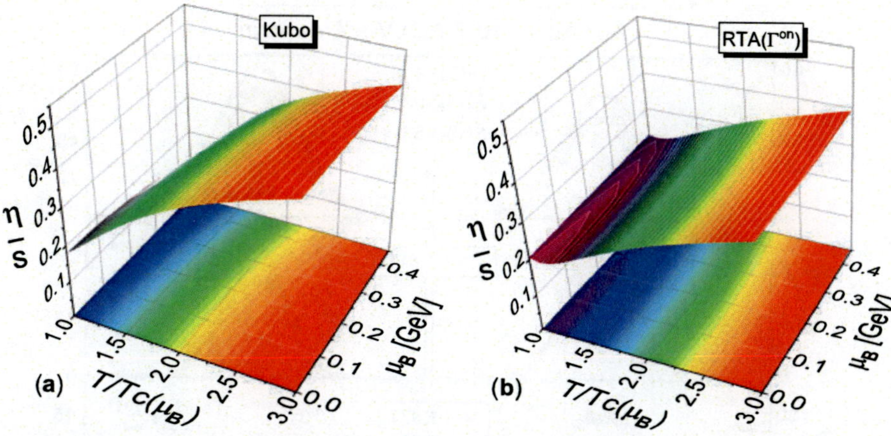

Fig. 10.3 The ratio of shear viscosity to entropy density η/s as a function of the scaled temperature $T/T_c(\mu_B)$ and baryon chemical potential μ_B calculated within the Kubo formalism (a) from (10.7) and in the Relaxation Time Approximation (RTA) (b) from (10.8) using the on-shell interaction rate Γ^{on}

In summarizing this section we find that the results for the ratio of shear viscosity over entropy density from the original DQPM and those from the microscopic calculations are similar and within error bars compatible with present results from lattice QCD. However, having the differential cross sections for each partonic channel at hand one might find substantial differences for non-equilibrium configurations as encountered in relativistic heavy-ion collisions where a QGP is formed initially out-off equilibrium.

10.4 Observables from Relativistic Nucleus-Nucleus Collisions

We briefly report on results from PHSD calculations at lower and intermediated energies covered experimentally by the AGS (BNL) and SPS (CERN) with a focus on central Au+Au or Pb+Pb collisions. In this energy range, the average baryon chemical potential μ_B is essentially finite—contrary to RHIC and LHC energies—and one might find some traces of the explicit μ_B dependence of the partonic cross sections in observables. To this end, we compare results for the rapidity distributions from the PHSD calculations based on the default DQPM parameters (PHSD4.0) [46] with the new PHSD5.0 including the differential partonic cross sections for the individual partonic channels at finite T and μ_B (cf. [37]). A comparison to the available experimental data is included (for orientation) but not discussed explicitly since this has been done in more detail in [46]. When implementing the differential cross sections and parton masses into the PHSD5.0 approach one has to specify the

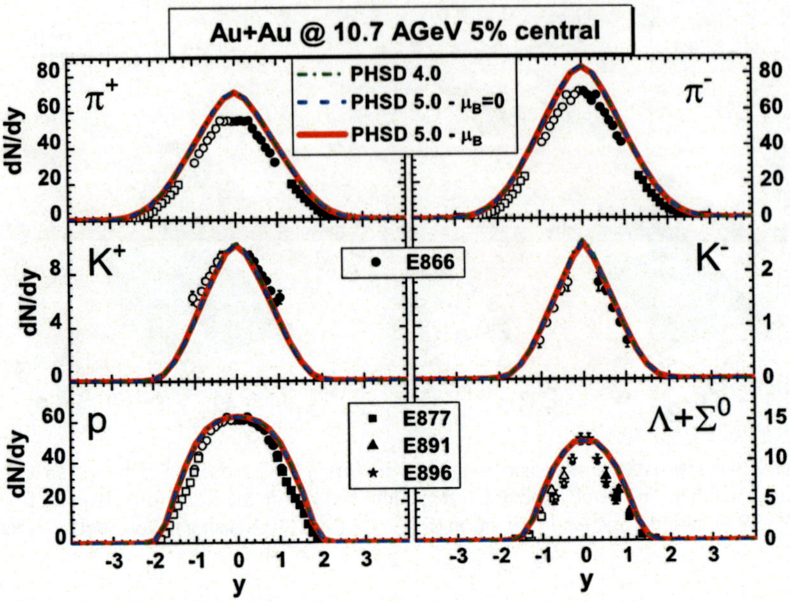

Fig. 10.4 The rapidity distributions for 5% central Au+Au collisions at 10.7 AGeV for PHSD4.0 (green dot-dashed lines), PHSD5.0 with partonic cross sections and parton masses calculated for $\mu_B = 0$ (blue dashed lines) and with cross sections and parton masses evaluated at the actual chemical potential μ_B in each individual space-time cell (red lines) in comparison to the experimental data from the E866 [47], E877 [48], E891 [49], E877 [50], and E896 [51] collaborations. All PHSD results are the same within the linewidth

Lagrange parameters T and μ_B in each computational cell in space-time. This has been done by employing the DQPM equation of state, which is practically identical to the lattice QCD equation of state, and a diagonalization of the energy-momentum tensor from PHSD as described in [37].

Figure 10.4 displays the actual results for hadronic rapidity distributions in case of 5% central Au+Au collisions at 10.7 AGeV for PHSD4.0 (green dot-dashed lines), PHSD5.0 with partonic cross sections and parton masses calculated for $\mu_B = 0$ (blue dashed lines), and with cross sections and parton masses evaluated at the actual chemical potential μ_B in each individual space-time cell (red lines) in comparison to the experimental data from the E866 [47], E877 [48], E891 [49], E877 [50], and E896 [51] collaborations. Here we focus on the most abundant hadrons, i.e., pions, kaons, protons, and neutral hyperons. We note in passing that the effects of chiral symmetry restoration are incorporated as in [46] since this was found to be mandatory to achieve a reasonable description of the strangeness degrees of freedom reflected in the kaon and neutral hyperon dynamics. As seen from Fig. 10.4 there is no difference in rapidity distributions for all the hadron species from the different versions of PHSD within linewidth which implies that there is no sensitivity to the new partonic differential cross sections and parton masses employed. One could argue that this result might be due to the low amount of QGP produced at this energy

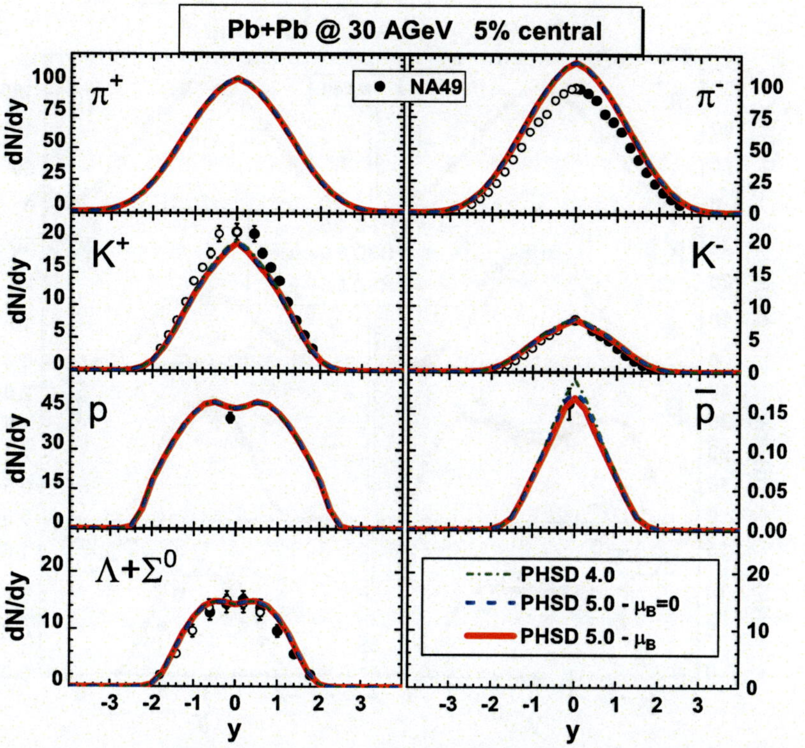

Fig. 10.5 The rapidity distributions for 5% central Pb+Pb collisions at 30 AGeV for PHSD4.0 (green dot-dashed lines), PHSD5.0 with partonic cross sections and parton masses calculated for $\mu_B = 0$ (blue dashed lines) and with cross sections and parton masses evaluated at the actual chemical potential μ_B in each individual space-time cell (red lines) in comparison to the experimental data from the NA49 Collaboration [52–54]. All PHSD results are practically the same within the linewidth

but the different PHSD calculations for 5% central Pb+Pb collisions at 30 AGeV in Fig. 10.5 for the hadronic rapidity distributions do not provide a different picture, too. Only when stepping up to the top SPS energy of 158 AGeV one can identify a small difference in the antibaryon sector ($\bar{p}$, $\bar{\Lambda} + \bar{\Sigma}^0$) in case of 5% central Pb+Pb collisions (cf. Fig. 10.6).

10.5 Summary

In this contribution, we have described the PHSD transport approach [33] and its recent extension to PHSD5.0 [37] to incorporate differential "off-shell" cross sections for all binary partonic channels that are based on the same effective propagators and couplings as employed in the QGP equation of state and the parton propagation. To this end we have recalled the extraction of the partonic masses and the coupling g^2

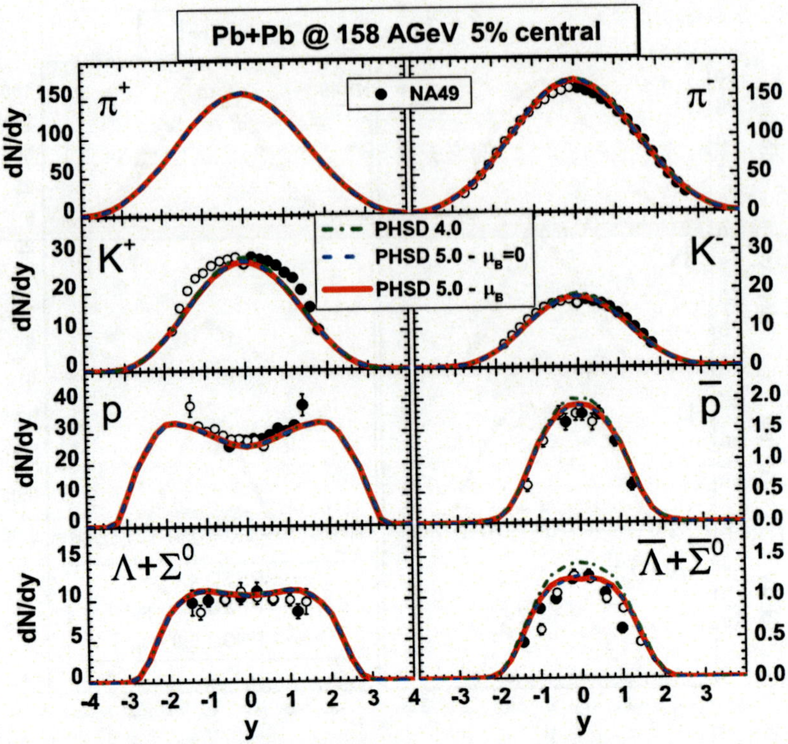

Fig. 10.6 The rapidity distributions for 5% central Pb+Pb collisions at 158 AGeV for PHSD4.0 (green dot-dashed lines), PHSD5.0 with partonic cross sections and parton masses calculated for μ_B = 0 (blue dashed lines) and with cross sections and parton masses evaluated at the actual chemical potential μ_B in each individual space-time cell (red lines) in comparison to the experimental data from the NA49 Collaboration [55–58]. All PHSD results are the same within the linewidth except for the antibaryons

from lattice QCD data (within the DQPM) and calculated the partonic differential cross sections as a function of T and μ_B for the leading tree-level diagrams (cf. Appendices of [37]). Furthermore, we have used these differential cross sections to evaluate partonic scattering rates $\Gamma_i(T, \mu_B)$ for fixed T and μ_B as well as to compute the ratio of the shear viscosity η to entropy density s within the Kubo formalism in comparison to calculations from lQCD. It turns out that the ratio η/s calculated with the partonic scattering rates in the relaxation-time approximation is very similar to the original result from the DQPM and to lQCD results such that the present extension of the PHSD approach does not lead to different partonic transport properties. We recall that the novel PHSD version (PHSD5.0) is practically parameter free in the partonic sector since the effective coupling (squared) is determined by a fit to the scaled entropy density from lQCD. The dynamical masses for quarks and gluons then are fixed by the HTL expressions. The interaction rate in the time-like sector is, furthermore, calculated in leading order employing the DQPM propagators and coupling.

When implementing the differential cross sections and parton masses into the PHSD5.0 approach one has to specify the Lagrange parameters T and μ_B in each computational cell in space-time. This has been done by employing the DQPM equation of state, which is practically identical to the lattice QCD equation of state, and a diagonalization of the energy-momentum tensor from PHSD as described in [37].

In Sect. 10.4 we then have calculated 5% central Au+Au (or Pb+Pb) collisions and compared the results for hadronic rapidity distributions from the previous PHSD4.0 with the novel version PHSD5.0 (with and without the explicit dependence of the partonic differential cross sections and parton masses on μ_B). No differences for all the hadron bulk observables from the various PHSD versions have been found at AGS and FAIR/NICA energies within linewidth which implies that there is no sensitivity to the new partonic differential cross sections employed. Only in case of the kaons and the antibaryons $\bar{p}$ and $\bar{\Lambda} + \bar{\Sigma}^0$, a small difference between PHSD4.0 and PHSD5.0 could be seen at top SPS energy, however, no clear difference between the PHSD5.0 calculations with partonic cross sections for $\mu_B = 0$ and actual μ_B in the local cells.

Our findings can be understood as follows: The fact that we find only small traces of the μ_B-dependence of partonic scattering dynamics in heavy-ion bulk observables—although the differential cross sections and parton masses clearly depend on μ_B—means that one needs a sizable partonic density and large space-time QGP volume to explore the dynamics in the QGP phase. These conditions are only fulfilled at high bombarding energies (top SPS, RHIC energies) where, however, μ_B is rather low. On the other hand, decreasing the bombarding energy to FAIR/NICA energies and, thus, increasing μ_B, leads to collisions that are dominated by the hadronic phase where the extraction of information about the parton dynamics will be rather complicated based on bulk observables. Further investigations of other observables (such as flow coefficients v_n of particles and antiparticles, fluctuations and correlations) might contain more visible μ_B−traces from the QGP phase.

Acknowledgements The authors acknowledge inspiring discussions with J. Aichelin, H. Berrehrah, C. Ratti, E. Seifert, A. Palmese, and T. Steinert. This work was supported by the LOEWE center "HIC for FAIR". Furthermore, P.M., L.O. and E.B. acknowledge support by the Deutsche Forschungsgemeinschaft (DFG, German Research Foundation) through the grant CRC-TR 211 'Strong-interaction matter under extreme conditions'—Project number 315477589—TRR 211. O.S. acknowledges support from HGS-HIRe for FAIR; L.O. and E.B. thank the COST Action THOR, CA15213. The computational resources have been provided by the LOEWE-Center for Scientific Computing.

References

1. J. Cugnon, Monte Carlo calculation of high-energy heavy-ion interactions. Phys. Rev. C **22**, 1885–96 (1980)
2. E.A. Uehling, G.E. Uhlenbeck, Transyort phenomena in Einstein-Bose and Fermi-Dirae gases. I. Phys. Rev. **43**, 552–561 (1932)
3. G.F. Bertsch, S.Das Gupta, Guide to microscopic models for intermediate-energy heavy ion collisions. Phys. Rep. **160**, 189 (1988)
4. W. Cassing, V. Metag, U. Mosel, K. Niita, Production of energetic particles in heavy ion collisions. Phys. Rep. **188**, 363 (1990)
5. J. Aichelin, H. Stöcker, Quantum molecular dynamics. A Novel approach to N body correlations in heavy ion collisions. Phys. Lett. B **176**, 14–19 (1986)
6. C. Hartnack, R.K. Puri, J. Aichelin, J. Konopka, S.A. Bass, H. Stöcker, W. Greiner, Modeling the many body dynamics of heavy ion collisions: present status and future perspective. Eur. Phys. J. A **1**, 151–169 (1998)
7. C.M. Ko, Q. Li, Ren-Chuan Wang, Relativistic Vlasov equation for heavy ion collisions. Phys. Rev. Lett. **59**, 1084–1087 (1987)
8. B. Blaettel, V. Koch, W. Cassing, U. Mosel, Covariant Boltzmann-Uehling-Uhlenbeck approach for heavy-ion collisions. Phys. Rev. C **38**, 1767–1775 (1988)
9. T. Maruyama, W. Cassing, U. Mosel, S. Teis, K. Weber, Covariant Boltzmann-Uehling-Uhlenbeck approach for heavy-ion collisions. Nucl. Phys. A **573**, 653–675 (1994)
10. H.-U. Bengtsson, T. Sjöstrand, The Lund Monte-Carlo for hadronic processes PYTHIA version 4.8. Comput. Phys. Commun. **46**, 43–82 (1987)
11. W. Ehehalt, W. Cassing, Relativistic transport approach for nucleus nucleus collisions from SIS to SPS energies. Nucl. Phys. A **602**, 449–486 (1996)
12. W. Cassing, E.L. Bratkovskaya, Hadronic and electromagnetic probes of hot and dense nuclear matter. Phys. Rep. **308**, 65–233 (1999)
13. S.A. Bass et al., Microscopic models for ultrarelativistic heavy ion collisions. Prog. Part. Nucl. Phys. **41**, 255–369 (1998)
14. E.L. Bratkovskaya, M. Bleicher, M. Reiter, S. Soff, H. Stöcker, M. van Leeuwen, S.A. Bass, W. Cassing, Strangeness dynamics and transverse pressure in relativistic nucleus-nucleus collisions. Phys. Rev. C **69**, 054907 (2004)
15. R. Marty, J. Aichelin, Molecular dynamics description of an expanding q/q plasma with the Nambu-Jona-Lasinio model and applications to heavy ion collisions at energies available at the BNL relativistic heavy ion collider and the CERN large hadron collider. Phys. Rev. C **87**, 034912 (2013)
16. R. Marty, E. Bratkovskaya, W. Cassing, J. Aichelin, Observables in ultrarelativistic heavy-ion collisions from two different transport approaches for the same initial. Phys. Rev. C **92**, 015201 (2015)
17. S. Juchem, W. Cassing, C. Greiner, Quantum dynamics and thermalization for out-of-equilibrium theory, Phys. Rev. D **69**, 025006; (2004). Nonequilibrium quantum-field dynamics and off-shell transport for -theory in dimensions. Nucl. Phys. A **743**, 92–126 (2004)
18. W. Cassing, S. Juchem, Nucl. Phys. A **665**, 377 (2000); ibid A **672**, 417 (2000)
19. Y.B. Ivanov, J. Knoll, D.N. Voskresensky, Resonance transport and kinetic entropy. Nucl. Phys. A **672**, 313–356 (2000)
20. E.L. Bratkovskaya, W. Cassing, Dilepton production and off-shell transport dynamics at SIS energies. Nucl. Phys. A **807**, 214–250 (2008)
21. E. Shuryak, Why does the quark-gluon plasma at RHIC behave as a nearly ideal fluid. Prog. Part. Nucl. Phys. **53**, 273–303 (2004)
22. M.H. Thoma, J. Phys. G **31**, L7 (2005); Nucl. Phys. A **774**, 307 (2006)
23. A. Peshier, W. Cassing, The hot nonperturbative gluon plasma is an almost ideal colored liquid. Phys. Rev. Lett. **94**, 172301 (2005)

24. I. Arsene, et al., Quark-gluon plasma and color glass condensate at RHIC? The perspective from the BRAHMS experiment. Nucl. Phys. A **757**, 1–27 (2005); B.B. Back et al., The PHOBOS perspective on discoveries at RHIC. Nucl. Phys. A **757**, 28–101 (2005); J. Adams et al., Experimental and Theoretical Challenges in the search for the quark-gluon plasma: the STAR collaboration's critical evidence from the RHIC collisions. Nucl. Phys. A **757**, 102 (2005); K. Adcox et al., Formation of dense partonic matter in relativistic nucleus-nucleus collisions at RHIC: experimental evaluation by the PHENIX collaboration. Nucl. Phys. A **757**, 184–283 (2005)
25. T. Hirano, M. Gyulassy, Perfect fluidity of the quark-gluon plasma core as seen through its dissipative hadronic corona. Nucl. Phys. A **769**, 71–94 (2006)
26. R.C. Hwa, C.B. Yang, Scaling behavior at high p_T and the p-p ratio. Phys. Rev. C **67**, 034902 (2003); V. Greco, C.M. Ko, P. Levai, Parton coalescence and the antiproton/pion anomaly at RHIC. Phys. Rev. Lett. **90**, 202302 (2003)
27. R.J. Fries, B. Müller, C. Nonaka, S.A. Bass, Parton coalescence and the antiproton/pion anomaly at RHIC. Phys. Rev. Lett. **90**, 202303 (2003)
28. Z.-W. Lin et al., Multiphase transport model for relativistic heavy ion collisions. Phys. Rev. C **72**, 064901 (2005)
29. W. Cassing, E.L. Bratkovskaya, Parton transport and hadronization from the dynamical quasiparticle point of view. Phys. Rev. C **78**, 034919 (2008)
30. W. Cassing, QCD thermodynamics and confinement from a dynamical quasiparticle point of view. Nucl. Phys. A **791**, 365–381 (2007)
31. W. Cassing, Dynamical quasiparticles properties and effective interactions in the sQGP. Nucl. Phys. A **795**, 70–97 (2007)
32. W. Cassing, From Kadanoff-Baym dynamics to off-shell parton transport E. Phys. J. ST **168**, 3–87 (2009)
33. O. Linnyk, E.L. Bratkovskaya, W. Cassing, Effective QCD and transport description of dilepton and photon production in heavy-ion collisions and elementary processes. Prog. Part. Nucl. Phys. **87**, 50 (2016)
34. W. Cassing, E.L. Bratkovskaya, Parton-Hadron-String Dynamics: an off-shell transport approach for relativistic energies. Nucl. Phys. A **831**, 215 (2009)
35. P. Moreau, O. Linnyk, W. Cassing, E. Bratkovskaya, Examinations of the early degrees of freedom in ultrarelativistic nucleus-nucleus collisions. Phys. Rev. C **93**, 044916 (2016)
36. V. Ozvenchuk, O. Linnyk, M.I. Gorenstein, E.L. Bratkovskaya, W. Cassing, Dynamical equilibration of strongly interacting infinite-parton matter within the parton-hadron-string dynamics transport approach. Phys. Rev. C **87**, 024901 (2013)
37. P. Moreau, O. Soloveva, L. Oliva, T. Song, W. Cassing, E.L. Bratkovskaya, Exploring the partonic phase at finite chemical potential within an extended off-shell transport approach. Phys. Rev. C **100**, 014911 (2019)
38. R. Kubo, Statistical-mechanical theory of irreversible processes. J. Phys. Soc. Jpn. **12**, 570 (1957)
39. V. Ozvenchuk, O. Linnyk, M.I. Gorenstein, E.L. Bratkovskaya, W. Cassing, Shear and bulk viscosities of strongly interacting infinite- parton-hadron matter within the parton-hadron-string dynamics transport approach. Phys. Rev. C **87**, 064903 (2013)
40. G. Aarts, J.M. Martinez Resco, Transport coefficients, spectral functions and the lattice. J. High Energy Phys. **04**, 053 (2002)
41. R. Lang, N. Kaiser, W. Weise, Shear viscosities from Kubo formalism in a large-Nc Nambu-Jona-Lasinio model. Eur. Phys. J. A **48**, 109 (2012)
42. C. Sasaki, K. Redlich, Bulk viscosity in quasiparticle models. Phys. Rev. C **79**, 055207 (2009)
43. J.-P. Blaizot, E. Iancu, A Boltzmann equation for the QCD plasma. Nucl. Phys. B **557**, 183–236 (1999)
44. P.K. Kovtun, D.T. Son, A.O. Starinets, Viscosity in strongly interacting quantum field theories from Black Hole physics. Phys. Rev. Lett. **94**, 111601 (2005)
45. N. Astrakhantsev, V. Braguta, A. Kotov, Temperature dependence of shear viscosity of SU(3)-gluodynamics within lattice simulation. J. High Energy Phys. **04**, 101 (2017)

46. A. Palmese, W. Cassing, E. Seifert, T. Steinert, P. Moreau, E.L. Bratkovskaya, Chiral symmetry restoration in heavy-ion collisions at intermediate energies. Phys. Rev. C **94**, 044912 (2016)
47. Y. Akiba et al., Dynamics of ultrarelativistic nuclear collisions with heavy beams: an experimental overview, (E802 Collaboration). Nucl. Phys. A **610**, 139–152 (1996)
48. R. Lacasse et al., Hadron yields and spectra in Au+Au collisions at the AGS, (E877 Collaboration). Nucl. Phys. A **610**, 153–164 (1996)
49. S. Ahmad et al., Lamda production by 11.6 A GeV/c Au beam on Au target. Phys. Lett. B **382**, 35 (1996); Lamda production by 11.6 A GeV/c Au beam on Au target. Phys. Lett. B **386**, 496(E) (1996)
50. J. Barrette et al., Lamda production and flow in Au+Au collisions at 11.5 A GeV/c, (E877 Collaboration). Phys. Rev. C **63**, 014902 (2001)
51. S. Albergo et al., Spectra in 11.6 A GeV/c Au-Au collisions. Phys. Rev. Lett. **88**, 062301 (2002)
52. C. Alt et al., Energy and centrality dependence of p and p production and the ratio in Pb+Pb collisions between 20A GeV and 158A GeV, (NA49 Collaboraion). Phys. Rev. C **73**, 044910 (2006)
53. C. Alt et al., Pion and kaon production in central Pb + Pb collisions at 20 A and 30 A GeV: Evidence for the onset of deconfinement, (NA49 Collaboration). Phys. Rev. C **77**, 024903 (2008)
54. C. Alt et al., Energy dependence of Lamda and Epsilon production in central Pb + Pb collisions at 20 A, 30 A, 40 A, 80 A, and 158 A GeV measured at the CERN Super Proton Synchrotron, (NA49 Collaboration). Phys. Rev. C **78**, 034918 (2008)
55. S.V. Afanasiev et al., Energy dependence of pion and kaon production in central Pb+Pb collisions, (NA49 Collaboration). Phys. Rev. C **66**, 054902 (2002)
56. T. Anticic et al., Lamda and Lamda bar production in central Pb-Pb collisions at 40, 80, and 158 AGeV, (NA49 Collaboration). Phys. Rev. Lett. **93**, 022302 (2004)
57. T. Anticic et al., Centrality dependence of proton and antiproton spectra in Pb + Pb collisions at 40 A GeV and 158 A GeV measured at the CERN super proton synchrotron (NA49 Collaboration). Phys. Rev. C **83**, 014901 (2011)
58. T. Anticic et al., System-size and centrality dependence of charged kaon and pion production in nucleus-nucleus collisions at 40 A GeV and 158 A GeV beam energy, (NA49 Collaboration). Phys. Rev. C **86**, 054903 (2012)

Chapter 11
Influence of the Neutron Skin of Nuclei on Observables

Christoph Hartnack, Arnaud Le Févre, Yvonne Leifels, and Jörg Aichelin

Abstract We use the Isospin-dependent Quantum Molecular Dynamics approach to analyze the influence of the neutron skin on hadronic observables in heavy-ion collisions at 400 AMeV incident energy. While the most observables on the behavior of nucleons seem to be only slightly affected, the isospin ratios of pions show a quite significant effect. It is found that the comparison of the centrality dependence at different energies allows to gain information on the neutron skin of the nuclei on one hand, but also on other ingredients like the nuclear equation of state of asymmetric matter.

11.1 Introduction

One of the main interests of the study of relativistic heavy-ion collisions is the investigation of the properties of nuclear matter at extreme densities and excitation energies. These investigations include dynamical observables of nucleons on one side but also the production of secondary particles on the other side, where the pion is the "cheapest" in production and thus the most prominent produced particle in heavy-ion collisions beyond 1 AGeV. For our calculations we use the isospin-dependent quantum molecular dynamics (IQMD) approach [1], a microscopic transport model calculating heavy-ion collisions on an event-by-event basis. In the first part we will concentrate on the production of pions, especially on their isospin ratios, where recently a strong dependence on the neutron skin has been reported [2]. For this we will first describe how the production of pions is done in IQMD and how the density

C. Hartnack (✉) · J. Aichelin
SUBATECH, UMR 6457, IMT Atlantique, IN2P3/CNRS, Université de Nantes, 4 rue Alfred Kastler, 44 307 Nantes, France
e-mail: hartnack@subatech.in2p3.fr

A. Le Févre · Y. Leifels
GSI Helmholtzzentrum für Schwerionenforschung GmbH, Planckstraße 1, 64291 Darmstadt, Germany

J. Aichelin
Frankfurt Institute for Advanced Studies, Ruth Moufang Str. 1, 60438 Frankfurt, Germany

R. K. Puri et al. (eds.), *Advances in Nuclear Physics*, Springer Proceedings in Physics 257, https://doi.org/10.1007/978-981-15-9062-7_11

profiles of protons and neutrons are described in IQMD model. Afterwards we will recall the influence of neutron skin on the isospin ratios of pions and study whether other observables might be affected as well.

11.1.1 Pion Production in IQMD

IQMD decomposes, as many other microscopic transport models [3–5] do similarly, the interaction of nucleons into a long-range part described by local two-body potentials and a short-range part related to the effect of stochastic collisions. The long-range term leads to nuclear potentials, which become important for the stability of the nucleus and also touch topics like the nuclear equation of state [6]. We will shortly summarize the important part of the collision term and refer for a detailed description of both parts and their application in IQMD on [7]. Particles in IQMD are described as Gaussian wave packages in coordinate and momentum space, where the centroides of these packages are propagated by Hamilton's equation of motion. Two particles collide if their minimum distance d, i.e., the minimum relative distance of the centroids of the Gaussians during their motion, in their centre of mass (CM) frame fulfills the requirement:

$$d \leq d_0 = \sqrt{\frac{\sigma_{\text{tot}}}{\pi}}, \qquad \sigma_{\text{tot}} = \sigma(\sqrt{s},\ \text{type}). \tag{11.1}$$

The total cross section is assumed to be the free cross section of the regarded collision type ($N - N, N - \Delta, \ldots$). The cross sections for elastic and inelastic collisions are obtained by a table lookup using experimentally measured cross sections (when available) or derived from available cross sections using symmetry assumptions and detailed balance.

The pion production in IQMD is done via the Δ-channel, where deltas can be produced in nucleon-nucleon (NN) collisions but also be reabsorbed in $N\Delta$ collisions. The Δ decays and produces a free pion, which can be reabsorbed in collisions with a nucleon and form a Δ again:

$$NN \leftrightarrow N\Delta \qquad \Delta \leftrightarrow N\pi. \tag{11.2}$$

These reactions have to comply with detailed balance and isospin effects have to be taken into account by the use of Clebsch-Gordon coefficients. For more details see [8].

The production of pions in IQMD has successfully been tested by various comparisons with experimental measurements performed by the FOPI collaboration at GSI [9]. Giving an example taken from [10], Fig. 11.1 compares the excitation function of the total pion yield in Au+Au collisions measured by FOPI to IQMD calculations (left-hand side). For this purpose, the events calculated by IQMD have undergone the same analysis procedures as the experimental data. The multiplicities calculated

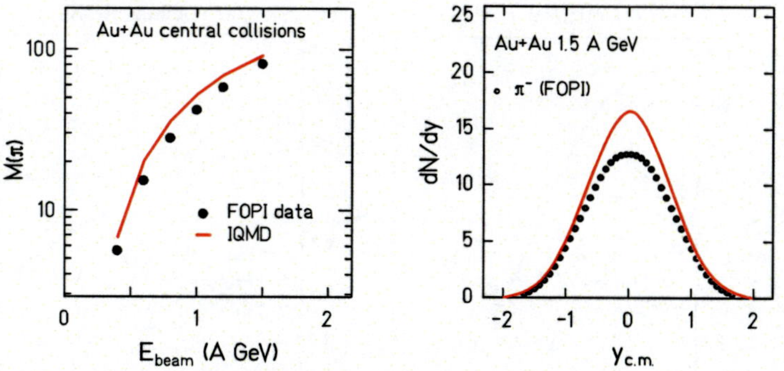

Fig. 11.1 Comparison between IQMD calculations and FOPI results on the absolute pion number as function of the incident energy (left) and of the rapidity distribution of π^- in collisions of Au+Au at 1.5 AGeV (right). (from [10])

by IQMD are slightly higher than those obtained by FOPI but the excitation functions show nicely the same behavior. Figure 11.1 compares on the right-hand side the rapidity distributions of negative pions in central collisions. As already stated before, IQMD shows slightly larger absolute pion yields; therefore, it is not astonishing that the absolute numbers of the rapidity distribution are also higher than the experimental points. However, the structure of the distribution is quite similar. The distribution is peaked at midrapidity underlining that most of the pions are produced by first collisions or in collisions of the stopped participant matter. Afterwards they will undergo rescattering. This will become important later on when we will discuss the influence of the neutron skin.

11.1.2 Density Profiles of Protons and Neutrons

In standard IQMD calculations the centroids of the Gaussians are distributed inside a sphere in the rest frame of the nucleus according to

$$(\vec{r}_i - \vec{r}_{CM})^2 \le R_A^2 \qquad R_A = R_0 A^{1/3}, \tag{11.3}$$

where $\vec{r}_i$ and $\vec{r}_{CM}$ are the position vectors of particle i and of the center-of-mass of the nucleus, respectively, and $A = Z + N$ is the number of nucleons of the nucleus. The radius parameter is chosen as $R_0 = 1.12$ fm. This initialization, which we will call "$R_P = R_N$ ", assures the same rms radius for protons and neutrons even for heavy isospin-asymmetric systems. Consequently, for nuclei with neutron numbers higher than those of the protons the neutrons have systematically a higher density than the protons in the whole nucleus. It should be noted that most other microscopic models with the exception of [11] use a similar procedure yielding the same rms radius for neutrons and protons. If we want to keep the same density of protons and neutrons

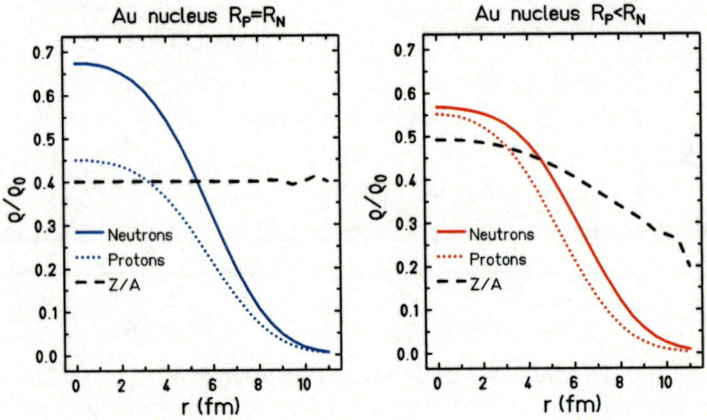

Fig. 11.2 Profiles of proton and neutron densities and for the charge ration $Z/A(r)$ for initialisations assuming $R_P = R_N$ (left) and $R_P < R_N$ (right)

at least in the centre of the nucleus we are forced to allow protons and neutrons to have different rms radii. This can be obtained by distributing the centroids of the Gaussians according to

$$R_P = R_0 \cdot (2Z)^{1/3}; \qquad R_N = R_0(2N)^{1/3}. \tag{11.4}$$

Here R_P and R_N denote the radii for protons and neutrons. This initialization, which we will call "$R_P < R_N$ ", yields, however, a difference of around 0.5 fm for the rms radii of protons and neutrons in a system like ^{197}Au.

In Fig. 11.2 the density profiles of protons (dotted lines) and neutrons (full lines) are presented for a ^{197}Au nucleus if we use both initializations $R_P = R_N$ (left-hand side) and $R_P < R_N$ (right-hand side). While for the first one the charge density $Z/A(R)$ (dashed line) remains constant over the whole nucleus, this ratio varies strongly for $R_P < R_N$. It should be noted that the rescattering cross sections of π^- with neutrons are higher than those with protons—and analogously those of π^+ with protons are higher than those with neutrons; therefore, this difference in density has an effect on the isospin ratios [2].

11.2 The Effect of the Neutron Skin on the Isospin Ratio

The study of the isospin ratio π^-/π^+ at 400 AMeV has been frequently proposed as a possibility for determining the nuclear equation of state of asymmetric matter [12–15]. The reason for this energy choice is due to the effect that at this energy the FOPI collaboration has measured quite high values of this ratios for the system Au+Au [9]. Since this ratio could not be explained by many simulation models (IQMD included) it was assumed that this might be related to the density dependence of the asymmetry potential. A very common description is to scale the density dependence of the asymmetry potential with an exponent γ. The (classical) linear dependence thus corresponds to $\gamma = 1$, a soft asy-eos to $\gamma < 1$ and a hard one to $\gamma > 1$. Here we want to compare its influence with the effect of the neutron skin of the nucleus and try to disentangle the different contributions.

11.2.1 Centrality Dependence at 400 AMeV

Figure 11.3 presents on the left-hand side the effect of the equation of state of asymmetric matter: for an incident energy of 400 AMeVwe see a significant effect for central collisions. A softer asy-eos (blue dashed line) enhances the π^-/π^+ ratio with respect to a hard one (black dotted line). However this effect gets weaker when going to peripheral collisions. Nevertheless all calculations show a strong rise of the ratio when going to high impact parameters. While the decreasing significance of the asy-eos when going to peripheral collisions is due to the smaller densities reached in such collisions the global rise of the ratios is due to the neutron skin, as it can be seen on the right-hand side. Here we compare the effects of the initializations $R_P < R_N$ (neutron skin, red full line) and $R_P = R_N$ (no neutron skin, blue dashed line). When applying a neutron skin the ratios become higher than without such a skin. Note that the calculations on the left-hand side were done with a neutron skin. The reason for this is the rescattering of the pions. We will shortly sketch the argument line—a more detailed study of this effect can be found in [2]. In case of a neutron skin the pions leaving the system will have their last interaction matter dominated by neutrons. Neutron rich matter acts differently on π^- and π^+: while for π^- rescattering will

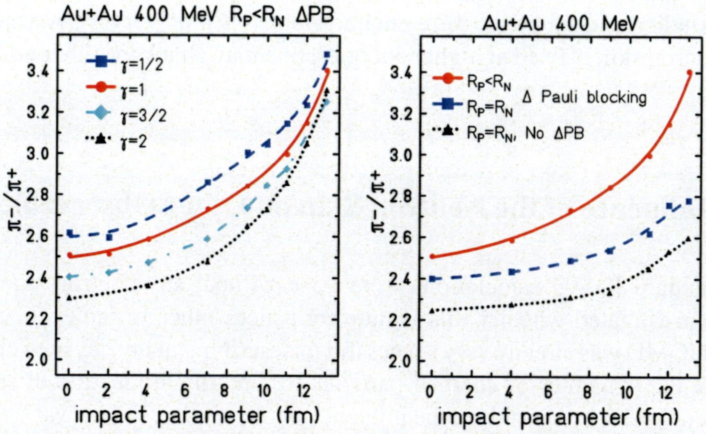

Fig. 11.3 Impact parameter dependence of the ratio π^-/π^+ in collisions of Au (400 AMeV)+Au. Left: influence of the equation of state of asymmetric matter, right: influence of neutron skin and delta Pauli blocking

lead to an intermediate Δ^- which decays only by $\Delta^- \rightarrow n\pi^-$ and thus does not influence the π^- yield, a rescattering of a π^+—yielding a Δ^+—will reproduce a π^+ with a probability of only one third. Thus the π^+ yield will be penalized when passing neutron dominated matter. This effect even is enhanced by the Pauli blocking of the delta decay: the high density of neutrons will add a penalty to the the $\Delta^+ \rightarrow n\pi^+$ channel and reduce the π^+ yield even more. The Pauli blocking also acts on the channel $\Delta^- \rightarrow n\pi^-$, but since there is no concurrent channel, this will only delay the decay of the delta, but it does not change the π^- yield. This effect can be seen when disabeling it: a calculation without Pauli blocking of the delta decay (black dotted line) yields even smaller ratios.

11.2.2 Dependence of the Ratio on the Incident Energy

How to disentangle these effects? The FOPI collaboration had shown that the isospin ratios in central collisions decrease when increasing the incident energy. Additionally at these energies the isospin ratios could be explained by IQMD (and other models) without any assumption on the density dependence of the asymmetry potential. Figure 11.4 presents the same analysis as in Fig. 11.3 but for now for Au (1200 AMeV)+Au. Indeed, for central collisions we see smaller values of the isospin ratios and the differences due to the equation of state of asymmetric matter vanish completely. However, the effect of the neutron skin becomes very prominent at peripheral collisions while the influence of the Pauli blocking in the delta decay disappears as well. Here we get a very good handle to test the neutron skin from π^-/π^+ ratios at high energies: from comparison of central and very peripheral collisions we may estimate the thickness of the neutron skin. A more detailed investigation of that procedure is presented in [16], where more refined parametrizations of the neutron skin are studied and applied to different nuclei like ^{48}Ca or ^{208}Pb. Once the question of the neutron skin is fixed at higher energies, one may attack the other effects at low energies.

11.3 Influence of the Neutron Skin on Other Observables

Since standard IQMD calculations were done without any neutron skin it should now be investigated whether this feature influences other variables, knowing that standard IQMD was already very successful in describing many physical observables including the dynamics of charged particles [6] and the production of strangeness [10]. Therefore, we will analyze the influence of the neutron skin on variables related to stopping and transverse pressure, to transverse and elliptic flow and to strangeness production. It should also be noted that the absolute total pion yield (see, e.g., Fig. 11.1) is not affected by the neutron skin. The rescattering of pions in neutron rich matter only changes their isospin flavors. We will compare dynamical observables

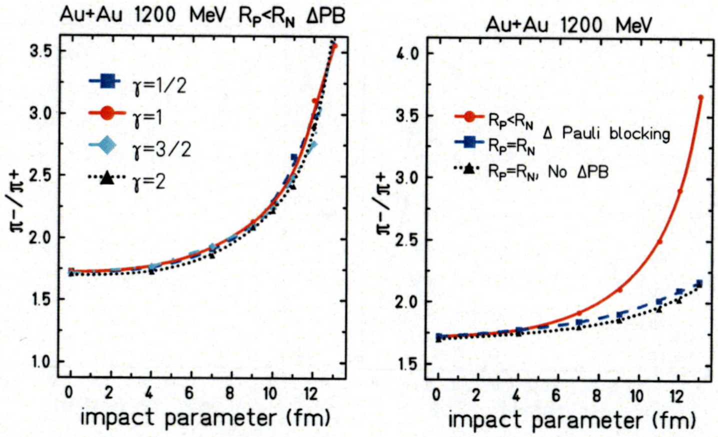

Fig. 11.4 Impact parameter dependence of the ratio π^-/π^+ in collisions of Au(400 AMeV)+Au. Left: influence of the equation of state of asymmetric matter, right: influence of neutron skin and delta Pauli blocking

for protons and neutrons using calculations with and without neutron skin. In the following figures we will mark protons with red full lines when using $R_P = R_N$ and with blue dashed lines for $R_P < R_N$. Neutrons are presented by magenta dashed-dotted lines for $R_P = R_N$ and by cyan dotted lines for $R_P < R_N$.

11.3.1 Observables Related to Stopping on Transverse Pressure

Let us first look on the rapidity distribution of protons and neutrons in a semiperipheral reaction of Au+Au at 400 AMeV incident energy. Figure 11.5 shows on the left-hand side the rapidity distribution of protons and neutrons. Up to the difference of absolute numbers there is no significant effect. The right-hand side presents the transverse energy ratio $E_{\text{rat}} = E_{\text{trans}}/E_{\text{long}}$ for protons and neutrons. An isotropic system would yield a value of 2, higher values correspond to the dominance of the transverse direction, lower to the dominance of the longitudinal direction, e.g., for incomplete stopping. We see very similar behavior for protons and neutrons, independently of the use of the neutrons skin. Similarly the transverse momentum distribution $dN/p_T dp_T$ at central rapidity shown on the left-hand side Fig. 11.6 is very similar for all cases. Also the mean transverse momentum show no significance. Only for very peripheral systems there seems to be a slight enhancement for neutrons, independent of the choice on the neutron skin.

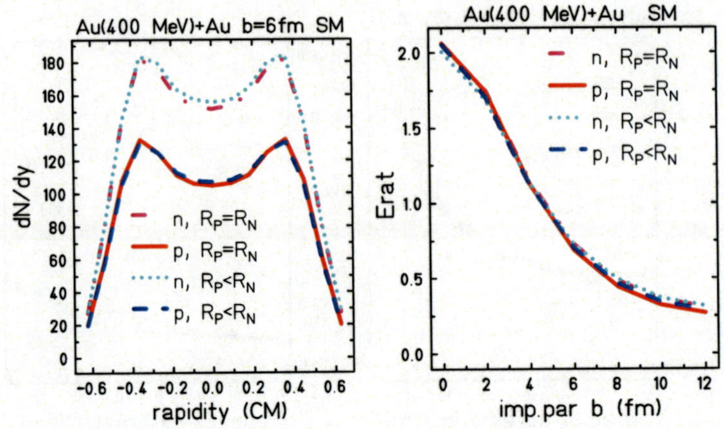

Fig. 11.5 Stopping in Au(400 AMeV)+Au. Left: rapidity distribution for b = 6 fm, right: centrality dependence of $E_{rat} = E_{trans}/E_{long}$

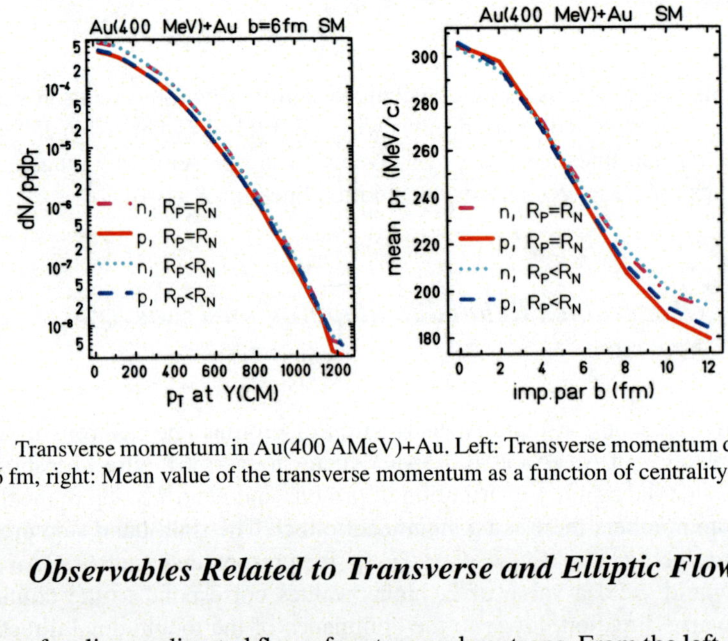

Fig. 11.6 Transverse momentum in Au(400 AMeV)+Au. Left: Transverse momentum distribution for b = 6 fm, right: Mean value of the transverse momentum as a function of centrality

11.3.2 Observables Related to Transverse and Elliptic Flow

Next let us focalize on directed flow of protons and neutrons. From the left-hand side of Fig. 11.7 we see that in calculations without neutron skin ($R_P = R_N$) we yield slightly higher flow in semiperipheral collisions, an effect that can also be seen in the centrality dependence of the directed flow parameter v_1 at the right-hand side. However, the effects are quite small. Similarly it seems that the elliptic flow is a bit stronger negative without neutron skin. Even if this can hardly be seen in the azimuthal distribution on the left-hand side of Fig. 11.8 the centrality dependence of v_2 on the right-hand side shows some indication of this effect for semiperipheral collisions.

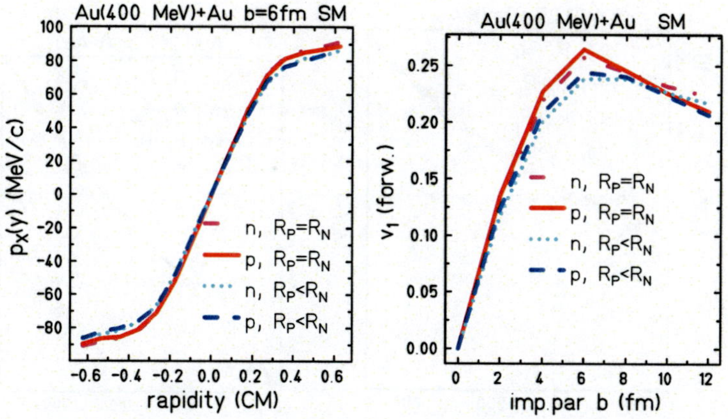

Fig. 11.7 Transverse flow Au(400 AMeV)+Au. Left: In plane momentum as a function of rapidity for = 6 fm, right: centrality dependence of v_1 in forward direction

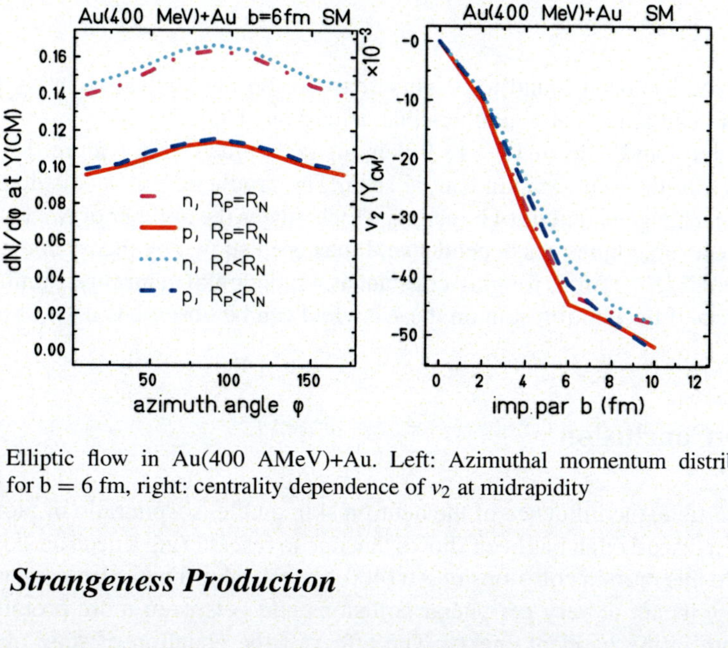

Fig. 11.8 Elliptic flow in Au(400 AMeV)+Au. Left: Azimuthal momentum distribution at midrapidty for b = 6 fm, right: centrality dependence of v_2 at midrapidity

11.3.3 Strangeness Production

Strangeness production below the energetic threshold of the elementary production is a process which is strongly related to the properties of the high-density region build up during the compression phase of the reaction. The basic key is a multistep process forming first a delta in a high energetic collision of two nucleons followed by a collision of this delta with another nucleon in order to form a kaon-hyperon pair. Since the delta decay is a process strongly concurring to that channel, short mean free pathes are needed. Thus strangeness production is very sensitive on the density reached in the reaction. For details see [10]. This is also the reason why kaon

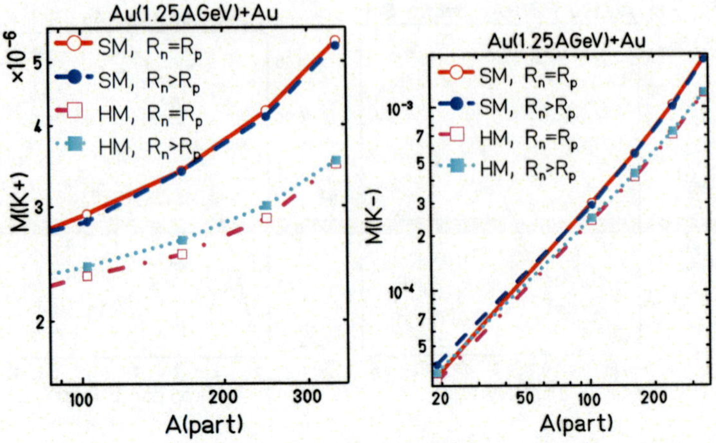

Fig. 11.9 Strangeness production in Au(1230 AMeV)+Au. Left: centrality dependence of K^+ for hard and soft eos. right: centrality dependence of K^- for hard and soft eos

production has been found to be very sensitive on the nuclear equation of state: a soft eos yield larger kaon numbers than a hard one [17].

The left-hand side of Fig. 11.9 demonstrates nicely this feature. However, no influence of the neutron skin can be seen. The production of K^- is dominated by charge exchange reactions of hyperons which relates the number of K^- to that of K^+ yielding an analogues eos dependence. It has been shown by theory and experiment that the K^-/K^+ ratio is roughly constant as a function of centrality. Similarly to K^+ no effects of the neutron skin on the K^- yield can be observed.

11.4 Conclusion

In this article, the influence of the neutron skin on the isospin ratio of pions π^-/π^+ has been discussed. It has been shown that due to rescattering a neutron skin is able to enhance this ratio in collisions of Au (400 AMeV)+Au. This enhancement becomes very significant at very peripheral collisions and gets even more prominent when going up in the incident energy. The effect of the equation of state of asymmetric matter only shows up at low-incident energy and dominate in central collisions. Therefore, a procedure of disentangling the effects can be proposed by measuring the full impact parameter dependence of π^-/π^+ at low and at high incident energies. From the comparison of high precision data to simulation one may thus reveal better information on the neutron skin and get another handle to attack the nuclear equation of state of asymmetric matter. Furthermore it was found that the neutron skin seems not to effect significantly to the majority of dynamical observables like rapidity distributions, transverse and elliptic flow, transverse momenta or strangeness production. Only slight effects may be seen at rather peripheral collisions. This means that we do not expect that assumptions on the neutron skin of nuclei might alter previous conclusions already taken from data comparisons using only the same radius of protons and neutrons on the simulation side.

References

1. C. Hartnack et al., Quantum molecular dynamics: a microscopic model from Unilac to Cern energies. Nucl. Phys. A **495**, 303–319 (1989)
2. C. Hartnack, A. Le Fèvre, Y. Leifels, J. Aichelin, CERN Proc. **1**, 211 (2019)
3. H. Stöcker, W. Greiner, High energy heavy-ion collisions- probing the equation of state of highly excited hadronic matter. Phys. Rep. **137**, 277 (1986)
4. W. Cassing, V. Metag, U. Mosel, K. Niita, Production of energetic particles in heavy-ion collisions. Phys. Rep. **188**, 361 (1990)
5. J. Aichelin, Quantum molecular dynamics-a dynamical microscopic n-body approach to investigate fragment formation and the nuclear equation of state in heavy ion collisions. Phys. Rep. **202**, 233–360 (1991)
6. A. Le Fèvre, Y. Leifels, W. Reisdorf, J. Aichelin, C. Hartnack, Constraining the nuclear matter equation of state around twice saturation density. Nucl. Phys. A **945**, 112–133 (2016)
7. C. Hartnack, R.K. Puri, J. Aichelin, J. Konopka, S.A. Bass, H. Stoecker, W. Greiner, Modelling the many-body dynamics of heavy ion collisions: present status and future perspective. Eur. Phys. J. A **1**, 151–169 (1998)
8. S.A. Bass, C. Hartnack, H. Stoecker, W. Greiner, Azimuthal correlations of pions in relativistic heavy ion collisions at 1-GeV/nucleon. Phys. Rev. C **51**, 3343–3356 (1995)
9. W. Reisdorf et al., [FOPI Collaboration], Systematics of pion emission in heavy ion collisions in the 1A- GeV regime. Nucl. Phys. A **781**, 459–508 (2007)
10. C. Hartnack, H. Oeschler, Y. Leifels, E.L. Bratkovskaya, J. Aichelin, Strangeness production close to threshold in proton-nucleus and heavy-ion collisions. Phys. Rep. **510**, 119–200 (2012)
11. G.F. Wei, B.A. Li, J. Xu, L.W. Chen, Influence of neutron-skin thickness on π^-/π^+ ratio in Pb+Pb collisions. Phys. Rev. C **90**, 014610 (2014)
12. Z. Xiao, B.A. Li, L.W. Chen, G.C. Yong, M. Zhang, Circumstantial Evidence for a Soft Nuclear Symmetry Energy at Suprasaturation Densities. Phys. Rev. Lett. **102**, 062502 (2009)
13. Z.Q. Feng, G.M. Jin, Probing high-density behavior of symmetry energy from pion emission in heavy-ion collisions. Phys. Lett. B **683**, 140–144 (2010)
14. J. Hong, P. Danielewicz, Subthreshold pion production within a transport description of central Au + Au collisions. Phys. Rev. C **90**, 024605 (2014)
15. M.D. Cozma, The impact of energy conservation in transport models on the π^-/π^+ multiplicity ratio in heavy-ion collisions and the symmetry energy. Phys. Lett. B **753**, 166–172 (2016)
16. C. Hartnack, A. Le Fèvre, Y. Leifels, J. Aichelin, The influence of neutron skin and asymmetry energy on π^-/π^+ ratio (2018), arxiv:1808.09868
17. C. Hartnack, H. Oeschler, J. Aichelin, Hadronic matter is soft. Phys. Rev. Lett. **96**, 012302 (2006). [arXiv:nucl-th/0506087]

Chapter 12
Nuclear Matter Properties at High Densities: Squeezing Out Nuclear Matter Properties from Experimental Data

Yvonne Leifels

Abstract The nuclear equation of state is a topic of highest current interest in nuclear physics and astrophysics. The nuclear equation of state governs the evolution of heavy-ion reactions as well as the characteristics of compact stellar objects like neutron stars, the explosions of supernovae, and the merging of two neutron stars. The symmetry energy is the part of the equation of state which is connected to the asymmetry in the neutron/proton content. During recent years a multitude of experimental and theoretical efforts on different fields have been undertaken to constraint its density dependence at low densities but also above saturation density ($\rho_0 = 0.16 \mathrm{fm}^{-3}$). Conventionally, the symmetry energy is described by its magnitude S_v and the slope parameter L, both at saturation density. Values of $L \approx$ 44–66 MeV and $S_v \approx$ 31–33 MeV have been deduced in recent compilations of nuclear structure, heavy-ion reaction, and astrophysics data. Apart from astrophysical data on mass and radii of neutron stars and the gravitational wave signal of neutron star mergers, heavy-ion reactions above incident energies of several 100 MeV are the only means to access the high-density behavior of the symmetry energy. In particular, meson production and collective flows up to about 1 GeV/nucleon are predicted to be sensitive to the slope of the symmetry energy as a function of density. From the measurement of elliptic flow of neutrons with respect to charged particles at GSI, a stringent constraint for the slope of the symmetry energy at supra-saturation densities has been deduced. Future options to reach even higher densities will be discussed.

Y. Leifels (✉)
GSI Helmholtzzentrum für Schwerionenforschung, Planckstr. 1, 64291 Darmstadt, Germany
e-mail: y.leifels@gsi.de

R. K. Puri et al. (eds.), *Advances in Nuclear Physics*, Springer Proceedings in Physics 257, https://doi.org/10.1007/978-981-15-9062-7_12

12.1 Introduction

The nuclear matter equation of state (EOS) is one of central topics in nuclear physics. It defines the evolution of nuclear reactions and the characteristics of compact stars and cataclysmic astrophysical events like supernova explosions and neutron star mergers. A theoretical determination of the nuclear EOS from first principles by microscopic calculations using realistic two and three-body nuclear interactions is highly non-trivial and a subject of current scientific research using different approaches, i.e., quantum many-body theory in the Brueckner-Hartree-Fock approximation or chiral perturbation theory (ChPT). From astrophysical observations meaningful constraints can be obtained to the nuclear EOS, however, heavy-ion collisions are the only means to study the characteristics of the nuclear EOS in the laboratory.

The nuclear EOS describes the relation between density, pressure, energy, temperature, and the isospin asymmetry $\delta = (\rho_n - \rho_p)/\rho$, where ρ_n, ρ_p, and ρ are neutron, proton, and nuclear matter densities, respectively. It is conventionally divided into a symmetric matter part independent of the isospin asymmetry and an isospin term, also quoted as symmetry energy $E_{sym}(\rho)$, that enters with a factor δ^2 into the equation of state. In this description, the symmetry energy is the difference in the energy between symmetric matter $\rho_n = \rho_p$ and pure neutron matter. Different density dependencies of $E_{sym}(\rho)$ can be described quantitatively by expanding the symmetry energy in terms of $(\rho - \rho_0)/\rho_0$ using the value of $E_{sym,0} = E_{sym}(\rho = \rho_0)$ and the slope parameter at normal nuclear matter density $L = 3\rho_0 \frac{\delta E_{sym}(\rho)}{\delta\rho}|_{\rho=\rho_0}$ leading to the following equation:

$$E_{sym}(\rho) = E_{sym,0} + \frac{L}{3}\left(\frac{(\rho - \rho_0)}{\rho_0}\right) + \frac{K_{sym}}{18}\left(\frac{(\rho - \rho_0)}{\rho_0}\right)^2 + \ldots,$$

where K_{sym} is referred to as curvature parameter.

Microscopic calculations of the energy functional of nuclear matter employing different approaches to the nucleon-nucleon interaction predict rather different forms of the EOS. Most calculations for the symmetry energy coincide at or slightly below normal nuclear matter density (compare Fig. 12.5 right panel), which demonstrates that constraints from finite nuclei are active for an average density smaller than saturation density and surface effects play a role. In contrast to that extrapolations to supra-normal densities diverge dramatically. The density dependence of the nuclear symmetry energy is an important constituent for drip lines, masses, densities, and collective excitations of neutron-rich nuclei [1], flows,m and multi-fragmentation in heavy-ion collisions at intermediate energies [2, 3], but also for astrophysical phenomena like supernovae, neutron stars [4], and the merging of two neutron stars [5], which have been recently observed and identified by the characteristic gravitational wave signal and the simultaneous emission of γ-rays.

Many results of nuclear structure and nuclear reaction measurements as well as astrophysical observations have been collected in [6]. The symmetry energy $E_{sym,0}$ and the slope parameter L at saturation density have been deduced to be

$E_{sym,0} = 31.6 \pm 2.7$ MeV and $L = 59 \pm 16$ MeV, despite the quite large variations in the individual measurements. However, L and $E_{sym,0}$ cannot be individually determined in most experiments, often a larger value of L is compensated by a smaller $E_{sym,o}$ or vice versa.The authors of [7] have extracted experimental constraints from nuclear physics and astrophysical measurements in the $E_{sym,0}$–L plane, which give a consensus region of 40 MeV $< L(\rho_0) <$ 60 MeV and 30 MeV $< E_{sym,0} <$ 32 MeV with 68 % confidence, which is in mutual agreement with the results quoted in [6]. However, in another publication [8] it was argued, while using isobaric analog states and isovector skins on neutron-rich nuclei, that both symmetry parameters may be larger than the commonly adopted values.

The symmetry energy at higher densities $\rho > \rho_0$ can be accessed by the determination of the mass and the radii of neutron stars [9] or by investigating observables in heavy-ion collisions which are related to the early, high-density phase of the reactions and its isospin content. Theoretical model calculations predict that for a short time period of 20 fm/c densities up to 3 ρ_0 are reached in the central zone of a heavy-ion collision even at moderate energies $\approx$ 1 GeV/nucleon [2]. At lower energies around 200 MeV/nucleon up to $2\rho/\rho_0$ may still be reached. However, one should be aware that the highest density reached during a heavy-ion collision is not necessarily equivalent to the density which is probed by a certain observable. This question has to be addressed when extracting constraints on the density dependence of the symmetry energy from experimental data.

A multitude of observables have been proposed to be sensitive to the symmetry energy (for a review see [2]): ratio of multiplicities or spectra of isospin partners (e.g., π^-/π^+, n/p or t/^{3}He) and the comparison of their flows: The ratio of positively and negatively charged pions measured close to or below the production threshold in the NN system ($E_{beam,thr} = 280$ MeV) is one of the observables discussed. It is predicted to be sensitive to the density dependence of the symmetry energy. Indeed, model predictions obtained with the transport code IBUU4 [11] could only reproduce existing experimental data on pion production around $E_{beam} = 400$ MeV/nucleon in various collisions systems [12] when a rather soft density dependence of the symmetry energy was applied. Incorporation of in-medium effects in addition to the symmetry energy, like, e.g., pion potentials, s-wave production of pions, and the properties of intermediate Delta resonances, may lead to different and even opposite conclusions [13, 14], while describing the experimental data equally well. It is still not settled how the symmetry energy influences the pion production ration. An experimental way out is to measure double ratios of pion production, i.e., compare pion production in a neutron-rich and proton-rich system having the same Z. Here, some input parameters to the models will drop out. Such experiments have been accomplished at the Riken facility with the BIGRIPS magnet and the SPIRIT TPC, recently.

Other observables sensitive to the symmetry energy at supra-normal densities are collective flows. At energies below 1 GeV/nucleon the reaction dynamics is largely determined by the nuclear mean field. The resulting pressure produces a collective motion of the compressed material whose strength will be influenced by the symmetry energy in isospin-asymmetric systems. The FOPI collaboration has measured a multitude of flow observables in many different collisions systems [15] in the energy regime between 200 and 2000 MeV/nucleon with FOPI setup at SIS18@GSI.

12.2 Neutron and Charged Particle Elliptic Flow

The strengths of collective flows in heavy-ion collisions are determined by a Fourier expansion of the azimuthal distributions of particles around the reaction plane:

$$\frac{d\sigma(y)}{d\phi} = C(1 + 2v_1(y)\cos\phi + 2v_2(y)\cos 2\phi.....) \tag{12.1}$$

The side flow of particles is characterized by the coefficient v_1 and the elliptic flow by v_2. The value of v_2 around mid-rapidity is negative at incident beam energies between 0.2 and 6 GeV/nucleon which signifies that matter is squeezed out perpendicular to the reaction plane. Elliptic flow at those energies is known to be sensitive to the stiffness of the symmetric part of the nuclear equation of state. In Fig. 12.1 data of (negative) elliptic flow of protons is shown for mid-central Au+Au collisions at 400 MeV/nucleon (left panel) and 1.2 GeV/nucleon (right panel). The bell-shaped experimental data are compared to IQMD predictions [16] employing a soft (SM) and hard (HM) density dependence of the nuclear matter equation of state. The particles within IQMD are interacting via a momentum dependent interaction which has been deduced from experimental data on proton scattering off nuclei [16]. A momentum dependent interaction is needed to describe the energy dependence of flow and pion

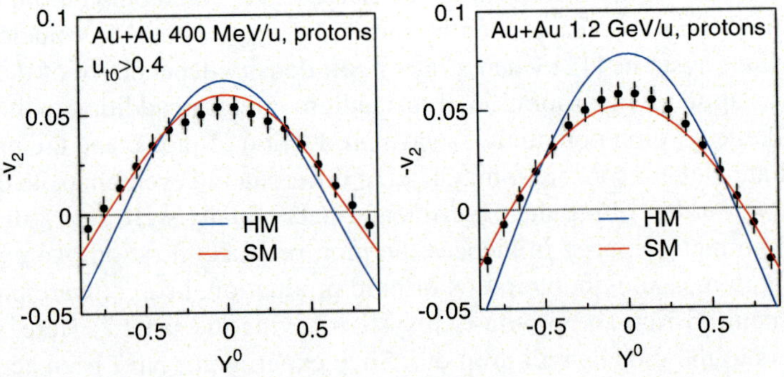

Fig. 12.1 The elliptic flow v_2 measured with the FOPI setup in semi-central Au+Au collisions as function of normalized rapidity $Y^0 = y/y_p$, where y is the particle rapidity in the center-of-mass system and y_p the rapidity of the projectile. The initial target-projectile rapidity gap always extends from $Y^0 = -1$ to $Y^0 = 1$. The experimental data (shown as block points) are compared to IQMD predictions [16] employing a soft EOS (denoted as red line) and a hard EOS (blue line). The calculations have been performed using a momentum dependent interaction. Data and calculations are taken from [15, 17]

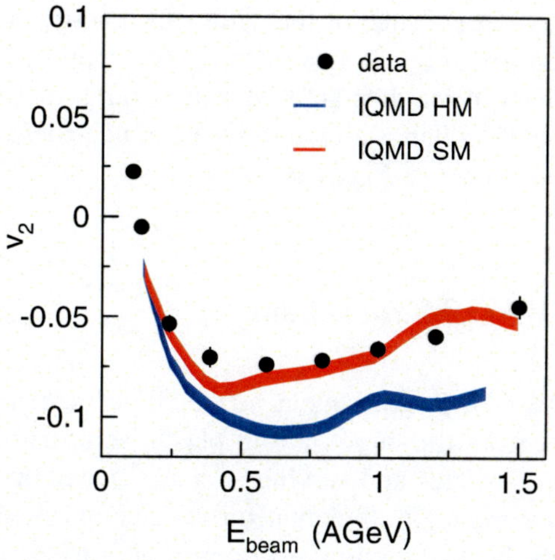

Fig. 12.2 Elliptic flow v_2 of protons as a function of beam energy for Au+Au collisions and semi-central collisions [15]. The data points are shown as full circles, predictions of the IQMD model for a soft and a hard nuclear equation of state with momentum dependent interactions by red and blue bands, respectively. Particles around mid-rapidity ($|Y^0| < 0.2$) and with high transverse momenta ($u_t^0 > 0.8$) are selected. u_t is the transverse component of the four-velocity u, $u_t = \beta_t \gamma$, where $u_t^0 = u_t / u_p$, with $u_p = \beta_p \gamma_p$, the index p referring to the incident projectile in the center-of-mass system [15, 16]

production data simultaneously. Clearly, a soft density dependence of the nuclear EOS is needed to describe the experimental data at both energies. Not only the size of the elliptic flow v_2 at mid-rapidity is described but also the rapidity dependence of v_2, which supports in addition the choice of the specific input parameters to the transport model, which is in addition constraint by other experimental results of the FOPI collaboration.

Data for the elliptic flow v_2 of protons emitted around mid-rapidity in semi-central Au+Au collisions [15] as a function of beam energy together with predictions of the IQMD model for HM and SM nuclear equation of state augmented by a momentum dependent interaction are shown in Fig. 12.2. Again, the experimental data are best described by calculations utilizing a soft (SM) equation of state. The larger negative v_2-values for a hard equation of state are almost completely due to the higher density gradients perpendicular to the reaction plane and, therefore, an effect of the acting potentials [17].

The ratio of elliptic flow strengths of neutrons and light charged particles has been proposed by using a version of the UrQMD transport code [18] as a robust observable sensitive to the EOS of asymmetric matter [19]. Neutron and proton emission has been already measured by a combination of the LAND neutron detector with a part of the FOPI setup [20]. A reanalysis of the data and comparison to the results of the UrQMD transport model [19] yielded a moderately soft symmetry energy dependence on density, $L = 86 \pm 15$ MeV. Despite the large error a particular soft or a very stiff density dependence of the symmetry energy could ruled out. These results have been confirmed by using a different transport code [21]. Nevertheless, the statistical error bars of the data are rather large, and, therefore, an attempt was started by the ASY-EOS collaboration to remeasure neutron/proton flow in Au+Au collisions at an incident energy $E_{beam} = 400A$ MeV.

12.2.1 The ASY-EOS Experiment

The setup of the ASY-EOS experiment at the SIS18 accelerator is presented in Fig. 12.3. Upstream from the target, a thin plastic scintillator foil was placed for measuring the beam particles and serving as a start detector for the time-of-flight systems. The granular and high-efficient neutron detector LAND was placed close to 45° with respect to the beam direction. A veto wall of thin plastic scintillator material in front of LAND was used to discriminate between neutrons and charged particles. The Krakow Tripple Telescope Array (KRATTA was placed opposite to LAND covering approximately the same angular acceptance. This device allows to identify light charged fragments. Three detector systems were used for event characterization and background suppression. The ALADIN Time-of-Flight (ATOF) detected

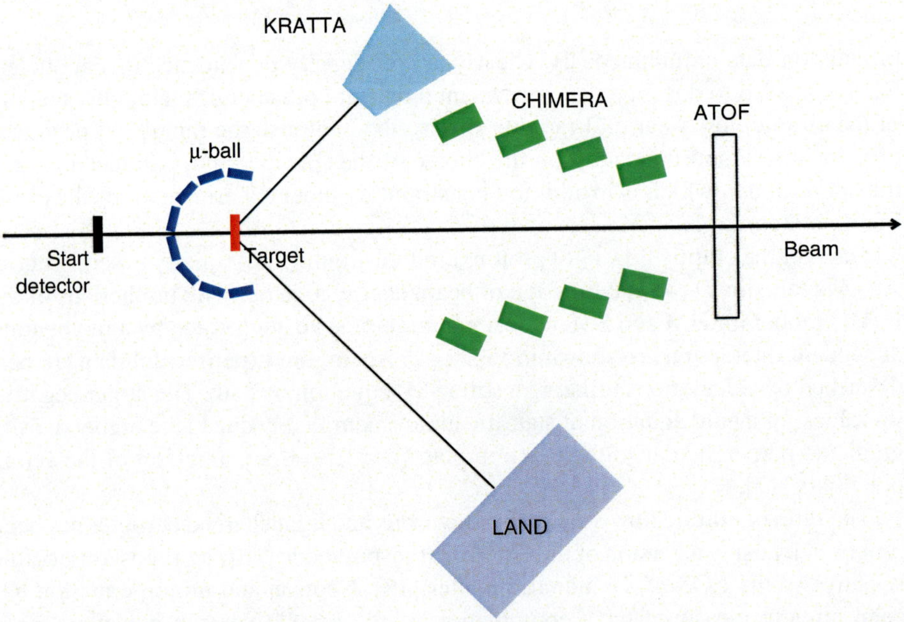

Fig. 12.3 Sketch of the setup of the ASY-EOS experiment (S394) at the SIS18 of GSI showing the six main detector systems and their positions relative to the beam direction. Dimensions and distances are not to scale (from [22])

charged particles and fragments emitted at very small polar angles $\Theta_{lab} < 7°$ close to thebeam. Four double rings of the CHIMERA multidetector consisting out of 352 CsI(Tl) scintillators were placed in forward direction, and the target was surrounded by four rings with 50 thin CsI(Tl) elements of the Washington University Microball array. Those detectors provided sufficient granularity and solid angle coverage to determine the orientation of the reaction plane and the impact parameter. A detailed description of the setup is available in [22].

12.2.2 Experimental Results

The data analysis procedure is described in detail in [22]. The v_2 ratios obtained for neutrons v_2^n and light charged particles v_2^{ch} after applying all corrections are shown in Fig. 12.4. Constraints to the symmetry energy were obtained by comparing the ratio v_2^n/v_2^{ch} with corresponding UrQMD predictions.

A soft iso-scalar EOS was chosen for the calculations and the following parameterization was used for the density dependence of the symmetry energy:

$$E_{sym}(\rho) = E_{sym}^{pot}(\rho) + E_{sym}^{kin}(\rho) = 22\text{MeV}(\rho/\rho_0)^{\gamma} + 12(\rho/\rho_0)^{2/3}\text{MeV}, \quad (12.2)$$

with $\gamma = 0.5$ and $\gamma = 1.5$ corresponding to a soft and a stiff density dependence, respectively.

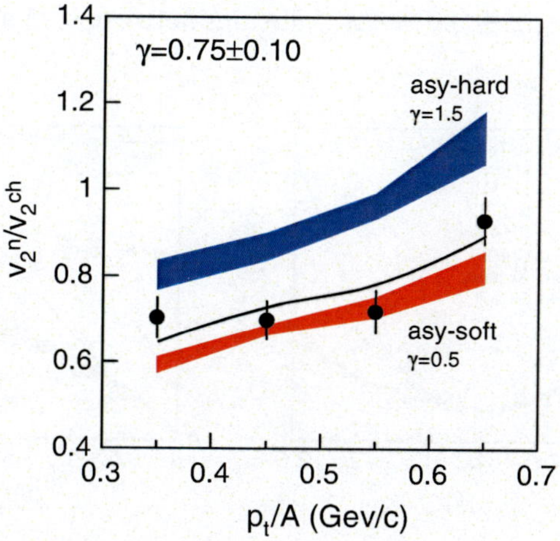

Fig. 12.4 Elliptic flow ratio of neutrons over charged particles measured in the same acceptance range for central ($b < 7.5$ fm) Au+Au collisions at 400 MeV/nucleon as a function of transverse momentum, p_t/A. The black circles represent the experimental data. The blue and red bands represent the results of UrQMD model calculations employing a hard and soft density dependence of the symmetry energy. Fitting a linear interpolation between the two extremes to the experimental data yields a γ-value of $\gamma = 0.75 \pm 0.10$. The black line shows the resulting predictions

The two predictions obtained under these conditions are presented in Fig. 12.4 together with the experimental results. The parameters which describe the data best are obtained by fitting a linear interpolation between the two predictions. The resulting power-law coefficient is $\gamma = 0.75 \pm 0.10$. The corresponding v_2 ratio is shown as black line in Fig.12.4. In comparison with the FOPI-LAND data, the statistical accuracy of the ASY-EOS experiments represents an improvement by a factor of 2. A systematic uncertainty arose from occasional malfunction of the electronic circuits for the time measurement with LAND. It prohibits extending the data evaluation into the region of large transverse momenta. For this reason this analysis is restricted to $p_t < 0.7$ GeV/c. Another uncertainty is related to the lower energy threshold for neutron detection. The necessary corrections and the methods used for estimating the remaining errors are described in detail in [22].

With all corrections and errors included, the acceptance-integrated elliptic flow values lead to a power-law coefficient $\gamma = 0.72 \pm 0.19$. The corresponding slope value is $L = 72 \pm 13$ MeV. Changing the absolute value of the symmetry energy at ground state nuclear matter density in the simulations to a lower value, $E_{sym}(\rho_0) =$ 31MeV, results in a lower γ-value $\gamma = 0.68 \pm 0.19$. This is still within the error bar of the above result.

This result is displayed in Fig. 12.5 (right panel) where the symmetry energy is shown as a function of reduced density ρ/ρ_0 together with various microscopic model predictions [23]. The experimental data are represented as colored bands. The results of the FOPI-LAND experiment are shown in yellow and the one of the ASY-Experiment in orange. For completeness the calculations from [23] for the EOS of symmetric matter are shown in the left panel of Fig. 12.5 together with the experimental constraint from the FOPI collaboration [17] utilizing elliptic flow of protons and light charged particles in Au+Au collisions between 0.25 and 1.5 GeV/nucleon.

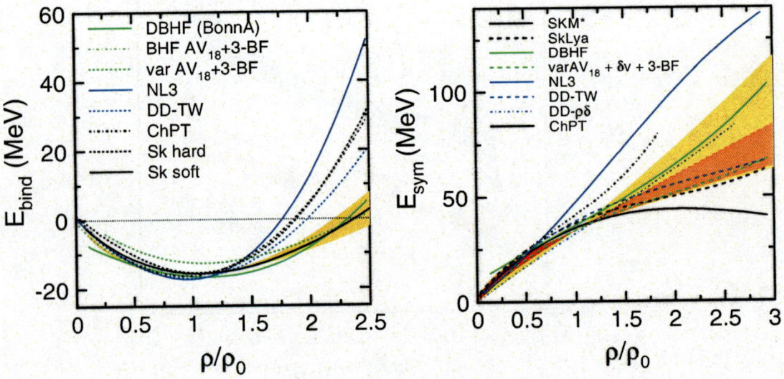

Fig. 12.5 Left panel: Constraints from experimental data of the FOPI collaboration as presented in [17] (yellow area) are shown together with predictions of microscopic model calculations employing different methods and different interactions [23]. Right panel: Result of the ASY-EOS experiment (orange area) and the FOPI-LAND analysis (yellow area) together with theoretical predictions for $E_{sym}(\rho)$ using different microscopic approaches in addition to Skyrme interactions. The calculations are taken from [23]

12.2.3 Density Tested in the ASY-EOS Experiment

Since microscopic transport models are reproducing reasonably well the experimental data it seems to be appropriate to deduce the density range relevant for the sensitivity to the symmetry energy in this reaction by using one of those codes. For this investigation the Tübingen Version of the QMD model, TüQMD, [21] was chosen (see [22] for more details). The two density dependencies of the symmetry energy, hard and soft (blue and red lines in Fig. 12.6, respectively), are shown in the left panel of Fig. 12.6 together with the linear form (black line) used as default parameterization. To quantify the results, the function $DERF$ (Difference of elliptic flow ratio) is defined:

$$DERF^{n,Z}(\rho) = \frac{v_2^n}{v_2^Z}(hard, \rho) - \frac{v_2^n}{v_2^Z}(soft, \rho). \tag{12.3}$$

An example of the function $DERF(\rho)$ is presented in the middle panel of Fig. 12.6. At zero density $DERF(0)$ is zero by definition and at high densities $DERF(\rho)$ is approaching the maximal value, which is the variation of v_2 for the two options of E_{sym}. The region in density where $DERF(\rho)$ is changing most rapidly is also the density regime which is most relevant for the determination of the symmetry energy. Hence, the value of the derivative $\mathrm{d}DERF(\rho)/\mathrm{d}\rho$ is a measure of the impact of the symmetry energy on the elliptic flow observables. In the right panel of Fig. 12.6 the derivative of $DERF(\rho)$ is shown for three choices of the elliptic flow ratio: For protons (n/p), hydrogen isotopes (n/H) and all charged particles (n/ch).

It can be seen in the figure that the maximum sensitivity achieved with the elliptic flow ratio of neutrons to charged particles is reached close to saturation density and extends beyond twice this value. This finding is in agreement with results of [17] which were obtained by analyzing also elliptic flow data of charged particles and

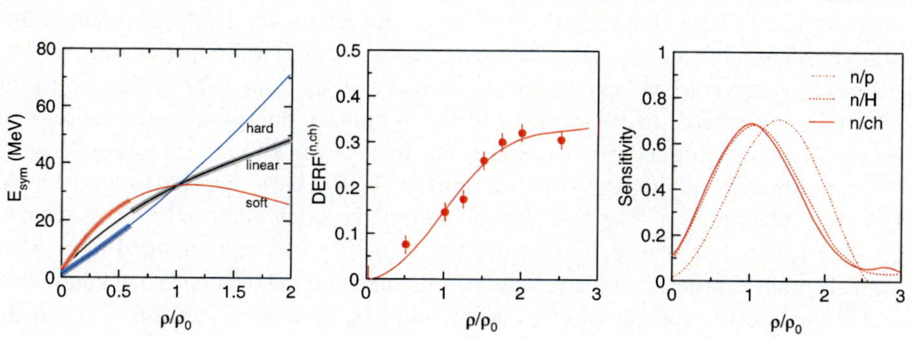

Fig. 12.6 Left panel: The different parameterizations describing the density dependence of the symmetry energy which were used for the investigation described in the text. The linear one was employed as a normal or default form of $E_{sym}(\rho)$. Middle panel: The function $DERF(\rho)$ for the ratio of elliptic flows of neutrons and charged particles. Right panel: Derivative of $DERF(\rho)$ as a function of density for different particle choices. These results are taken from [22]

using the IQMD model [16]. For $^{197}Au + Au$ collisions at 400 MeV/u the force-weighted density, defined by the authors, is spread over a broad density regime extending from 0.8$< \rho/\rho_0 <$ 1.6.

The derivative of $DERF(\rho)$ for the neutron-proton ratio peaks at a higher value, 1.4 to 1.5ρ_0. This observation gives rise to the expectation that with sufficient isotope separation one would be able not only to constraint the slope of the symmetry energy but also its curvature.

12.3 Conclusions and Outlook

The symmetry energy at supra-saturation densities can be effectively probed with the elliptic flow ratio of neutrons and charged particles. The ASY-EOS collaboration measured data for Au+Au collisions at 400A MeV, and a slope for the of $L = 72 \pm 13$ for the density dependence of the symmetry energy could be deduced from the v_2 ratio of neutrons and charged particles. According to model predictions, similar observables might also be sensitive not only to the slope of the symmetry energy but also to its curvature at high densities. Hence, it would be interesting to measure n and light charged particle flow also at higher energies and for different systems. In collisions of Sn-isotopes one could investigate the evolution of the v_2 ratio as a function of N/Z of the colliding system.

How the sensitivity of the various observables to the symmetry energy develops with rising beam energy and larger densities is by no means obvious and more calculations using different models and prescriptions need to be performed. Particle production is predicted to be most sensitive to modifications in the neutron/proton density close to threshold. Kaon production below and close to threshold ($E_{thr,NN} =$ 1.6 GeV has proven to be sensitive to the density reached in the course of the collision. It has been shown in particular that Kaon yields are robust observables to constrain the iso-scalar EOS [24]. At sub-threshold energies Kaons are produced in the central high-density region and, because they are not re-absorbed by the surrounding nuclear matter, they are true messengers of the dense overlap zone. The production ratio (K^+/K^0) is predicted to be sensitive to the symmetry energy and, therefore, may be an interesting observable accessing densities beyond twice saturation density. An experiment studying Au+Au reactions at 1.25A GeV was performed by the HADES collaboration. The HADES detector setup allows not only for measuring charged Kaons but also reconstructing K^0 by their decay into charged pions with high accuracy. Sufficient statistic was accumulated and the data have been published and compared to predictions of transport models. However, the sensitivity to the symmetry energy has not yet been tested.

Symmetry effects are very small, because the contribution of the symmetry energy enters with a factor δ^2 into the equation of the state of nuclear matter. Since δ is rather small even for the most neutron-rich stable isotopes it is evident that those studies have to be extended to radioactive beams with enhanced neutron-proton asymmetry. Experimental facilities, e.g., RIKEN/RIBF, MSU/FRIB, GANIL/SPIRAL2 are or will become available, which allow further studies of the symmetry energy by different experimental methods. But the highest energies and consequently highest densities will be accessed with FAIR/Super-FRS facility. The super fragment separator Super-FRS will have a magnetic rigidity of 20 Tm and will provide radioactive beams up to 1 GeV/nucleon. This will offer the possibility to study the symmetry energy at high densities. Nevertheless, the neutron-proton asymmetry which may be accessed in such experiments is not significantly larger than the one reached in Au+Au or Pb+Pb collisions with the concomitant disadvantage that the collision systems are getting smaller and the densities reached in the course of the reactions are not as large as in the heavy Au-system. Hence, for flow and particle production observables it may not be necessary to utilize radioactive beams, and stable beams and targets would be sufficient. In such case, it would be impossible to generate double ratios in order to diminish systematic errors in the experiment and uncertainties due to unknown input parameters to the models. This would require sophisticated transport codes which have been benchmarked intensively to experimental data (see [25] for an ongoing effort in comparing different transport codes).

Acknowledgements Results presented in this contribution have been obtained within the ASY-EOS collaboration. See [22] for a complete list of authors.

References

1. B.A. Brown, Neutron radii in nuclei and the neutron equation of state. Phys. Rev. Lett. **85**, 5296 (2000); X. Roca-Maza, M. Centelles, X. Viñas, M. Warda, Neutron skin of ^{208}Pb, nuclear symmetry energy, and the parity radius experiment. Phys. Rev. Lett. **106**, 252501 (2011)
2. B.-A. Li, L.-W. Chen, C.M. Ko, Recent progress and new challenges in isospin physics with heavy-ion reactions. Phys. Rep. **464**, 113–281 (2008)
3. B. Tsang et al., Constraints on the density dependence of the symmetry energy. Phys. Rev. Lett. **102**, 122701 (2009)
4. A.W. Steiner et al., Isospin asymmetry in nuclei and neutron stars. Phys. Rep. **411**, 325–375 (2005)
5. A. Bauswein et al. Identifying a first-order phase transition in neutron-star mergers through gravitational waves. Phys. Rev. Lett. **122**, 061102 (2019)
6. B.-A. Li, X. Han, Constraining the neutron-proton effective mass splitting using empirical constraints on the density dependence of nuclear symmetry energy around normal density. Phys. Lett. B **727**, 276–281 (2013)
7. I. Tews et al., Symmetry parameter constraints from a lower bound on neutron-matter energy. Astro. Phys. J. **848**, 1 (2017)
8. P. Danielewicz et al., Symmetry energy III: Isovector skins. Nucl. Phys. A **958**, 147–186 (2017)
9. J.M. Lattimer, A. Steiner, Constraints on the symmetry energy using the mass-radius relation of neutron stars. Eur. Phys. J. A **50**, 40 (2014)
10. B.A. Brown, Constraints on the Skyrme equations of state from properties of doubly magic nuclei. Phys. Rev. Lett. **111**, 232502 (2013)
11. Z. Xiao et al., Circumstantial evidence for a soft nuclear symmetry energy at suprasaturation densities. Phys. Rev. Lett. **102**, 062502 (2009)
12. W. Reisdorf et al., Systematics of pion emission in heavy ion collisions in the 1AGeV regime. Nucl. Phys. A **781**, 459–508 (2007)

13. Z.Q. Feng, G.M. Jin, Probing high-density behavior of symmetry energy from pion emission in heavy-ion collisions. Phys. Lett. B **683**, 140–144 (2010)
14. J. Hong, P. Danielewicz, Subthreshold pion production within a transport description of central Au + Au collisions. Phys. Rev. C **90**, 024605 (2014)
15. W. Reisdorf et al., Systematics of azimuthal asymmetries in heavy ion collisions in the 1AGeV regime. Nucl. Phys. A **876**, 1–60 (2012)
16. Ch. Hartnack et al., Modelling the many-body dynamics of heavy ion collisions: present status and future perspective. Eur. Phys. J. A **1**, 151–169 (1998)
17. A. Le Fèvre et al., Constraining the nuclear matter equation of state around twice saturation density. Nucl. Phys. A **945**, 112–133 (2016)
18. Q. Li, M. Bleicher, H. Stocker, The effect of pre-formed hadron potentials on the dynamics of heavy ion collisions and the HBT puzzle. Phys. Lett. B **659**, 525–530 (2008)
19. P. Russotto et al., Importance of nuclear effects in the measurement of neutrino oscillation parameters. Phys. Lett. B **697**, 471–476 (2011)
20. Y. Leifels et al., Exclusive studies of neutron and charged particle emission in collisions of ^{197}Au + ^{197}Au at 400 MeV/nucleon. Phys. Rev. Lett. **71**, 963–966 (1993)
21. M.D. Cozma, Neutron-proton elliptic flow difference as a probe for the high density dependence of the symmetry energy. Phys. Lett. B **700**, 139–144 (2011)
22. P. Russotto et al., Results of the ASY-EOS experiment at GSI: The symmetry energy at suprasaturation density. Phys. Rev. C **94**, 034608 (2016)
23. C. Fuchs, H.H. Wolter, Modelization of the EOS. Eur. Phys. J. A **30**, 5–21 (2006)
24. Ch. Hartnack et al., Hadronic matter is soft. Phys. Rev. Lett **96**, 012302 (2006)
25. O. Akira et al., Comparison of heavy-ion transport simulations: collision integral with pions and δ resonances in a box (2019), arXiv:1904.02888v2 [nucl-th]

Chapter 13
Elliptic Flow in Relativistic Heavy-Ion Collisions

Madan M. Aggarwal

Abstract The basic aim of the heavy-ion physics is to investigate matter at extreme densities and temperatures where quarks and gluons are no longer confined inside hadrons. Such a state of matter, that may have existed a few microseconds after the Big Bang, is created in the laboratory by colliding nuclei at the Relativistic Heavy-Ion Collider (RHIC), Brookhaven, and at the Large Hadron Collider (LHC), CERN at top center of mass energies $\sqrt{s_{\mathrm{NN}}} = 200\,\mathrm{GeV}$ and $5.02\,\mathrm{TeV}$, respectively. The large elliptic flow and number of constituent quark (NCQ) scaling observed at the RHIC and similar observations at the LHC with some deviation indicate the formation of de-confined state in relativistic heavy-ion collisions. The elliptic flow measurement and its dependence on collision centrality, transverse momentum, particle species, etc., will be presented.

13.1 Introduction

The Quantum Chromodynamics (QCD) predicted a phase transition from normal nuclear matter into the de-confined plasma phase of quarks and gluons at an energy density of $\sim 1\,\mathrm{GeV/fm^3}$ or for a critical temperature $T_c \sim 170$ MeV [1, 2]. The Ultra-relativistic heavy-ion collisions provide a very good opportunity to create and observe Quark-Gluon Plasma (QGP) in the laboratory. The QCD phase diagram, baryon chemical potential (μ_B) versus temperature (T), is shown in Fig. 13.1. A solid line separates the hadronic region (at comparatively low temperature or chemical potential) from the quark-gluon plasma. One expects the phase transition line to end at a critical point as indicated in the figure. The RHIC beam energy scan programme searches for a critical point in the phase diagram [3]. The Present and future nucleus-nucleus experiments are depicted in the diagram. The LHC and the RHIC are investigating the baryon free QGP at high temperature (low μ_B) whereas experiments at the FAIR will look for the baryon rich QGP at low T (high μ_B).

M. M. Aggarwal (✉)
Department of Physics, Panjab University, Chandigarh, India
e-mail: aggarwal@pu.ac.in

R. K. Puri et al. (eds.), *Advances in Nuclear Physics*, Springer Proceedings in Physics 257, https://doi.org/10.1007/978-981-15-9062-7_13

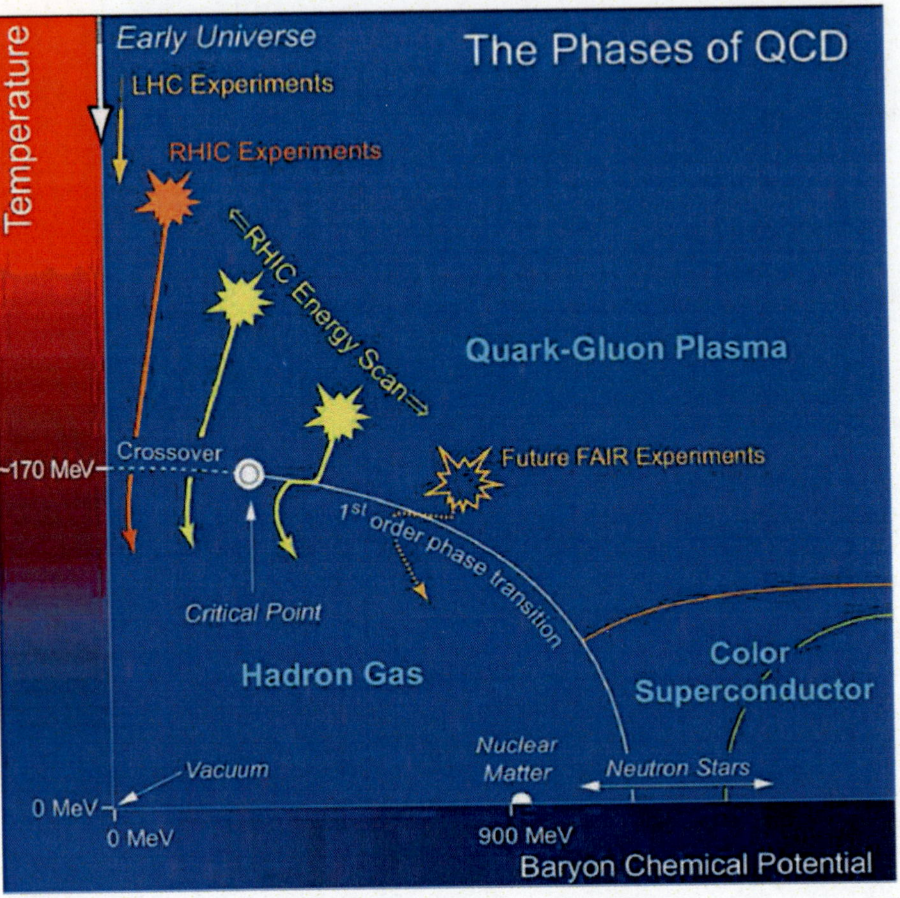

Fig. 13.1 The schematic phase diagram of the quantum chromodynamics (QCD) in terms of temperature (T) and baryon chemical potential (μ_B)

A space-time picture of the nucleus-nucleus collision is shown in Fig. 13.2. Two nuclei approach each other nearly at the velocity of light. At the instant of collision (i.e., t = 0, z = 0), the two nuclei strike each other and in the initial stage of the collision, i.e., pre-equilibrium stage, processes of parton-parton hard scatterings may predominantly occur, depositing a large amount of energy in the overlap region of two colliding nuclei. At a time ∼1 fm/c, the QGP phase would be formed, in which parton-parton interactions reach thermal equilibration state. Hadrons continue to collide elastically and inelastically. The chemical freeze-out occurs at T_{ch} when inelastic processes stop. The kinetic freeze-out occurs at T_{fo} when elastic scatterings stop and hadrons freely stream out from the medium.

The experimental programme started with the fixed target experiments at CERN and BNL in 1986. The CERN announced in the year 2000 the formation of Quark-Gluon Plasma (QGP) in the heavy-ion collisions based on the results from

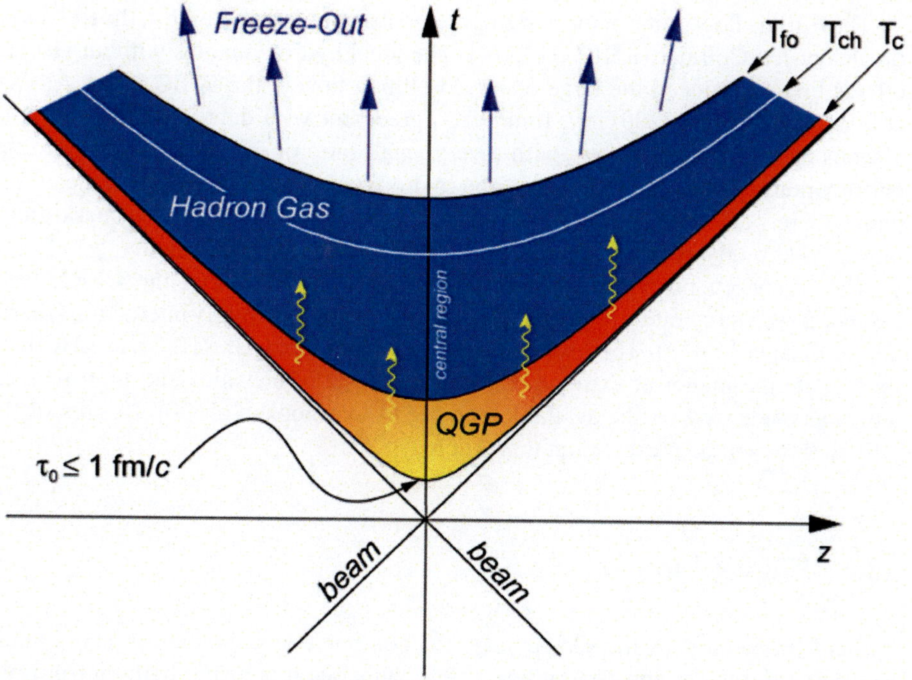

Fig. 13.2 The space-time picture of nucleus-nucleus collision. Two nuclei shown as beams, approach each other nearly at the velocity of light and collide at $t = 0$, $z = 0$

seven experiments, viz., NA44, NA45/CERES, NA49, NA50, NA52/NEWMASS, WA97/NA57, and WA98, at the CERN SPS [4]. The detailed investigations of the QGP formation in heavy-ion collisions became possible with the commissioning of the Relativistic Heavy-Ion Collider (RHIC) at BNL. Four experiments, viz., STAR (Solenoidal Tracker at RHIC), PHENIX (Pioneering High Energy Nuclear Interaction Experiment), PHOBOS, and BRAHMS (Broad Range Hadron Magnetic Spectrometer), were installed at different interaction points at the RHIC. Ollitrault [5] proposed the elliptic flow as a signature of hydrodynamic behaviour of the nuclear matter produced in high energy nuclear collisions. Since then lot of activities are observed both experimentally and theoretically in investigating collective behaviour in the heavy-ion collisions as evidenced by number of review articles [6–11]. The STAR collaboration defined the QGP as a (locally) thermally equilibrated state of matter in which quarks and gluons are de-confined from hadrons, so that colour degrees of freedom become manifest over nuclear, rather than merely nucleonic volumes [12]. The observation of large elliptic flow in Au+Au collisions at the RHIC points to the production of thermally equilibrated state of matter and opacity to jets in Au+Au collisions at the RHIC means quarks and gluons are de-confined or dense matter formed at early times is partonic.

The Elliptic flow is one of the first important observables measured at the Relativistic Heavy-Ion Collider (RHIC) [13, 14]. The ideal hydrodynamics without viscous effects first reproduced the large observed elliptic flow at the RHIC in the Au+Au collisions at $\sqrt{s_{\mathrm{NN}}} = 200\,\mathrm{GeV}$. Improved agreement with data was achieved with viscous hydrodynamic models with a very small ratio of the shear viscosity to the entropy density. The applicability of the hydrodynamics requires a short mean free path with respect to the system size. This led to conclude that the created quark-gluon plasma is strongly interacting and behaves like a nearly perfect liquid.

The heavy-ion programmes at the Large Hadron Collider at the CERN have extended the energy range for measuring the properties of strongly interacting Quark-Gluon Plasma (sQGP). The goal of heavy-ion experiments at RHIC and LHC is to investigate the matter at extreme conditions of energy densities and high temperatures as evidenced by the available number of text books [15–20]. Results of the elliptic flow will be presented in this article.

13.2 Elliptic Flow

Particle production in the elementary nucleon-nucleon collisions is azimuthally isotropic whereas in non-central heavy-ion collisions it is azimuthally anisotropic. Figure 13.3 (left) exhibits the spatial anisotropy of the almond-shaped overlap zone in the initial state of non-central heavy-ion collision. The initial momenta of particles in the overlap zone are predominantly longitudinal and transverse momenta, if any, distributed isotropically. If the particles interact amongst themselves, the probability of interaction and getting scatter is larger for the particles moving along the long axis than those moving along the short axis. This results in the large pressure gradient in the direction of the short axis. Therefore, flow velocity is larger along the short axis leading to the emission of more particles along the short axis than along the long axis. This leads to an anisotropic distribution of particles in the transverse plane

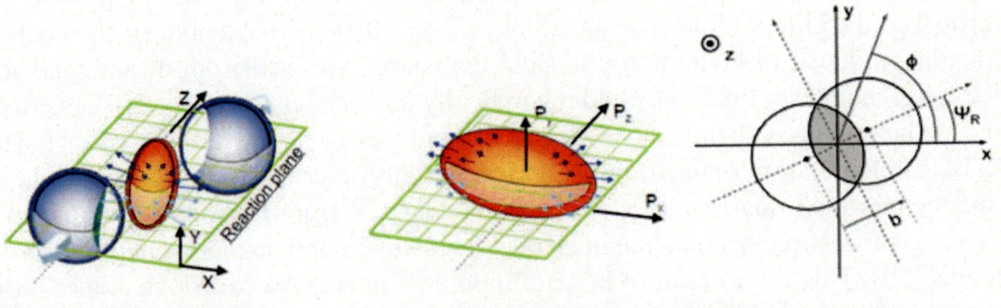

Fig. 13.3 Left: Depicting the spatial anisotropy of the almond-shaped overlap zone in non-central heavy-ion collision. Center: Spatial anisotropy resulting in the momentum anisotropy. Right: Picture of a non-central heavy-ion collision in the transverse (x versus y) plane. Z (beam axis) is perpendicular to the plane of the figure. Φ is the azimuthal angle of one of the outgoing particles. Ψ_R is the reaction plane angle and b is the impact parameter

(Fig. 13.3 (center)). So, it is believed that the elliptic flow helps in understanding the interactions between the constituents at an early time in the evolution of the produced system and hence sensitive to the equation of state of the system.

The invariant triple differential distribution of particles emitted in the collision is given by the following Fourier expansion [21]:

$$E\frac{d^3N}{dp^3} = \frac{1}{2\pi}\frac{d^2N}{p_t dp_t dy}\left(1 + \sum_{n=1}^{\infty} 2v_n cos(n(\phi - \Psi_{RP}))\right) \tag{13.1}$$

where p is the momentum of the particle, p_t is the transverse momentum, E is the energy, y is the rapidity, ϕ is the azimuthal angle and ψ_{RP} is the reaction plane angle (Fig. 13.3 (right)). Sine terms vanish in the above Fourier expansion due to the reflection symmetry with respect to the reaction plane and are not included in the (13.1). The v_2 for n = 2 represents the elliptic flow and is obtained as

$$v_2 = \langle cos(2(\phi - \Psi_{RP}))\rangle \tag{13.2}$$

here, angular brackets represent an average over the particles, summed over all events in a given sample.

13.2.1 Elliptic Flow Methods

The elliptic flow is found to be very useful for understanding relativistic nuclear collisions but its value spreads over a range of 20% determined by different analysis methods. It is well known that non-flow correlations not related to the reaction plane and fluctuations affect the measured v_2 values. We will discuss the standard event plane method [22, 23], Q-cumulants [24] and probability $p(v_2)$ [25]. A higher accuracy is needed in determining v_2 to compare with the relativistic viscous hydrodynamic calculations to extract the ratio of the shear viscosity to entropy.

13.2.1.1 Event Plane Method

The reaction plane (Ψ_{RP}) cannot be determined experimentally for an event. So one uses the event plane which is good approximation to the reaction plane. The nth order event plane is defined as

$$\Psi_n^{EP} = tan^{-1}(Y_n/X_n)/n \tag{13.3}$$

where $X_n = \sum cos(n\phi_i)$, $Y_n = \sum sin(n\phi_i)$, and ϕ_i is the azimuthal angle of the ith particle in an event and summation runs over all particles used to determine the event plane. The event plane determined is not isotropic due to detector non-uniformity. To obtain uniform event plane angle distributions following corrections are applied as discussed in [26]:

- Recenter the distributions of X_n and Y_n by subtracting the $\langle X_n \rangle$ and $\langle Y_n \rangle$ values averaged over all events, i.e.,

$$\begin{aligned} X_n^{'} &= X_n - \langle X_n \rangle \\ Y_n^{'} &= Y_n - \langle Y_n \rangle \end{aligned} \tag{13.4}$$

- After re-centering, correction for twist of the vector is applied as

$$\begin{aligned} X_n^{''} &= \frac{(X_n^{'} - \lambda_{2n}^{s-} Y_n^{'})}{(1 - \lambda_{2n}^{s+} \lambda_{2n}^{s-})} \\ Y_n^{''} &= \frac{(Y_n^{'} - \lambda_{2n}^{s+} X_n^{'})}{(1 - \lambda_{2n}^{s+} \lambda_{2n}^{s-})} \end{aligned} \tag{13.5}$$

where

$$\begin{aligned} \lambda_{2n}^{s\pm} &= \langle S_{2n} \rangle / A_{2n}^{\pm} \\ A_{2n}^{\pm} &= 1 \pm \langle C_{2n} \rangle \\ C_{2n} &= \sum cos(2n\phi) \\ S_{2n} &= \sum sin(2n\phi) \end{aligned} \tag{13.6}$$

here average is taken over all particles in an event and over all events in a sample.
- Finally re-scale the vector as

$$\begin{aligned} X_n^{'''} &= X^{''} / A_{2n}^{+} \\ Y_n^{'''} &= Y^{''} / A_{2n}^{-} \end{aligned} \tag{13.7}$$

One gets 2nd order event plane as

$$\Psi_2^{EP} = tan^{-1}\left(\frac{Y_2^{'''}}{X_2^{'''}}\right)/2 \tag{13.8}$$

The v_2^{obs} can be obtained using (13.2) as

$$
\begin{aligned}
v_2^{obs} &= \langle cos2(\phi - \Psi_2^{EP}) \rangle \\
&= \langle cos(2(\phi - \Psi_2^{EP}) + 2(\Psi_{RP} - \Psi_{RP})) \rangle \\
&= \langle cos(2(\phi - \Psi_{RP}) + 2(\Psi_{RP} - \Psi_2^{EP})) \rangle \\
&= \langle cos(2(\phi - \Psi_{RP}) cos(2(\Psi_{RP} - \Psi_2^{EP})) \rangle \\
&= v_2^{true} \langle cos(2(\Psi_{RP} - \Psi_2^{EP})) \rangle
\end{aligned} \tag{13.9}
$$

$$
v_2^{true} = v_2^{obs} / \langle cos(2(\Psi_{RP} - \Psi_2^{EP})) \rangle \tag{13.10}
$$

Here Ψ_{RP} is still unknown. $\langle cos(2(\Psi_{RP} - \Psi_2^{EP})) \rangle$ is the event plane resolution correction factor which one needs to apply to v_2^{obs} to get v_2^{true}. This is obtained by dividing an event into two sub-events a and b randomly having equal number of particles. The $\langle cos(2(\Psi_a^{EP} - \Psi_b^{EP})) \rangle$ can be written as:

$$
\begin{aligned}
\langle cos(2(\Psi_a^{EP} - \Psi_b^{EP})) \rangle &= \langle cos(2(\Psi_a^{EP} - \Psi_{RP}) - 2(\Psi_b^{EP} - \Psi_{RP})) \rangle \\
&= \langle cos(2(\Psi_a^{EP} - \Psi_{RP})) \langle cos(2(\Psi_b^{EP} - \Psi_{RP}) \rangle
\end{aligned} \tag{13.11}
$$

The resolution correction factor for sub-events is given as

$$
\langle cos(2(\Psi_a^{EP} - \Psi_{RP}) \rangle = \sqrt{(\langle cos(2(\Psi_a^{EP} - \Psi_b^{EP})) \rangle} \tag{13.12}
$$

and for full event as

$$
\langle cos(2(\Psi_2^{EP} - \Psi_{RP}) \rangle = \sqrt{2} \sqrt{(\langle cos(2(\Psi_a^{EP} - \Psi_b^{EP})) \rangle} \tag{13.13}
$$

13.2.1.2 Q-Cumulants

2- and 4-particle cumulants are defined in terms of multi-particle azimuthal correlations for the detectors with uniform azimuthal acceptance as [27]

$$
C_2\{2\} = \langle \langle 2 \rangle \rangle \tag{13.14}
$$

$$
C_2\{4\} = \langle \langle 4 \rangle \rangle - 2 \langle \langle 2 \rangle \rangle^2 \tag{13.15}
$$

The multi-particle azimuthal correlations $\langle 2 \rangle$ and $\langle 4 \rangle$ are calculated either using the generating function technique given by Borghini et al., [27] or from the following equations for each event [24]:

$$\langle 2\rangle_{n|n} \equiv \langle e^{\iota n(\phi_1-\phi_2)}\rangle \equiv \frac{1}{P_{M,2}} \sum_{\substack{i,j=1\\(i\neq j)}}^{M} e^{\iota n(\phi_i-\phi_j)} \quad (13.16)$$

$$\langle 4\rangle_{n,n|n,n} \equiv \langle e^{\iota n(\phi_1+\phi_2-\phi_3-\phi_4)}\rangle \equiv \frac{1}{P_{M,4}} \sum_{\substack{i,j,k,l=1\\(i\neq j\neq k\neq l)}}^{M} e^{\iota n(\phi_i+\phi_j-\phi_k-\phi_l))} \quad (13.17)$$

where $P_{M,n} = M!/(M-n)!n!$, M is the event multiplicity and ϕ is azimuthal angle of the particle.
2nd order cumulant is obtained by decomposing $|Q_n|^2$,

$$| Q_n |^2 = \sum_{i,j=1}^{M} e^{\iota n(\phi_i-\phi_j)} \quad (13.18)$$

where $\phi_i(\phi_j)$ is the azimuthal angle of the particle. Indices i and j can be either the same or different. The decomposition of $| Q_n |^2$ contains contributions from 2-particle correlations when indices are different and auto-correlations when indices are same, as given below:

$$| Q_n |^2 = \langle 2\rangle_{n,n} P_{M,2} + 1.M \quad (13.19)$$

So one gets

$$\langle 2\rangle_{n,n} = \frac{(| Q_n |^2 - M)}{M(M-1)} \quad (13.20)$$

Here Q_n is Q-vector:

$$Q_n = \sum_{i=1}^{M} \exp(n\phi_i) \quad (13.21)$$

Here summation runs over all the particles in an event. Similarly one can decompose $|Q_n|^4$ as

$$\begin{aligned} | Q_n |^4 = \langle 4\rangle_{n,n|n,n} M(M-1)(M-2)(M-3) \\ +(\langle 3\rangle_{2n|n,n} + \langle 3\rangle_{n,n|2n}) M(M-1)(M-2) \\ +\langle 2\rangle_{n|n}[M(M-1)2!(M-2)2! + M(M-1)2!2!] \\ +\langle 2\rangle_{2n|2n} M(M-1) + 1.[M(M-1)2! + M] \end{aligned} \quad (13.22)$$

thus for $\langle 3\rangle_{2n|n,n}$ and $\langle 3\rangle_{n,n|2n}$ one has to decompose $Q_{2n}Q_n^*Q_n^*$ and $Q_nQ_nQ_{2n}^*$.

$$\langle 4\rangle_{n,n|n,n} = \frac{|Q_n|^4 + |Q_{2n}|^2 - 2\Re[Q_{2n}Q_n^*Q_n^*] - 4(M-2)|Q_n|^2}{M(M-1)(M-2)(M-3)} + \frac{2}{(M-1)(M-2)} \tag{13.23}$$

For details see [24]. The average is taken over all events in a given sample using multiplicity weights $w_2 = M(M-1)$ and $w_4 = M(M-1)(M-2)(M-3)$ corresponding to different 2- and 4-particle combinations, one can form with multiplicity M.

$$\langle\langle 2\rangle\rangle_{n|n} \equiv \frac{\sum_{events}(W_2)_i(\langle 2\rangle_{n|n})_i}{\sum_{events}(W_2)_i} \tag{13.24}$$

$$\langle\langle 4\rangle\rangle_{n,n|n,n} \equiv \frac{\sum_{events}(W_4)_i(\langle 4\rangle_{n,n|n,n})_i}{\sum_{events}(W_4)_i} \tag{13.25}$$

Integrated flow estimates from cumulants is given as [27]

$$v_2^2\{2\} = C_2\{2\} \tag{13.26}$$

$$v_2^4\{4\} = -C_2\{4\} \tag{13.27}$$

13.2.1.3 Event by Event $p(v_2)$

The event-by-event v_2 coefficients and phases can be estimated [25] with

$$\begin{aligned} v_{2,x}^{obs} &= |\vec{v}_2^{obs}|\, cos(2\Psi_2^{obs}) = \langle cos(2\phi)\rangle = \frac{\sum_i w_i cos(2\phi_i)}{\sum_i w_i}, \\ v_{2,y}^{obs} &= |\vec{v}_2^{obs}|\, sin(2\Psi_2^{obs}) = \langle sin(2\phi)\rangle = \frac{\sum_i w_i sin(2\phi_i)}{\sum_i w_i}, \end{aligned} \tag{13.28}$$

$$|\vec{v}_2^{obs}| = \sqrt{(v_{2,x}^{obs})^2 + (v_{2,y}^{obs})^2} \tag{13.29}$$

$$p(v_2^{obs}) = p(v_2^{obs} \mid v_2) * p(v_2) \tag{13.30}$$

The v_2^{obs} is the magnitude of the observed EbyE per particle flow vector. The response function $p(v_2^{obs}|v_2)$ is needed to determine the EbyE v_2. In order to determine $p(v_2^{obs}|v_2)$ each event is divided into two sub-events with symmetric η range, i.e., $\eta > 0$ and $\eta < 0$ and labelled as a and b sub-events, respectively. The graph is obtained $(v_{2,x}^{obs})^a - (v_{2,x}^{obs})^b$ versus $(v_{2,y}^{obs})^a - (v_{2,y}^{obs})^b$ in which physical flow signal gets cancel and it contains mainly the effects of statistical smearing and non-flow. One can obtain the response functions from this graph as discussed in [25].

13.2.2 *Elliptic Flow Measurements and Comparison with Hydrodynamic Models*

In the heavy-ion collisions, events are characterized by the impact parameter "b" which is, however, not a direct observable. Experimentally, it is determined from the number of produced particles in a collision assuming that the multiplicity is a monotonic function of the impact parameter. Figure 13.4 (left) exhibits the charged particle multiplicity distribution obtained in Pb-Pb collisions at $\sqrt{s_{NN}} = 2.76\,\text{TeV}$ by the ALICE collaboration [28]. It also shows the centrality classifications, i.e., $0{-}5\%, 5{-}10\%, \ldots\ldots$ etc., based on the fraction, $\pi b^2/\pi(2R_a)^2$, of the geometrical cross section with R_a the nuclear radius. Figure 13.4 (right) displays the transverse energy distribution measured in Pb-Pb collisions at 5.02 TeV by the ATLAS collaboration showing a classification in centrality percentiles [29]. In Monte-Carlo models centrality is characterized by the number of participating nucleons in a collision or by the number of binary collisions, which can be related to the impact parameter "b" using proper nuclear geometry in Glauber model.

Figure 13.5 (left) displays the measurement of the elliptic flow in Au+Au collisions at $\sqrt{s_{NN}} = 200\,\text{GeV}$ obtained using various methods as indicated in figure by the STAR collaboration [13]. It is seen that v_2 values obtained by standard method, scalar product, η sub, random sub, and $v_2\{2\}$ cumulants agree within 5% whereas v_2 values from q-distribution and $v_2\{4\}$ values are about 25% less. The v_2 values fall in two bands. It is worth mentioning that same data cuts were used for different methods. It is argued that higher value of v_2 is because of the contribution of non-flow effects whereas smaller value indicates that non-flow effects are removed, i.e., in $v_2\{4\}$, q-distribution. Further, it is shown that event-by-event fluctuations in eccentricity, lead two-particle ($v_2\{2\}$) cumulants overestimate true value of v_2 and four/six $(v_2\{4\})/(v_2\{6\})$ particles azimuthal correlations underestimate true value of v_2. Sometimes one uses the mean of $v_2\{2\}$ and $v_2\{4\}$ for the true value of v_2 [30]. The v_2 reaches the maximum value for 40–50% collision centrality. Figure 13.5 (right)

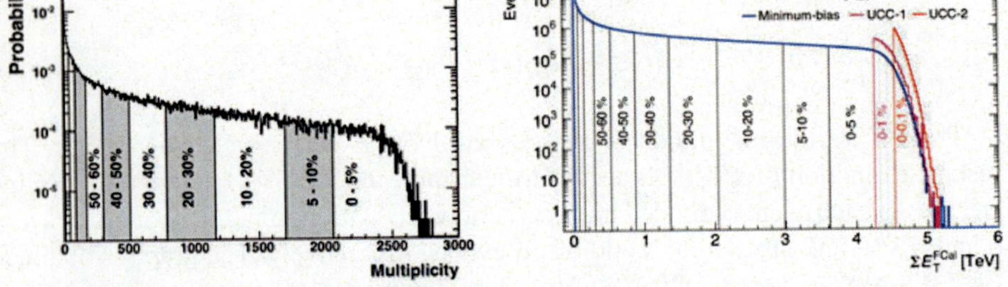

Fig. 13.4 Left: Multiplicity distribution of charged particles in the TPC ($\eta < 0.8$). The cumulative percentage of total events is indicated. Right: Total transverse energy distribution measured by the ATLAS. Several centrality intervals are marked with vertical lines and labelled on the plot

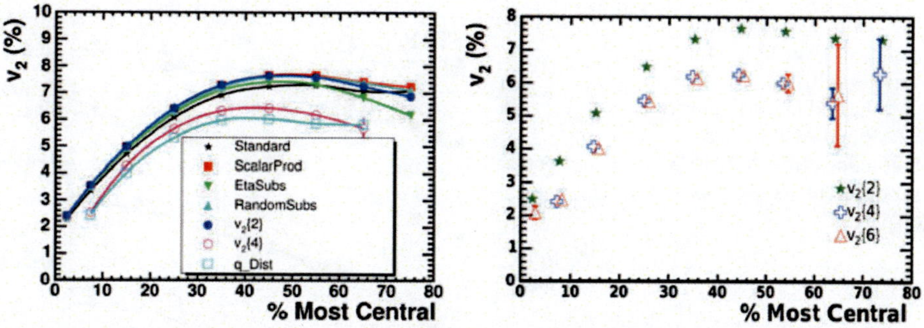

Fig. 13.5 Left: $v_2(\%)$ versus collision centrality. $v_2(\%)$ determined from various methods as indicated in the figure. Right: compares $v_2(\%)$ determined from $v_2\{2\}$,$v_2\{4\}$and $v_2\{6\}$

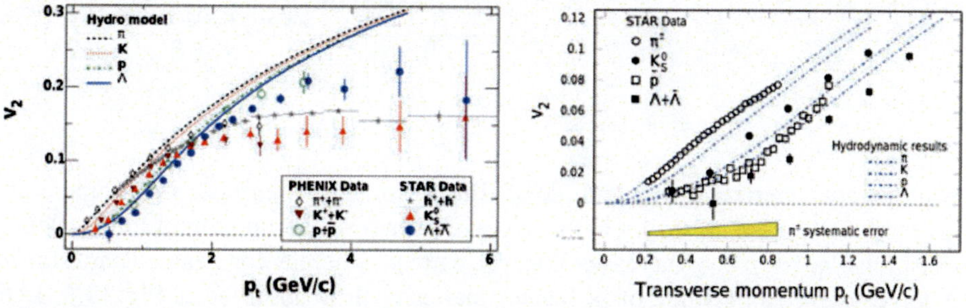

Fig. 13.6 Left: v_2 versus transverse momentum p_t for different identified particles as indicated in the figure. Predictions of Hydro are also shown. The PHENIX data points are also displayed. Right: Comparison of the STAR data with Hydro for $p_t < 1.6$

compares the $v_2\{2\}$, $v_2\{4\}$ and $v_2\{6\}$; and it is observed that $v_2\{4\}$ and $v_2\{6\}$ agree indicating the non-flow effects are removed already with four particles cumulants.

Huovinen et al., [31] and Ollitrault [11] predicted the transverse momentum dependence of elliptic flow of identified particles using a hydrodynamical model. Transverse momentum dependences of v_2 for identified particles [13] are shown in Fig. 13.6 (left). It is seen that there is a good agreement between PHENIX and the STAR results. Figure 13.6 (right) compares the STAR data with the hydrodynamical calculations. The v_2 is in broad agreement with the hydrodynamical model. This observation led to the claim of formation of an almost perfect liquid at the RHIC. Further, one notices the mass ordering of v_2, i.e., lighter the particle higher the v_2 at same transverse momentum (at low p_t) indicating that as if all the particles are emitted from the same expanding thermal source [11, 31]. The dependence of v_2 on p_t for fast particles with $p_t > m_\mu$ and $m_t > m_{\mu^0}$ is obtained as [11]:

$$v_2 = \frac{\alpha}{T}(p_t - vm_t) \tag{13.31}$$

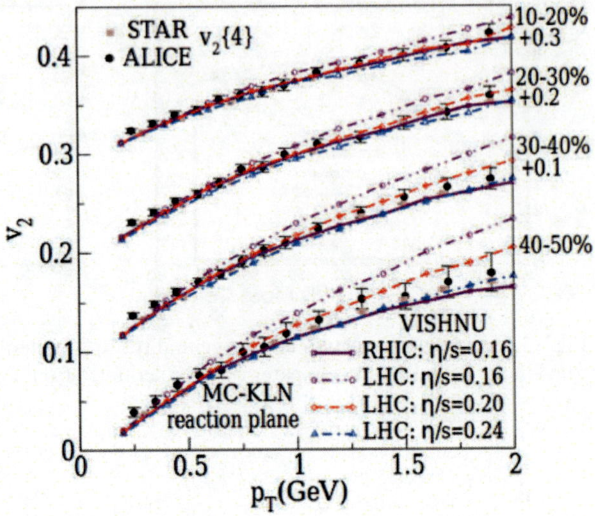

Fig. 13.7 v_2 determined from $v_2\{4\}$ versus p_T for various collision centralities as indicated in the figure for STAR and ALICE data. Predictions from VISHNU for different η/s are also displayed

where α characterizes the magnitude of elliptic flow, T is the freeze-out temperature, m_t the transverse mass and v is the average over ϕ of the maximum fluid velocity. For pions, $m_t \sim p_t$, v_2 increases linearly with p_t whereas for heavier particles m_t is larger at the same value of p_t hence smaller v_2. It is observed in Fig. 13.6 (Left) pions have larger v_2 than kaons and protons at the same p_t. In Fig. 13.6 (Right), a deviation from ideal hydrodynamics is seen for $p_t \sim 2\,\mathrm{GeV}/c$.

Since fluid dynamics can describe the elliptic flow quite well, so efforts were made to study the effects of the shear viscosity and on extracting the value of shear viscosity over entropy. Song et al., [32], used the hybrid model VISHNU which describes the expansion of the QGP using viscous hydrodynamics and successive evolution of hadronic matter with a microscopic transport model. Authors extracted the η/s values using data on the Au+Au collisions at 200 GeV and Pb-Pb collisions at 2.76 TeV. Figure 13.7 exhibits p_T dependence of $v_2\{4\}$ for different centralities for both Au+Au at 200 GeV [33] and Pb-Pb at 2.76 TeV [34]. Theoretical lines are from VISHNU calculations with different constant η/s. It is seen that the STAR data agree reasonably well with η/s = 0.16 whereas ALICE data is significantly larger for $p_T >$ 0.5 GeV. However, the ALICE data agree well for η/s = 0.20–0.24 at higher p_T but still model underestimates the ALICE data at $p_T <$ 0.5 GeV. Other hydrodynamical model calculations [35, 36] also under predict the ALICE low p_T data. Figure 13.8 shows the p_T integrated $v_2\{4\}$ dependence on centrality for both STAR and ALICE data. It is observed that the STAR data fits with η/s = 0.16 whereas the ALICE data fits better for η/s = 0.20. The STAR results for Cu+Au at $\sqrt{s_{\mathrm{NN}}} = 200$ GeV [37] agree with hydrodynamics calculations using Glauber initial conditions with η/s = 0.08–0.16 (Fig. 13.9). It seems that shear viscosity increases with temperature.

Figure 13.10 compares the ALICE Pb-Pb data at 2.76 TeV [34] for different particle species for centralities 10–20% and 40–50% with VISHNU having η/s = 0.16. It is seen that VISHNU exhibits qualitatively similar v_2 mass ordering as seen in the

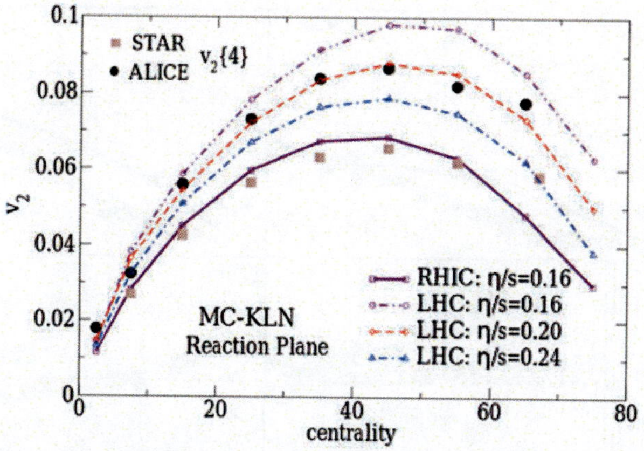

Fig. 13.8 Integrated v_2 determined from $v_2\{4\}$ versus collision centrality for STAR and ALICE data. Predictions from VISHNU for different η/s are also displayed

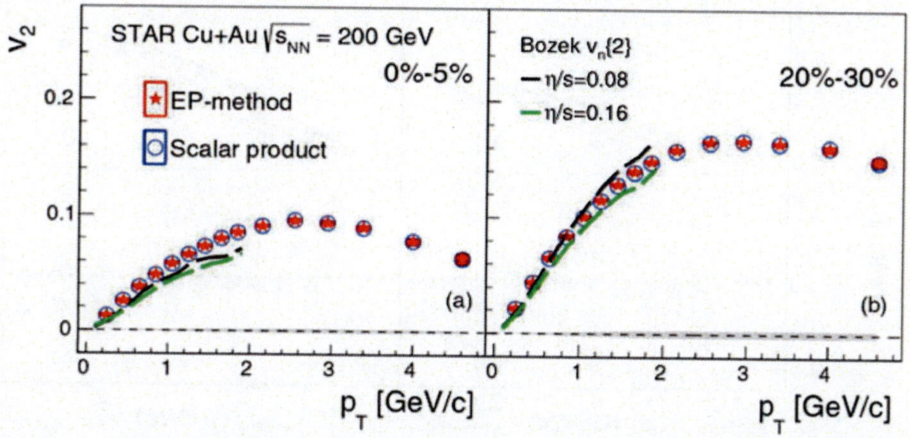

Fig. 13.9 v_2 versus p_T for 0–5% and 20–30% collision centralities for Cu+Au collisions at $\sqrt{s_{NN}} =$ 200 GeV. v_2 is determined using event plane or scalar product is indicated in the figure. Predictions of hydrodynamics calculations using Glauber initial conditions with $\eta/s = 0.08$–0.16 are also displayed

data. Ratio of v_2/v_2(hydro fit) in the central collisions reproduces K results whereas deviations are seen for other particle species. The measured $v_2(p_T)$ for pions is more than those of theoretically calculated for $p_T < 1\,\text{GeV}/c$ and also proton elliptic flow is significantly underestimated. It overestimates $v_2(p_T)$ for multi-strange baryons. The ϕ meson does not follow mass ordering in case of VISHNU. This may be due to small hadronic interaction cross section.

Figure 13.11 compares the v_2 probability distribution for Pb-Pb collisions at 2.76 TeV (ATLAS data) for 20–25% central events [25, 38] with the eccentricity distribution, $P(\epsilon_2/\langle\epsilon_2\rangle)$, obtained using IP-Glasma [39]. Reasonably good agreement is seen. The IP-Glasma model [40, 41] describes well the initial conditions for

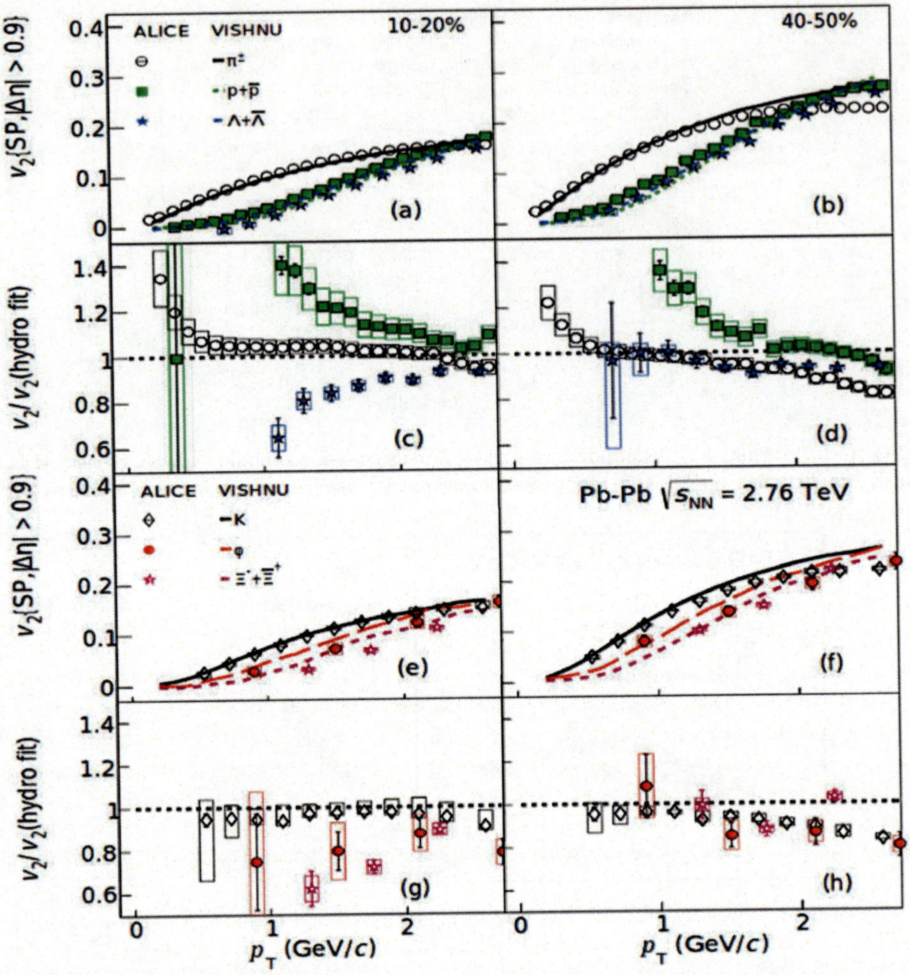

Fig. 13.10 Comparison of v_2 versus p_T for the ALICE Pb-Pb data at 2.76 TeV for different particle species for centralities 10–20% and 40–50% with VISHNU having η/s = 0.16. v_2/v_2(hydro fit) versus p_T are also shown as indicated in figure

Fig. 13.11 v_2 distribution for 20–25% central events for Pb-Pb collisions at $\sqrt{s_{NN}} = 2.76$ TeV ATLAS data compared to IP-Glasma and P-Glasma+MUSIC as indicated in the figure

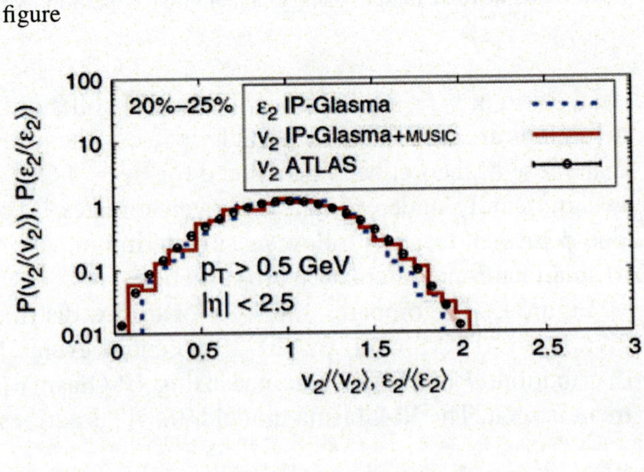

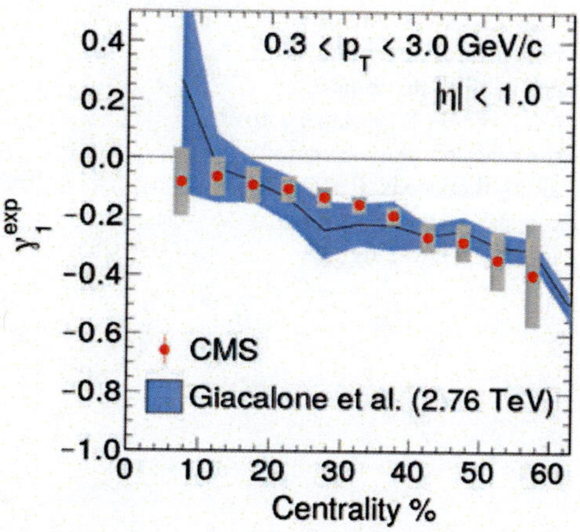

Fig. 13.12 γ_1^{exp} as a function of collision centrality for Pb-Pb collisions at $\sqrt{s_{NN}} = 5.02\,\text{TeV}$ CMS data. Prediction of Hydro at $\sqrt{s_{NN}} = 2.76\,\text{TeV}$ is also shown [50]

systematic flow studies. The IP-Glasma model incorporates IP-Sat (Impact Parameter Saturation Model) model [42, 43] of nucleon and nuclear wave-functions, and classical Yang–Mills (CYM) dynamics of the glasma fields produced in a heavy-ion collision [44–46]. A very good agreement is seen for, if IP-Glasma couples with the hydrodynamic evolution described by MUSIC, a 3+1 dimensional relativistic viscous hydrodynamic simulation [47] which uses the Kurganov-Tadmor algorithm [48]. The CMS [49] also used probability p(v_2) method and obtained skewness as

$$\gamma_1^{exp} = -6\sqrt{2}v_2\{4\}^2 * (v_2\{4\} - v_2\{6\})/(v_2\{2\}^2 - v_2\{4\}^2)^{2/3} \tag{13.32}$$

The good agreement is seen between the Pb-Pb data at 5.02 TeV and Hydro predictions at 2.76 TeV as shown in Fig. 13.12.

13.2.3 Elliptic Flow Fluctuations

It is noticed that v_2 measured by different methods differs by about 20%. In non-central heavy-ion collisions, initial eccentricity of the almond shape overlap zone results in the elliptic flow. This eccentricity fluctuates from event to event due to fluctuations in the impact parameter (positions of participating nucleons). This leads to fluctuations in the elliptic flow from one event to other within a given sample. Flow fluctuations depend on the initial geometry fluctuations of the system created in the collision. So flow develops relative to the participant plane instead of the reaction plane. The elliptic flow and its fluctuations are very useful for understanding the initial conditions of the expansion phase of heavy-ion collisions.

Agakishiev et al., [51] and Kumar [52], investigated the eccentricity and eccentricity fluctuations using three monte-carlo models, viz., Monte-Carlo Glauber model with nucleons as participants (MCG-N) [53, 54], a Monte-Carlo Glauber model

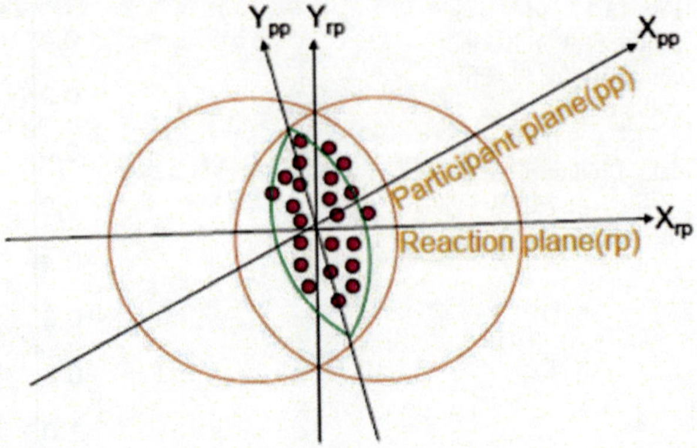

Fig. 13.13 Picture of a non-central heavy-ion collision in the transverse (X_{rp} versus Y_{rp}) plane with the reaction plane oriented along the x-axis. Participant plane X_{pp} and Y_{pp} are also indicated in the figure

with quarks as participants (MCG-Q), and the factorized Kharzeev, Levin, and Nardi Colour Glass Condensate model (fKLN-CGC) [55]. In the MCG-N it is assumed: (i) Au nuclei are spherical, (ii) nucleons move in a straight line along the beam direction, and (iii) $\sigma_{\rm NN}$ inelastic cross section as the transverse interaction range. The Woods–Saxon distribution of nucleons was used. The minimum distance (d) between nucleons inside nuclei before colliding is taken as $\sqrt{(\sigma_{\rm NN}/\pi)}$. A negative binomial distribution was used to generate events with width k = 2.1 for each participant and mean multiplicity given as (for details see [51, 52]):

$$n = (0.5933 \ln \sqrt{s_{\rm NN}} - 0.4153)((1 - x_{hard}) + 2x_{hard} N_{bin}/N_{part}) \tag{13.33}$$

where N_{bin} and N_{part} represent number of binary collisions and number of participants, respectively. x_{hard} is the fraction of multiplicity proportional to N_{bin}. In the MCG-Q, quarks are distributed inside the nucleon using another Woods–Saxon density distribution with R = 0.865 fm, skin depth a = 0.108 fm and $\sigma_{\rm QQ} = \sigma_{\rm NN}/7$. The fKLN-CGC model gives multiplicity and eccentricity. The eccentricity in the reaction and participant planes (Fig. 13.13) are given as:

$$\epsilon_{RP} = (\sigma_y^2 \sigma_x^2)/(\sigma_y^2 + \sigma_x^2) \tag{13.34}$$

$$\epsilon_{part} = \sqrt{((\sigma_y^2 \sigma_x^2)^2 + 4\sigma_{xy}^2)/(\sigma_y^2 + \sigma_x^2)} \tag{13.35}$$

where $\sigma_x^2 = \langle x^2 \rangle - \langle x \rangle^2$, $\sigma_y^2 = \langle y^2 \rangle - \langle y \rangle^2$ and $\sigma_{xy} = \langle xy \rangle$ - $\langle x \rangle \langle y \rangle$, here (x, y) denote position of nucleon in the reaction plane. The two- and four-particle cumulants of ϵ_{part} [56]:

$$\epsilon\{2\}^2 = \langle \epsilon_{part}^2 \rangle \tag{13.36}$$

$$\epsilon\{4\}^4 = 2\langle \epsilon_{part}^2 \rangle - \langle \epsilon_{part}^4 \rangle \tag{13.37}$$

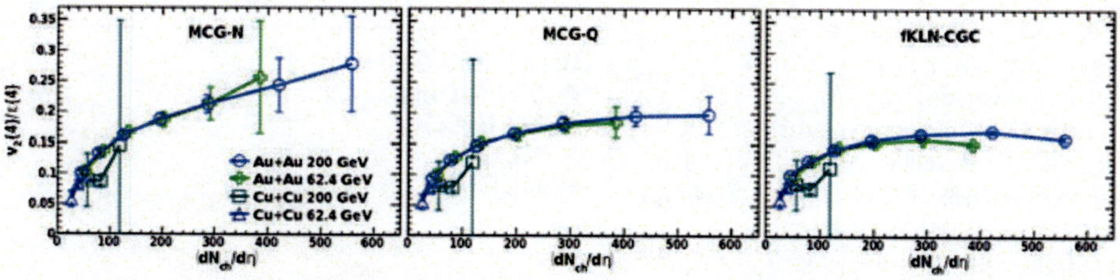

Fig. 13.14 The eccentricity scaled $v_2\{4\}$ i.e.,$v_2\{4\}/\epsilon\{4\}$, versus $dN_{ch}/d\eta$ with eccentricity ($\epsilon\{4\}$) taken from the MCG-N (left), MCG-Q (middle), or fKLN-CGC (right) model at $\sqrt{s_{\rm NN}} = 200$ and 62.4 GeV for Au+Au and Cu+Cu collisions as indicated in the figure

Assuming $v_2 \propto \epsilon$, then $\langle v_2 \rangle / \langle \epsilon \rangle$ in the reaction plane is given by $v_2\{4\}/\epsilon\{4\}$ [56]. The mean multiplicity versus $v_2\{4\}/\epsilon\{4\}$ displayed in Fig. 13.14 for Au+Au and Cu+Cu collisions at 200 and 62.4 GeV for above three models. The data points for different system and energies seem to fall on top of each other as expected if $v_2 \propto \epsilon$. The MCG-N exhibits the sharp increase of $v_2\{2\}/\epsilon$ violating the approximation $v_2 \propto \epsilon$ whereas the fKLN-CGC shows a saturation with $v_2 \propto \epsilon$. The MCG-Q displays intermediate trend between the sharp rise and the saturation. The v_2 saturation or nearly saturation trend exhibited in central Au+Au collisions at 200 GeV is consistent with a nearly perfect liquid behaviour.

New Q-cumulants method used to determine experimentally flow fluctuations does not require the knowledge of either reaction plane or participant plane. The estimation of $v_2\{2\}$ can be biased by correlations such as resonance decay, jets, etc., which are not related to the reaction or participant plane and denoted as δ. Voloshin et al., [57], have shown that $v_2\{2\}$ and $v_2\{4\}$ for the Gaussian form of flow fluctuations in the participant plane are correlated as

$$v_2\{2\}^2 = \langle v \rangle^2 + {\sigma_v}^2 + \delta \tag{13.38}$$

$$v_2\{4\}^2 \sim \langle v \rangle^2 - {\sigma_v}^2 \tag{13.39}$$

$$v_2\{2\}^2 - v_2\{4\}^2 \sim \delta + 2{\sigma_v}^2 \tag{13.40}$$

where v_2 is the elliptic flow, δ is the non-flow contribution and σ_v represent flow fluctuations. This is because fluctuations increase $v_2\{2\}$ and decrease $v_2\{4\}$ and the approximations are valid for $\sigma_{v_2}/\langle v_2 \rangle \ll 1$. It is not possible to determine the value of σ_{v_2} from the measurements of $v_2\{2\}$ and $v_2\{4\}$. However, we can find the upper limit on the flow fluctuations taking $\delta = 0$ in [(13.38)–(13.40)], i.e.,

$$R_{v(2-4)} = \sigma_{v_2}/\langle v_2 \rangle = \sqrt{\frac{(v_2\{2\}^2 - v_2\{4\}^2)}{(v_2\{2\}^2 + v_2\{4\}^2)}} \tag{13.41}$$

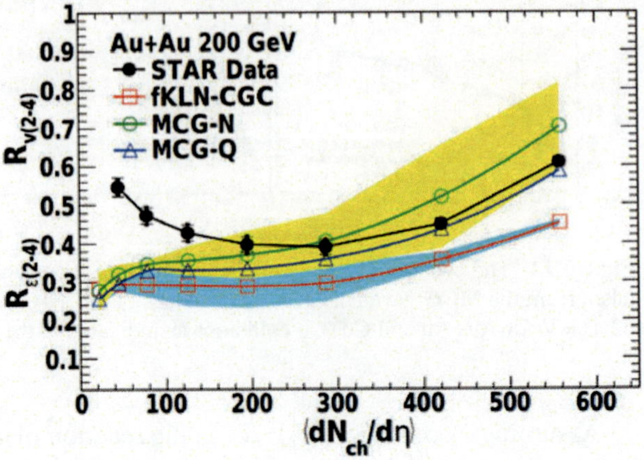

Fig. 13.15 Upper limit on v_2 fluctuations, i.e., $R_{v(2-4)}$ ($R_{\epsilon(2-4)}$) versus $dN_{ch}/d\eta$, for the STAR data on Au+Au at $\sqrt{s_{NN}} = 200$ GeV. Predictions from MCG-N, MCG-Q, and fKLN-CGC are indicated in the figure. The shaded bands display the uncertainties for the MCG-N and fKLN-CGC models

Figure 13.15 displays $R_{v(2-4)}$ versus mean multiplicity for Au+Au collisions at 200 GeV measured by the STAR collaboration [51]. Here, STAR collaboration used $v_2\{2\}$ for the like sign charged particles to reduce non-flow. The normalized elliptic flow variance $\sigma_{v_2}/\langle v_2 \rangle$ is large (~40% in mid central events). For non-central collisions the magnitude of elliptic flow is proportional to the eccentricity (ϵ) of the initial nuclear overlap zone, one expects that normalized variance of v_2 should be close to that of eccentricity. $R_{\epsilon(2-4)}$ for different eccentricity models is defined as

$$R_{\epsilon(2-4)} = \sqrt{\frac{(\epsilon\{2\}^2 - \epsilon\{4\}^2)}{(\epsilon\{2\}^2 + \epsilon\{4\}^2)}} \tag{13.42}$$

where $\epsilon\{2\}$ and $\epsilon\{4\}$ are the second and fourth cumulants of ϵ_{part}. Predictions from above three eccentricity models are also shown in Fig. 13.15. The shaded bands represent the uncertainties on the models which are dominated by uncertainty on the distribution of nucleons inside the nucleus. The MCG-N underestimates for $dN_{ch}/d\eta < 150$ whereas it overshoots for higher multiplicities and exceeds the upper limit. The MCG-Q approaches the upper limit in central collisions but never exceeds the data. However, the fKLN-CGC points always remain below the upper limit obtained from the data. The comparison of the STAR like sign charged particles and the PHOBOS all charged particles [58] for Au+Au data at $\sqrt{s_{NN}} = 200$ GeV on upper limit for σ_{v_2}/v_2 is displayed in Fig. 13.16. It is seen that the PHOBOS data points are slightly lower than those of the STAR data points. This may be due to the different procedures used in both papers. Also the PHOBOS has subtracted narrow $d\eta$ correlations. It seems that some residual non-flow is left in STAR like sign charged particles results. Similar observations were made in the Pb-Pb collisions at 2.76 TeV [59]. The CMS collaboration observed a non-Gaussian behaviour in the event-by-event fluctuations of the elliptic flow (v_2) in Pb-Pb collisions at 5.02 TeV [49].

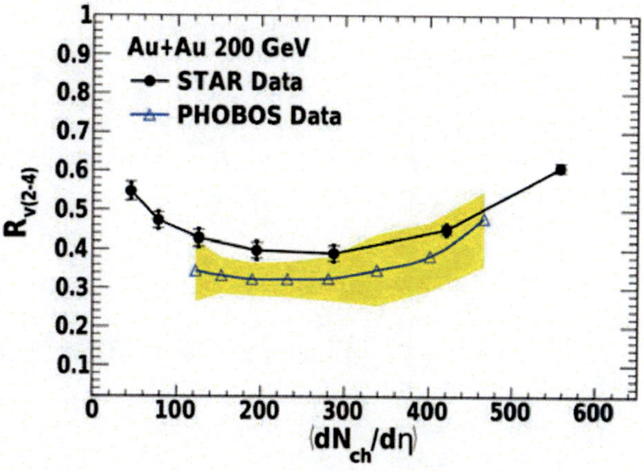

Fig. 13.16 Comparison of the STAR data with those of the PHOBOS data on v_2 fluctuations. The shaded band indicates errors on the PHOBOS data

13.2.4 Number of Constituent Quark (NCQ) Scaling

The hadron production in the intermediate p_t range is well described by coalescence or recombination of constituent quark models [60–62] and models predict the scaling of v_2 with the number of constituent quarks (n), i.e., universal curve for v_2/n versus p_t/n for all hadrons. Figure 13.17 (top) displays v_2/n versus p_t/n for K_s^0, $K^{\pm}$, $p + \bar{p}$ and Λ+$\bar{\Lambda}$ for minimum bias Au+Au collisions at $\sqrt{s_{NN}} = 200$ GeV [13]. It is seen that all points fall on one curve. In Fig. 13.17 (bottom), data points scaled by the fitted polynomial function are plotted. It is observed that all data points lie on a universal curve within error bars except the pion points deviating significantly which may be due to the pions coming from resonance decay. Similar trend is noticed for 30−70% and 5−30% collision centralities. The PHENIX data [62] for 10–40% central Au+Au collisions at $\sqrt{s_{NN}} = 62.4$ GeV is shown in Fig. 13.18. Here, the scaling seems to break. The better scaling is observed when plotted v_2/n_q versus $KE_T/n_q = (m_t - m)/n_q$. STAR data on Cu+Au at 200 GeV also exhibit similar scaling behaviour [63].

The PHENIX data [64] on Au+Au and Cu+Cu for different centralities at 200 GeV and Au+Au at 62.4 GeV for π/k/p are displayed in Fig. 13.19 (a). A large spread is seen for v_2 versus p_T. The KE_T scaled by n_q and v_2 scaled by n_q, eccentricity, $N_{part}^{1/3}$, i.e., $v_2/(n_q * \epsilon_{part} * N_{part}^{1/3})$ found to exhibit universal scaling displayed in the Fig. 13.19 (b) for $0.1 < KE_T/n_q < 1.0$ GeV. The better χ^2/NDF = 2.11 for the third-order polynomial was found for the $N_{part}^{1/3}$ scaling than that of the $N_{coll}^{1/3}$ scaling (χ^2/NDF = 5.39). The v_2 values exhibited some spread as the energy dependence of v_2 was not taken into account.

Figure 13.20 (left) presents p_T/n_q versus v_2/n_q for identified particles measured by the ALICE for the Pb-Pb collisions at $\sqrt{s_{NN}} = 2.76$ TeV for 40–50% collision centrality [65]. Here, the scaling seems to break again but when plotted v_2/n_q versus $(m_t - m)/n_q$ scaling seems to work (Fig. 13.20 (right)). In Fig. 13.21 double ratio $(v_2/n_q)/(v_2/n_q)_{Fitp}$ is plotted versus $(m_t - m)/n_q$. It is observed that scaling is

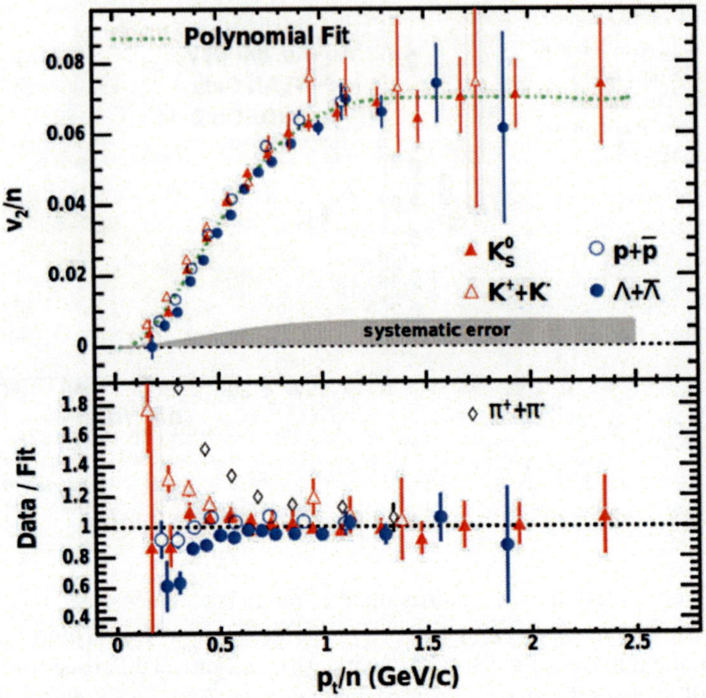

Fig. 13.17 Top: Scaled v_2/n versus scaled p_t/n for identified particles for minimum bias collisions. n is number of constituent quarks in the particle. A polynomial curve is fit to the data. Bottom: Ratio of v_2/n to the fitted curve versus p_t/n

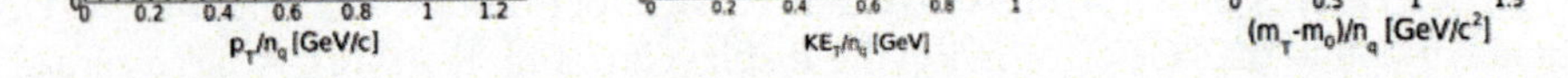

Fig. 13.18 Left: v_2/n_q versus p_T/n_q for different identified particles for the PHENIX data for 10–40% collision centrality for Au+Au collisions at $\sqrt{s_{NN}} = 62.4\,\text{GeV}$. Middle: v_2/n_q versus KE_T/n_q, Right: v_2/n_q versus $(m_T - m_0)/n_q$ for 0–40% collision centrality for Cu+Au collisions at $\sqrt{s_{NN}} = 200\,\text{GeV}$ the STAR data

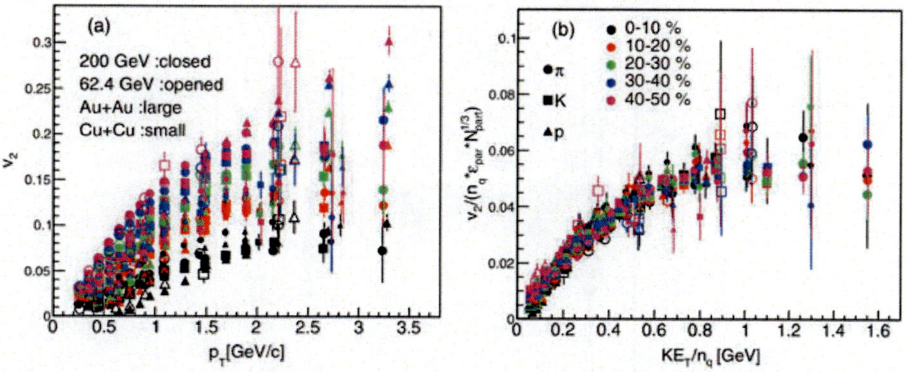

Fig. 13.19 **a** v_2 versus p_T the PHENIX data [64] on Au+Au and Cu+Cu for different centralities and for different energies indicated in the figure. **b** Scaled v_2 as discussed in the text as a function of KE_T/n_q

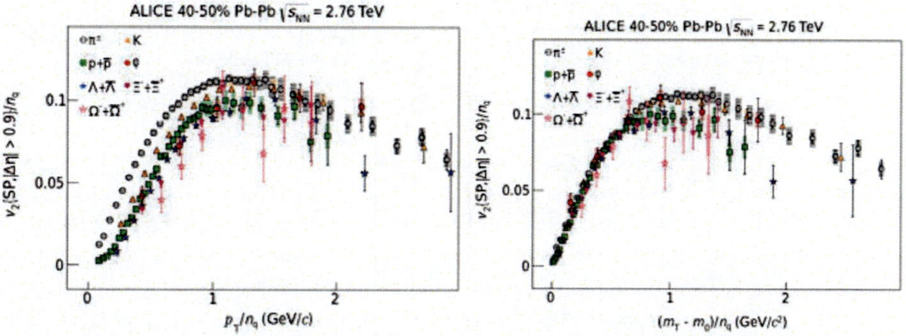

Fig. 13.20 Left: v_2/n_q versus p_T/n_q for identified particles measured by the ALICE for Pb-Pb collisions at $\sqrt{s_{\rm NN}} = 2.76$ TeV for 40–50% collision centrality [65]. Right: same as in left plot but v_2/n_q versus $(m_T - m_0)/n_q$

violated for $(m_t - m)/n_q \sim -0.6-0.8$ GeV/c^2 but for higher values there are deviations at the level of 20% with respect to the reference ratio. This is found to be true for all centralities.

13.2.5 Learning from Simple Scaling Behaviour

Figure 13.22 (left) presents the dependence of v_2 on p_T in Pb-Pb collisions at $\sqrt{s_{\rm NN}} = 5.02$ TeV for various centralities obtained by the ATLAS [66]. It is observed that v_2 increases with decreasing centrality and the p_T value at which v_2 reaches its maximum value changes from one centrality to another. The ATLAS collabora-

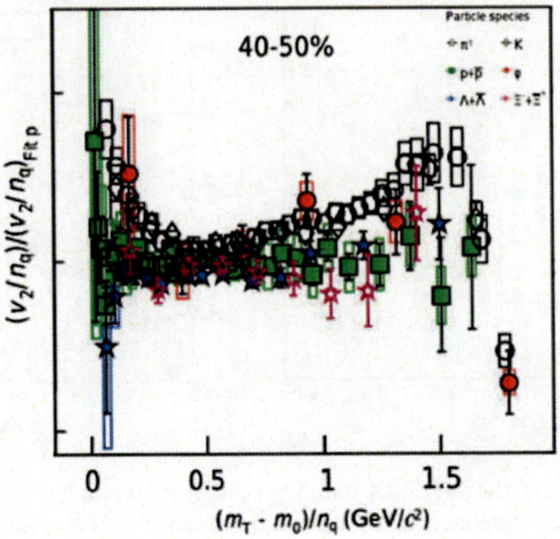

Fig. 13.21 Double ratio $(v_2/n_q)/(v_2/n_q)_{Fitp}$ versus $(m_T - m_0)/n_q$

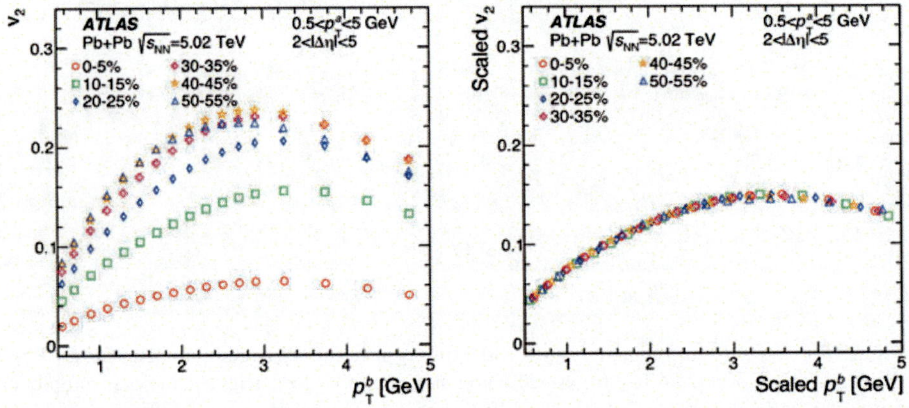

Fig. 13.22 Left) The dependence of v_2 on p_T in Pb-Pb collisions at $\sqrt{s_{\rm NN}} = 5.02\,$TeV for various centralities obtained by the ATLAS [66]. Right) Scaled v_2 versus scaled p_T. Details can be found in the text

tion [67] found that their findings regarding the similar p_T dependence in p-Pb and Pb-Pb collisions are also valid for different centralities in the Pb-Pb collisions at $\sqrt{s_{\rm NN}} = 5.02\,$TeV if we scale p_T to account for difference in $\langle p_T \rangle$ between different collision centralities and scale v_2 to account for the difference in the collision geometry between the centralities. Here, the 0–60% centrality bin is chosen as the reference, and the $v_2(p_T)$ for the individual 5% wide centralities are scaled to match best the $v_2(p_T)$ in the 0–60% centrality interval over the 0.5–5 GeV p_T range. Figure 13.22 (right) displays the scaled v_2 versus scaled p_T for different centralities. It is seen that all data points fall on the same curve supporting the observation made in p-Pb and Pb-Pb collisions. Details of p_T and v_2 scaled factors can be found in [66].

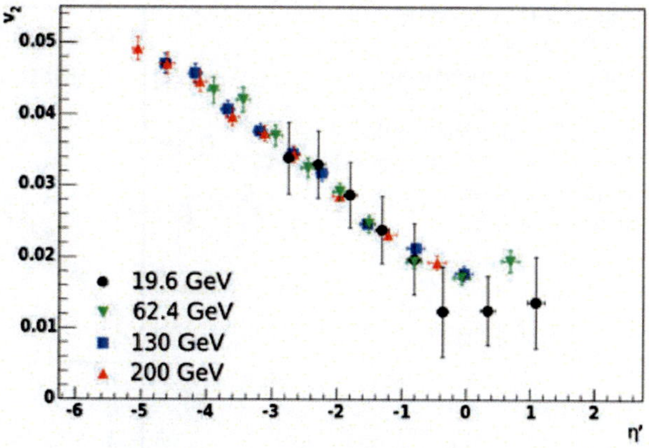

Fig. 13.23 v_2(EP) as a function of η' for Au +Au collisions at different energies as indicated in the figure

Back et al., [68] observed that the integrated elliptic flow varies linearly with the pseudo-rapidity η' $(\eta - y_{beam})$, measured with respect to the beam rapidity y_{beam}, in Au+Au collisions at $\sqrt{s_{\rm NN}} = 200$, 130, 62.4 and 19.6 GeV. Figure 13.23 displays the plot of v_2(EP) versus η'. It is seen that elliptic flow exhibits longitudinal scaling behaviour over the whole range of η'. The extrapolation of this to LHC energy, predicted an increase in v_2 of 50%. However, the ALICE collaboration reported an increase in $v_2 \sim 30\%$ from $\sqrt{s_{\rm NN}} = 200$ GeV to $\sqrt{s_{\rm NN}} = 2.76$ TeV [28].

13.2.6 Energy Dependence

Figure 13.24 (top) displays the STAR $v_2\{4\}$ obtained using the 4-particle cumulant method versus p_T for 20–30% centrality for Au+Au at $\sqrt{s_{\rm NN}} = 39$, 62.4. and 200 GeV [69]. The ALICE Pb-Pb data points for similar centrality at $\sqrt{s_{\rm NN}} =$ 2.76 TeV are also shown. All data points fall on the same curve. Ratio of v_2 to the polynomial fitted value at 200 GeV versus p_T is plotted in Fig. 13.24 (bottom). The good agreement within ~10% is seen for wide range of energies for $p_T > 500$ MeV/c. It is seen that $v_2\{4\}$ gets saturated for $p_T > 1$GeV but for $p_T < 1$ GeV/c increase in v_2 is seen for higher energy Pb-Pb collisions at 2.76 TeV and decrease in v_2 for lower energy taking 200 GeV as a reference which is consistent with the overall increase in the v_2 with increase in beam energy.

The comparison of p_T integrated v_2 for 20–30% centrality at different energies is shown in Fig. 13.25 [28]. There is continuous increase in the magnitude of elliptic flow for this centrality from RHIC to LHC. About 30% increase in v_2 is observed with increasing energy from $\sqrt{s_{\rm NN}} = 200$ GeV to $\sqrt{s_{\rm NN}} = 2.76$ TeV. The larger v_2 is due to the increase in the mean p_T. Hydrodynamical [70] and hybrid [71–73] models infact predicted an increase of ~10–30% in v_2 at the LHC energy.

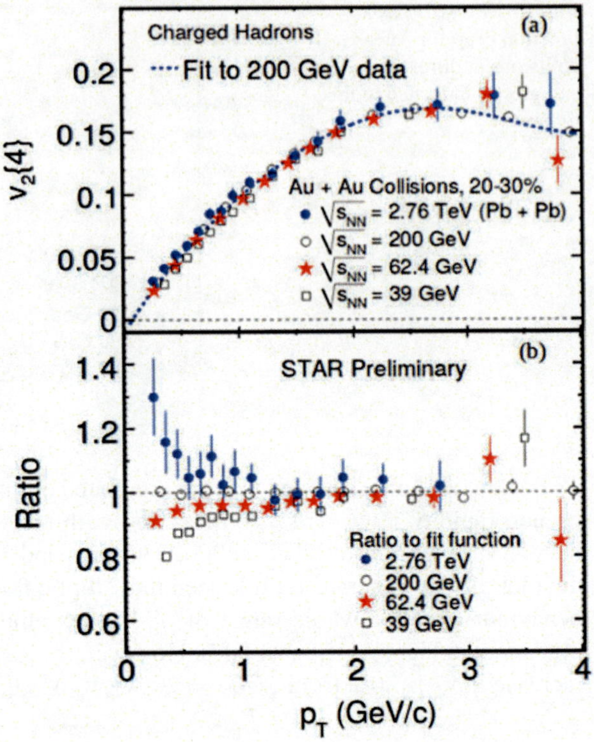

Fig. 13.24 Top: $v_2\{4\}$ versus p_T for 20–30% collision centrality for Au+Au and Pb-Pb collisions at different energies as indicated in the figure. Bottom: Plot of ratio of $v_2\{4\}$ to the fit value as a function p_T

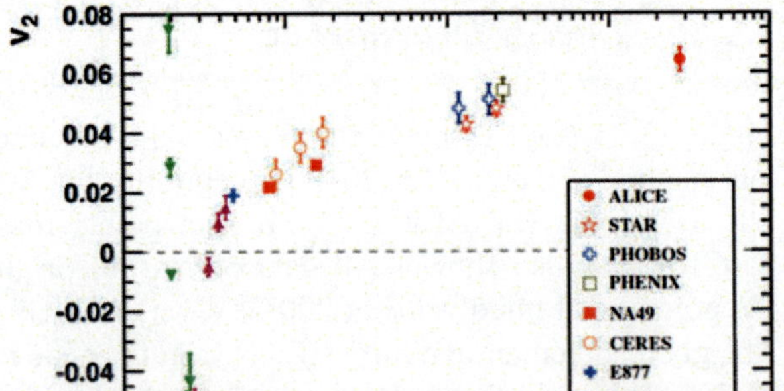

Fig. 13.25 Energy dependence of v_2 for 20–30% collision centrality for various experiments as indicated in the figure

13.3 Summary

The elliptic flow (v_2) measured from two-particle correlations is larger than that from multi-particle correlations. This can be because of either non-flow effects or fluctuations. The dependence of v_2 on p_T is well described by hydrodynamical models. The mass ordering of v_2 at low p_T follows the predictions of hydrodynamical models. RHIC and LHC data agree with VISHNU predictions and yielded $\eta/s = 0.16$ and 0.20 at $\sqrt{s_{\rm NN}} = 200$ GeV and 2.76 TeV energies, respectively. The v_2 scales with the number of constituent quarks, n_q, within $\sim$10% at the RHIC energies, above $p_T/n_q \sim 0.6$ GeV/c indicating hadron production via coalescence of constituent quarks. However, pions deviates from this which may be due to large contribution from resonance decay. There are deviations in the NCQ scaling at the level of 20% at the LHC energies. The v_2 fluctuations are well described by fKNL-CGC and IP-Glasma models. The v_2 versus $\eta' = \eta - y_{beam}$ follows linear scaling behaviour. The v_2 versus p_T for different collision centralities follows universal curve if v_2 and p_T are properly scaled for collision geometry and mean p_T changes, respectively. It is observed that v_2 increases with increase in energy due to increase in mean p_T.

Acknowledgements Financial assistance from the Department of Science & Technology and University Grants Commission is gratefully acknowledged.

References

1. E.V. Shuryak, Quantum chromodynamics and the theory of superdense matter. Phys. Rep. **61**, 71–158 (1980)
2. S. Borsanyi et al., The QCD equation of state with dynamical quarks. JHEP **11**, 077 (2010)
3. L. Kumar, D. Keane, Experimental studies of the quantum chromodynamics phase diagram at the STAR experiment. Pramana **84**, 773–786 (2015)
4. U. Heinz, M. Jacob, Evidence for a new state of matter: an assessment of the results from the CERN lead beam program nucl-th/0002042, pp. 1–7
5. J.Y. Ollitrault, Anisotropy as a signature of transverse collective flow. Phys. Rev. D **46**, 229–245 (1992)
6. U. Heinz, R. Snellings, Collective flow and viscosity in relativistic heavy-ion collisions. Ann. Rev. Nucl. Part. Sci. **63**, 123 (2013)
7. C. Gale, S. Jeon, B. Schenke, Hydrodynamic modelling of heavy-ion collisions. Int. J. Mod. Phys. A **28**, 1340011 (2013)
8. P. Huovinen, Hydrodynamics at RHIC and LHC: what have we learned? Int. J. Mod. Phys. E **22**, 1330029 (2013)
9. R.S. Bhalerao, (AEPSHEP 2012), Relativistic heavy-ion collisions, in *Proceedings on 1st Asia-Europe-Pacific School of High-Energy Physics TIFR-TH-14-11*, arXiv:1404.3294 [nucl-th]
10. R. Snellings, Elliptic flow: a brief review. New. J. Phys. **13**, 055008 (2011)
11. J.Y. Ollitrault, Relativistic hydrodynamics for heavy-ion collisions. Eur. J. Phys. **29**, 275–302 (2008)
12. J. Admas et al., STAR Collaboration, Experimental and theoretical challenges in the search for the quark-gluon plasma: the STAR Collaboration's critical assessment of the evidence from RHIC collisions. Nucl. Phys. A **757**, 102–183 (2005)

13. J. Admas et al., STAR Collaboration, Azimuthal anisotropy in Au+Au collisions at $\sqrt{s_{NN}}$ = 200 GeV. Phys. Rev. C **72**, 0149040 (2005)
14. S.S. Adler et al., PHENIX Collaboration, Elliptic flow of identified hadrons in Au + Au collisions at $\sqrt{s_{NN}}$ = 200 GeV. Phys. Rev. Lett. **91**, 182301 (2003)
15. H. Satz, *Extreme States of Matter in Strong Interaction Physics*. Lecture Notes in Physics, 1st edn., vol. 841 (Springer, Heidelberg, 2012)
16. W. Florkowski, *Phenomenology of Ultra-relativistic Heavy-Ion Collisions* (World Scientific, Singapore, 2010)
17. J. Bartke, *Relativistic Heavy-Ion Physics* (World Scientific, Singapore, 2008)
18. J. Letessier, J. Rafelski, *Hadrons and Quark-Gluon Plasma* (Cambridge University Press, Cambridge, 2002)
19. C.Y. Wong, *Introduction to High-Energy Heavy-Ion Collisions* (World Scientific, Singapore, 1994)
20. B. Muller, *The Physics of the Quark-Gluon Plasma*. Lecture Notes in Physics, 1st edn., vol. 225 (Springer, Heidelberg, 1985)
21. S.A. Voloshin, Y. Zhang, Flow study in relativistic nuclear collisions by Fourier expansion of Azimuthal particle distributions. Phys. Rev. C **70**, 665–672 (1996)
22. A.M. Poskanzer, S.A. Voloshin, Methods for analyzing anisotropic flow in relativistic nuclear collisions. Phys. Rev. C **58**, 1671–78 (1998)
23. J. Barrette et al., E877 Collaboration, Energy and charged particle flow in 10.8 A GeV/c Au+Au collisions. Phys. Rev. C **55**, 1420–1430 (1997)
24. A. Bilandzic, R. Snellings, Flow analysis with cumulants: direct calculations. Phys. Rev. C **83**, 044913 (2011)
25. G. Aad et al., Measurement of the distribution of event-by-event flow harmonics in lead-lead collisions at $\sqrt{s_{NN}}$ = 2.76 TeV with the ATLAS detector at the LHC. JHEP **11**, 183–205 (2013)
26. I. Selyuzhenkov, S. Voloshin, Effects of nonuniform acceptance in anisotropic flow measurements. Phys. Rev. C **77**, 034904 (2008)
27. N. Borghini, P.M. Dinh, J.Y. Ollitrault, Flow analysis from multiparticle Azimuthal correlations. Phys. Rev. C **64**, 054901 (2001)
28. K. Aamodt et al., ALICE Collaboration, Elliptic flow of charged particles in Pb-Pb collisions at $\sqrt{s_{NN}}$=2.76 TeV. Phys. Rev. Lett. **105**, 252302 (2010)
29. K. Burka et al., ATLAS Collaboration, Measurement of Azimuthal flow of soft and high-p_T charged particles in 5.02 TeV Pb+Pb collisions with the ATLAS detector. Nucl. Part. Phys. Proc. **289**, 441 (2017)
30. B. Abelev et al., ALICE Collaboration, Charge separation relative to the reaction plane in Pb-Pb. Phys. Rev. Lett. **110,** 012301 (2013)
31. P. Huovinen et al., Radial and elliptic flow at RHIC: further predictions. Phys. Lett. B **503**, 58–64 (2001)
32. H. Song, S.A. Bass, U. Heinz, Elliptic flow in $\sqrt{s_{NN}}$=200 GeV Au+Au collisions and $\sqrt{s}$=2.76 TeV Pb+Pb collisions: insights from viscous hydrodynamics+hadron cascade hybrid model. Phys. Rev. C **83**, 054912 (2011)
33. B.I. Abelev et al., STAR Collaboration, Centrality dependence of charged hadron and strange hadron elliptic flow from $\sqrt{s_{NN}}$=200 GeVAu+Au collisions. Phys. Rev. C **77**, 054901 (2008)
34. B.B. Abelev et al., ALICE Collaboration, (Anti-)deuteron production and anisotropic flow measured with ALICE at the LHC. JHEP **06**, 190 (2015)
35. P. Bozek, Component of elliptic flow in Pb-Pb collisions at 2.76 Tev. Phys. Lett. B **699**, 283 (2011). arXiv:1101.1791 [nucl-th]
36. T. Hirano, P. Huovinen, Y. Nara, Elliptic flow in Pb+Pb collisions at $\sqrt{s_{NN}}$ = 2.76 TeV: hybrid model assessment of the first data. Phys. Rev. C **84**, 011901 (2011). arXiv:1012.3955 [nucl-th]
37. L. Adamczyk et al., STAR Collaboration, Azimuthal anisotropy in Cu+Au collisions at $\sqrt{s_{NN}}$=200 GeV. Phys. Rev. C **98**, 014915 (2018)
38. ATLAS Collaboration, Measurement of the distributions of event-by-event flow harmonics in Pb-Pb collisions at $\sqrt{s_{NN}}$=2.76 TeV. ATLAS-CONF-2012-114, https://cdsweb.cern.ch/record/1472935

39. C. Gale et al., Event-by-event anisotropic flow in heavy-ion collisions from combined Yang-Mills and viscous fluid dynamics. Phys. Rev. Lett. **110**, 012302 (2013)
40. B. Schenke, P. Tribedy, R. Venugopalan, Fluctuating glasma initial conditions and flow in heavy ion collisions. Phys. Rev. Lett. **108**, 252301 (2012)
41. B. Schenke, P. Tribedy, R. Venugopalan, Event-by-event gluon multiplicity, energy density, and eccentricities in ultrarelativistic heavy-ion collisions. Phys. Rev. C **86**, 034908 (2012)
42. J. Bartels, K.J. Golec-Biernat, H. Kowalski, Modification of the saturation model: DGLAP evolution. Phys. Rev. D **66**, 014001 (2002)
43. H. Kowalski, D. Teaney, Impact parameter dipole saturation model. Phys. Rev. D **68**, 114005 (2003)
44. A. Kovner, L.D. McLerran, H. Weigert, Gluon production from non-Abelian Weizsacker-Williams fields in nucleus-nucleus collisions. Phys. Rev. D **52**, 6231–6237 (1995)
45. A. Krasnitz, R. Venugopalan, Initial energy density of gluons produced in very-high-energy nuclear collisions. Phys. Rev. Lett. **84**, 4309–4312 (2000)
46. T. Lappi, Production of gluons in the classical field model for heavy ion collisions. Phys. Rev. C **67**, 054903 (2003)
47. W. Israel, J.M. Stewart, Transient relativistic thermodynamics and kinetic theory. Ann. Phys. **118**, 341–372 (1979)
48. A. Kurganov, E. Tadmor, New high-resolution central schemes for nonlinear conservation laws and convection-diffusion equations. J. Comput. Phys. **160**, 241–282 (2000)
49. A.M. Sirunyan et al., CMS Collaboration, Non-Gaussian elliptic-flow fluctuation in Pb-Pb collisions at $\sqrt{s_{NN}}$=5.02 TeV. Phys. Lett. B **789**, 643–665 (2019)
50. G. Giacalone, L. Yan, J.N. Hostler, J.Y. Ollitrault, Skewness of elliptic flow fluctuations. Phys. Rev. C **95**, 014913 (2017)
51. G. Agakishiev et al., STAR Collaboration, Energy and system-size dependence of two- and four-particle v_2 measurements in heavy-ion collisions at $\sqrt{s_{NN}}$ = 62.4 and 200 GeV and their implications on flow fluctuations and nonflow. Phys. Rev. C **86**, 014904 (2012)
52. N. Kumar, Ph.D. thesis, Panjab University, Chandigarh, India (2011)
53. B. Alver et al., Importance of correlations and fluctuation on the initial source eccentricity in high-energy nucleus collisions. Phys. Rev. C **77**, 014906 (2008)
54. M.L. Miller, K. Reygers, S.J. Sanders, P. Steinberg, Glauber modeling in high energy nuclear collisions. Ann. Rev. Nucl. Part. Sci. **57**, 205–248 (2007)
55. H.J. Drescher, A. Dumitru, A. Hayashigaki, Y. Nara, The Eccentricity in heavy-ion collisions from color glass condensate initial conditions. Phys. Rev. C **74** 044905 (2006); H.J. Drescher, Y. Nara, Eccentricity in heavy-ion collisions from color glass condensate initial conditions. Phys. Rev. C **76**, 041903 (2007)
56. R.S. Bhalerao, J.Y. Ollitrault, Ecentricity fluctuations and elliptic flow at RHIC. Phys. Lett. B **641**, 260–269 (2006)
57. S.A. Voloshin, A.M. Poskanzer, A. Tang, G. Wang, Elliptic flow in the Gaussian model of eccentricity fluctuations. Phys. Lett. B **659**, 537–541 (2008)
58. B. Alver et al., Non-flow correlations and elliptic flow fluctuations in Au + Au collisions at $\sqrt{s_{NN}}$ = 200 GeV PHOBOS Collaboration. Phys. Rev. C **81**, 034915 (2010)
59. R. Snellings et al., ALICE Collaboration, Anisotropic flow at the LHC measured with the ALICE detector. J. Phys. G **38**, 124013 (2011)
60. J. Adams et al., STAR Collaboration, Particle-type dependence of Azimuthal anisotropy and nuclear modification of particle production in Au + Au collisions at $\sqrt{s_{NN}}$ = 200 GeV. Phys. Rev. Lett. **92**, 052302 (2004)
61. D. Molnar, S.A. Voloshin, Elliptic flow at large transverse momenta from quark coalescence. Phys. Rev. Lett. **91**, 092301 (2003); R.C. Hwa, C.B. Yang, Scaling distributions of quarks, mesons, and proton for all p_T, energy, and centrality. Phys. Rev. C **67**, 064902 (2003); R.C. Hwa, C.B. Yang, Measurement of event-plane correlations in $\sqrt{s_{NN}}$= 2.76 TeV lead-lead collisions with the ATLAS detector. Phys. Rev. C **70**, 024905 (2004); R.C. Hwa, C.B. Yang, Recombination of shower partons in fragmentation processes. Phys. Rev. C **70**, 024904 (2004)

62. S.S. Adler et al., PHENIX Collaboration, Scaling properties of proton and anti-proton production in $\sqrt{s_{NN}}$ 200 GeV Au+Au collisions. Phys. Rev. Lett. **91**, 172301 (2003)
63. L. Adamczyk et al., STAR Collaboration, Azimuthal anisotropy in Cu+Au collisions at $\sqrt{s_{NN}}$= 200 GeV. Phys. Rev. C **98**, 014915 (2018)
64. A. Adare et al., PHENIX Collaboration, Systematic study of Azimuthal anisotropy in Cu + Cu and Au + Au collisions at $\sqrt{s_{NN}}$=62.4 and 200 GeV. Phys. Rev. C **92**, 34913 (2015)
65. B.B. Abelev et al., ALICE Collaboration, Elliptic flow of identified hadrons in Pb-Pb collisions at $\sqrt{s_{NN}}$=2.76 TeV. JHEP **06**, 190 (2015)
66. M. Aaboud et al., ATLAS Collaboration, Measurement of the Azimuthal anisotropy of charged particles produced in $\sqrt{s_{NN}}$ = 5.02 TeV Pb+Pb collisions with the ATLAS detector. Eur. Phys. J. C **78**, 997 (2018)
67. G. Aad et al., ATLAS Collaboration, Measurement of long-range pseudorapidity correlations and Azimuthal harmonics in $\sqrt{s_{NN}}$ = 5.02 TeV proton-lead collisions with the ATLAS detector. Phys. Rev. C **90**, 044906 (2014)
68. B.B. Back et al., PHOBOS Collaboration, Energy dependence of elliptic flow over a large pseudorapidity range in Au + Au collisions at RHIC. Phys. Rev. Lett. **94**, 122303 (2005)
69. L. Kumar, for the STAR Collaboration, Results from the STAR beam energy scan program. Nucl. Phys. A **862**, 125–131 (2011)
70. G. Kestin, U.W. Heinz, Hydrodynamic radial and elliptic flow in heavy-ion collisions from AGS to LHC energies. Eur. Phys. J. C **61**, 545–552 (2009)
71. H. Masui, J.Y. Ollibrault, R. Snellings, A. Tang, The centrality dependence of v(2)/epsilon: the ideal hydro limit and eta/s Nucl. Phys. A **830**, 463C (2009)
72. M. Luzum, P. Romatschke, Viscous hydrodynamic predictions for nuclear collisions at the LHC. Phys. Rev. Lett. **103**, 262302 (2009)
73. T.H. Hirano, U.W. Heinz, D. Kharzeev, R. Lacey, Y. Nara, Hadronic dissipative effects on elliptic flow in ultrarelativistic heavy-ion collisions. Phys. Lett. B **636**, 299–304 (2006)

Chapter 14
Particle Production and Collective Phenomena in Heavy-Ion Collisions at STAR and ALICE

Lokesh Kumar

Abstract The recent results from STAR and ALICE experiments are presented on particle production and collective phenomena. The STAR results suggest that particle production at lower energies exhibit different features compared to that at higher energies. The lower energy results hint that there may not be Quark-Gluon Plasma formation at these energies. Small system results from ALICE experiment exhibit features similar to those associated generally with heavy-ion collisions. The results presented here constitute the heavy-ion collision energy range from lowest energy of $\sqrt{s_{NN}} = 7.7$ GeV to the highest energy of 5.02 TeV.

14.1 Introduction

The high energy heavy-ion collisions recreate the conditions suitable for the formation of Quark-Gluon Plasma (QGP), a state believed to have existed just after the big bang [1–4]. The Solenoidal Tracker at RHIC (STAR) experiment at Brookhaven National Laboratory (BNL), USA and A Large Ion Collider Experiment (ALICE) at CERN, Switzerland are the main experiments using the heavy-ion collisions to study the QGP formation [1, 5].

The schematic diagram of Quantum Chromo Dynamics is shown in Fig. 14.1 [6]. The x-axis represents the baryon chemical potential μ_B and y-axis represents the temperature T. The phase diagram has many structures, and mainly two phases—QGP and Hadron gas phase. The phase transition from hadron gas to QGP at $\mu_B \sim 0$ is a crossover while at large μ_B, a first-order phase transition is expected [7, 8]. In between, the point where the first-order phase transition line ends is called the QCD critical point [9, 10]. The x-axis and y-axis of the QCD phase diagram can be accessed through experiments by varying the collision energy.

The main goals of high energy heavy-ion collisions are as follows. Understand the particle production and freeze-out dynamics, study the QGP properties, and explore the QCD phase diagram to search for the signals of phase boundary, critical point,

L. Kumar (✉)
Department of Physics, Panjab University, Chandigarh 160014, India
e-mail: lokesh@pu.ac.in

R. K. Puri et al. (eds.), *Advances in Nuclear Physics*, Springer Proceedings in Physics 257, https://doi.org/10.1007/978-981-15-9062-7_14

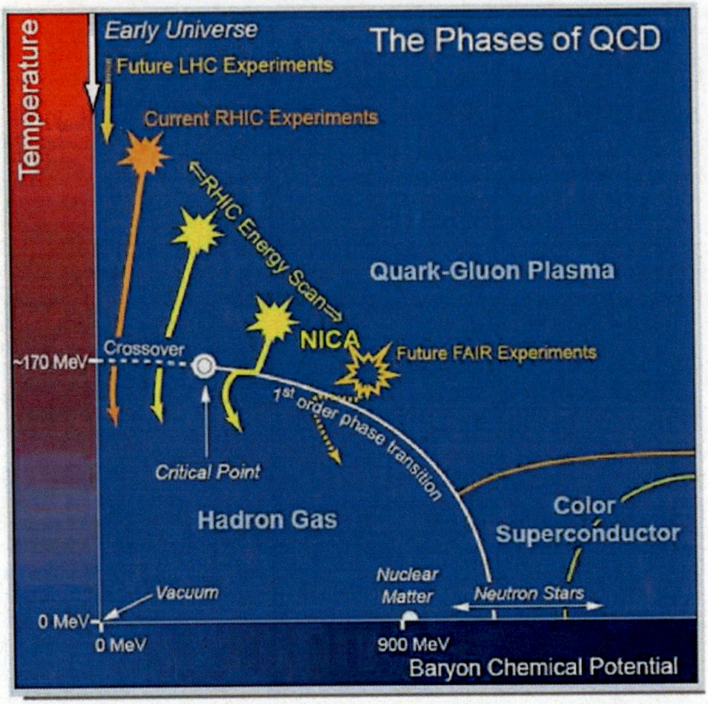

Fig. 14.1 The conjectured phase diagram of Quantum Chromo Dynamics showing different phase structures [6]

and first-order phase transition line. Exploring the QCD phase diagram and search for QCD critical point has been one of the main focus of STAR experiment. For this purpose, a dedicated program called the Beam Energy Scan (BES) program has been operational since 2010 [11, 12].

Recent results from high energy collisions of small systems (pp and $p-$Pb) from ALICE experiment suggest interesting features similar to those generally associated with heavy-ion collisions [13–15]. Understanding of these results has been one of the main focus in ALICE experiment recently.

14.2 Particle Production

Figure 14.2 shows the invariant yields of pions ($\pi^{\pm}$), kaons ($K^{\pm}$), protons (p), and anti-protons ($\bar{p}$) at midrapidity $|y| < 0.1$ as a function of transverse momentum p_T in Au+Au collisions at $\sqrt{s_{NN}} = 7.7$ GeV, the lowest collision energy achieved at RHIC so far [16]. Results are presented for various collision centralities from 0–5% (the most central) to the 70–80% (the most peripheral) collisions. The yield of particle increases from peripheral to central collisions. The slope of the distributions

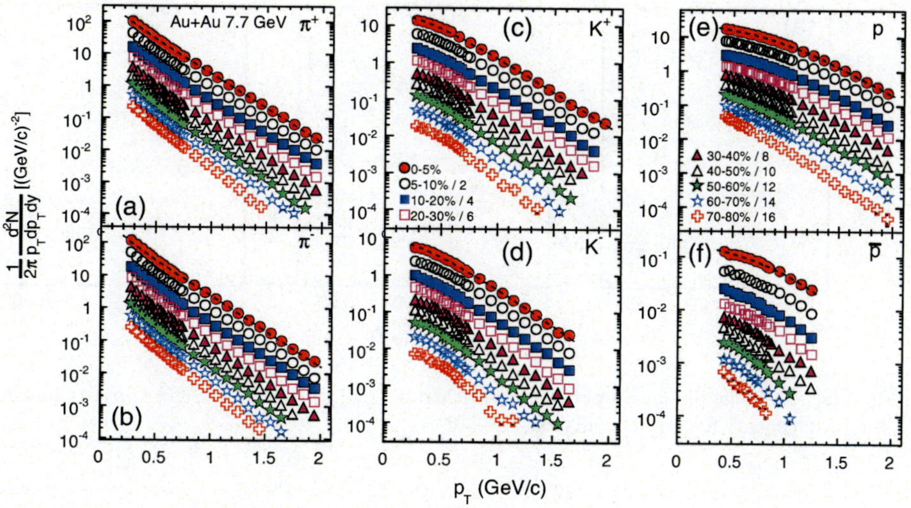

Fig. 14.2 Invariant yields of pions ($\pi^{\pm}$), kaons ($K^{\pm}$), protons (p) and anti-protons ($\bar{p}$) as a function of transverse momentum p_T in Au+Au collisions at $\sqrt{s_{NN}} = 7.7$ GeV for various collision centralities [16]

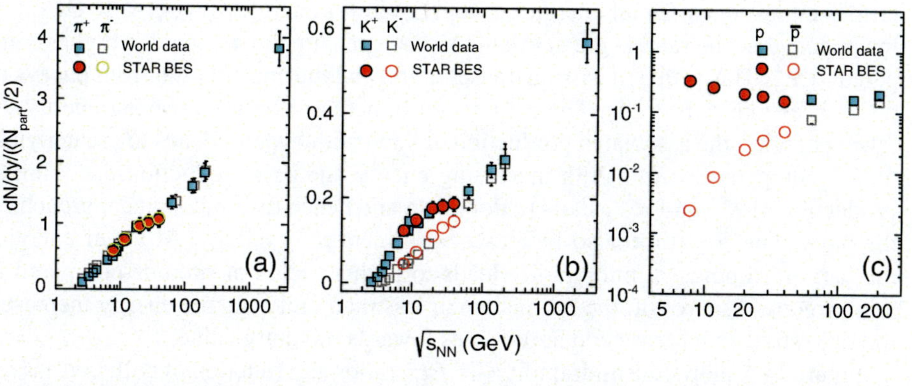

Fig. 14.3 The midrapidity dN/dy of $\pi^{\pm}$, $K^{\pm}$, p, and $\bar{p}$ per participant nucleon pair as a function of collision energy for central collisions [16–20]

decreases with increasing mass of the particle suggesting the effect of radial flow. These distributions are integrated over the full p_T range to obtain the particle yield dN/dy.

Figure 14.3 shows the midrapidity dN/dy of $\pi^{\pm}$, $K^{\pm}$, p, and $\bar{p}$ per participant nucleon pair as a function of collision energy for central collisions [16–20]. The pion yields increase with increasing energy and shows a kink behavior around $\sqrt{s_{NN}} =$ 19.6 GeV. At lower energies, the π^- yield is little higher than π^+. The kaon yields also increase with increasing energy. At lower energies, the K^+ yield is higher than K^-. The proton yields decrease with increasing energy and then starts to increase slightly towards higher energies. The anti-proton yields increase with increasing energy. The

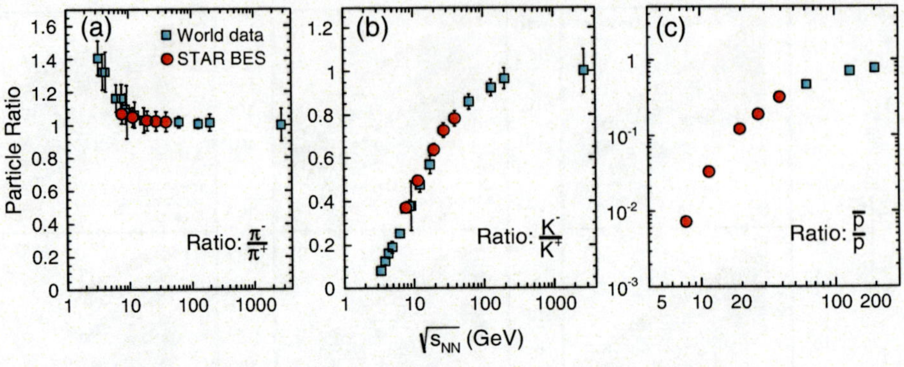

Fig. 14.4 The midrapidity anti-particle to particle ratios (π^-/π^+, K^-/K^+, and $\bar{p}/p$) as a function of collision energy for central collisions [16–20]

behavior of these particle yields as a function of collision energy is quantified in the anti-particle to particle ratios.

Figure 14.4 shows the midrapidity anti-particle to particle ratios (π^-/π^+, K^-/K^+, and $\bar{p}/p$) as a function of collision energy for central collisions [16–20]. The π^-/π^+ ratio is almost unity for all energies except the lower energies $\sqrt{s_{NN}} < 10$ GeV. For lower energies, the ratio is greater than unity suggesting the π^- yield is higher compared to π^+. This is due to isospin conservation and the contribution from decays of resonances like Δ baryons. The π^-/π^+ ratio increases with increasing energy. At lower energies, the associated production of kaon with hyperon leads to greater yield of K^+ compared to K^-. With increasing energy, the pair production, i.e., similar production of K^+ and K^-, starts to dominate and hence the ratio starts approaching the unity. The $\bar{p}/p$ ratio also increases with increasing energy. At lower energies the baryon stopping at midrapidity leads to higher yields of protons compared to anti-protons. As a result, the ratio is much less than unity. As the energy increases, the ratio starts to increase and approaches towards the unity value.

Figure 14.5 shows the midrapidity $K^\pm/\pi^\pm$ ratios as a function of collision energy for central collisions [16–20]. The ratio of K^-/π^- shows an increasing behavior from lower to higher energy. The K^+/π^+ ratio increases with energy at first, then shows a maximum value around $\sqrt{s_{NN}} = 7.7$ GeV, and then decreases with energy and become flat at higher energies to match with K^-/π^- ratio. The observation of maximum value of K^+/π^+ ratio as a function of collision energy has been suggested as the signature of phase transition from hadron gas to QGP [18]. However, various models that do not include such a phase transition could explain this type of energy dependence. The maximum of K^+/π^+ ratio is observed at the energy where maximum baryon density at midrapidity is predicted for heavy-ion collisions [21, 22].

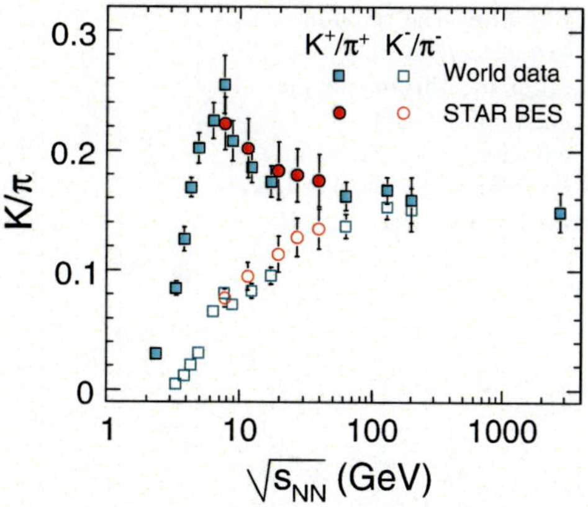

Fig. 14.5 The midrapidity $K^{\pm}/\pi^{\pm}$ ratios as a function of collision energy for central collisions [16–20]

14.2.1 Freeze-Out

The collisions of heavy-ions lead to the formation of new particles. The point in time when the inelastic collisions among the particles cease the particle numbers get fixed. This stage is referred to as the chemical freeze-out. At this point, the elastic interactions among the particles still continue which leads to the change in momenta of the particles. A stage comes when the elastic collisions also cease and the momenta of the produced particles.

The chemical freeze-out conditions in the heavy-ion collisions can be obtained using the statistical thermal model [23, 24]. Using this model, the particle multiplicities are given (in grand canonical ensemble) by

$$N_i = \frac{g_i V}{2\pi^2} \sum_{k=1}^{\infty} (\mp 1)^{k+1} \frac{m_i^2 T}{k} K_2 \left(\frac{k m_i}{T} \right) e^{\beta k \mu_i}, \tag{14.1}$$

where K_2 is the Bessel function of second order, g_i and μ_i are degeneracy and chemical potential of hadron species i, respectively, $\beta = 1/T$, and m_i is the mass of particle. If the number of particles is small, the conservation laws are implemented exactly and hence the strangeness conservation is considered exactly [25, 26]. So, the particle multiplicities are estimated accordingly.

The produced particle yields are fitted using the model and two main parameters are extracted—the chemical freeze-out temperature T_{ch} and the baryon chemical potential μ_B. In this way, the chemical freeze-out parameters are obtained at various collision energies. Figure 14.6 shows the variation of T_{ch} with μ_B at different collision energies [27]. The figure represents the phenomenological phase diagram and gives the phenomenological boundary between hadron gas and QGP. At higher energies (low μ_B) the temperature seems to be showing a constant or limiting behavior. As the energy decreases (μ_B increases), the temperature decreases and converges towards the value for ground state matter at $\mu_B = 931$ MeV. The band in the figure

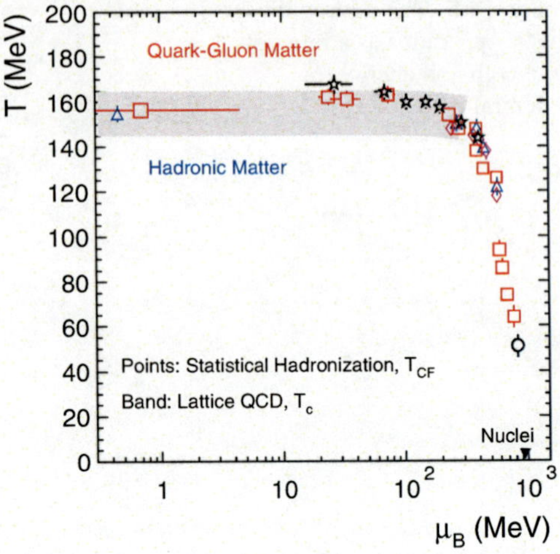

Fig. 14.6 The variation of extracted T_{ch} with μ_B at different collision energies in central heavy-ion collisions [27]. The band represents predictions from the Lattice QCD [28, 29]

represents the predictions from the Lattice QCD [28, 29]. The lattice QCD results are consistent with the observed dependence of T_{ch} on μ_B. This suggests that the heavy-ion collisions can probe the QCD phase boundary between hadron gas and QGP. It is also interesting to note that the thermal model works well for the vast energy range from 7.7 GeV to 2.76 TeV.

The kinetic freeze-out conditions can be obtained using the blast wave (BW) model [16, 30]. This is a hydrodynamics-based model which assumes that the particles are locally thermalized at a kinetic freeze-out temperature T_{kin} and are moving with a common transverse radial flow velocity β. The p_T distribution of the particles is given as

$$\frac{dN}{p_T\, dp_T} \propto \int_0^R r\, dr\, m_T I_0\left(\frac{p_T \sinh\rho(r)}{T_{\mathrm{kin}}}\right) K_1\left(\frac{m_T \cosh\rho(r)}{T_{\mathrm{kin}}}\right), \qquad (14.2)$$

where $m_T = \sqrt{p_T^2 + m^2}$ is the transverse mass of a hadron of mass m, $\rho(r) = \tanh^{-1}\beta$, and I_0 and K_1 are the modified Bessel functions. The radial flow velocity profile of the form $\beta = \beta_S (r/R)^n$ is used. The particle spectra of $\pi^{\pm}$, $K^{\pm}$, $\bar{p}$, and p are fitted simultaneously with the blast wave model and the kinetic freeze-out parameters (T_{kin}, $\langle\beta\rangle$, and n) are extracted.

Figure 14.7 shows the energy dependence of extracted kinetic freeze-out parameters the kinetic freeze-out parameter T_{kin} and average radial flow velocity $\langle\beta\rangle$, along with the chemical freeze-out temperature T_{ch} for central heavy-ion collisions. The kinetic and chemical freeze-out temperatures are similar around $\sqrt{s_{NN}} = 4$–$5\,\mathrm{GeV}$. With increasing collision energy, the chemical freeze-out temperature increases and becomes constant after $\sqrt{s_{NN}} = 11.5\,\mathrm{GeV}$. The T_{kin} is almost constant around the 7.7–39 GeV and then decreases up to the LHC energies. The difference between T_{ch} and T_{kin} increases with increasing energy. This could be related to the increasing

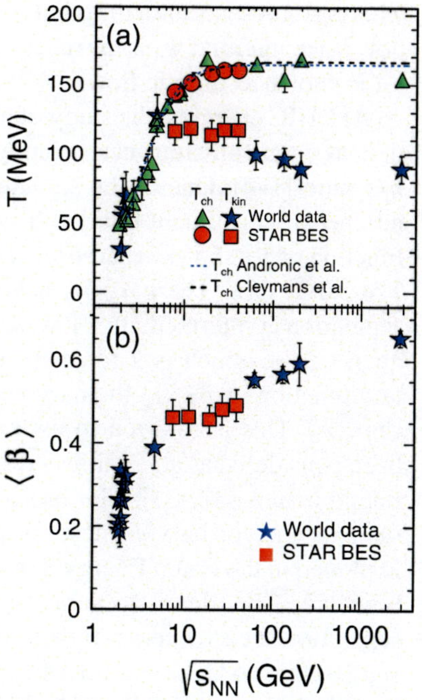

Fig. 14.7 The energy dependence of extracted kinetic freeze-out parameters the kinetic freeze-out parameter T_{kin} and average radial flow velocity $\langle \beta \rangle$, along with the chemical freeze-out parameter for central collisions [16, 30]

hadronic interactions between chemical and kinetic freeze-out at higher energies. The average transverse radial flow velocity exhibits a rapid increase at very low energies, and then a steady rise up to LHC energies. There is also a region of constant $\langle \beta \rangle$ around 7.7–19.6 GeV energies. It is interesting to note that the simple blast wave model is quite successful in describing the spectra of identified particles from the low energy of $\sqrt{s_{NN}} = 7.7\,\text{GeV}$ up to the energy of $\sqrt{s_{NN}} = 2.76\,\text{TeV}$ which is more than about 350 times higher.

14.3 Azimuthal Anisotropy

Anisotropic flow which measures the momentum anisotropy of the final-state particles can give information about the properties such as shear viscosity to entropy density (η/s) of the system formed in heavy-ion collisions [31]. In heavy-ion collisions, the reaction plane is defined as the plane formed by the impact parameter and the beam axis. For non-central heavy-ion collisions, an initial spatial anisotropy is created which gets transformed into the anisotropy in momentum space due to the interactions among the produced particles. The anisotropy in momentum space is characterized by the Fourier (flow) coefficients given as [31, 32]

$$v_n = \langle cos[n(\phi - \Psi_n)] \rangle, \tag{14.3}$$

where ϕ represents the azimuthal angle of produced particles, n represents the order of flow harmonic, and Ψ_n is the reaction plane of order n. The second Fourier coefficient, v_2, is known as elliptic flow.

At RHIC energies, the study of anisotropic flow suggested that the system formed in heavy-ion collisions is a strongly coupled QGP (sQGP) with a small value of the η/s ratio [1–4]. One of the signatures of the formation of QGP was the scaling of number of constituent quarks. It was observed that elliptic flow when plotted as a function of p_T shows separation between baryons and mesons at intermediate p_T (> 1.5–2.0 GeV/c). The baryons have larger v_2 than the mesons. The baryon-meson separation at intermediate p_T was associated with the difference in their constituent quarks (n_q), which is 2 for mesons and 3 for baryons. When the v_2/n_q is plotted as a function of p_T/n_q the baryons and mesons follow single behavior, i.e., exhibit same v_2. This is referred to as the number of constituent quarks scaling and has been considered as an established signature of QGP formation at high energy heavy-ion collisions [32, 33]. The basic requirement for this signature to be observed is the existence of baryon-meson separation at intermediate p_T. This information is exploited in the Beam Energy Scan program at RHIC to look for the signals of phase boundary. The idea is to vary the collision energy and observe the baryon-meson separation at each energy. If there is no separation observed at some energy, it would suggest that no QGP was formed at that energy and hence one could locate the phase boundary between QGP and hadron gas.

Figure 14.8 shows the v_2 of various (anti-) baryons and mesons as a function of $m_T - m_0$ for various energies in Au+Au collisions from $\sqrt{s_{NN}} = 7.7$–62.4 GeV for 0–80% central collisions [34]. Here, m_T represents the transverse mass and m_0 is

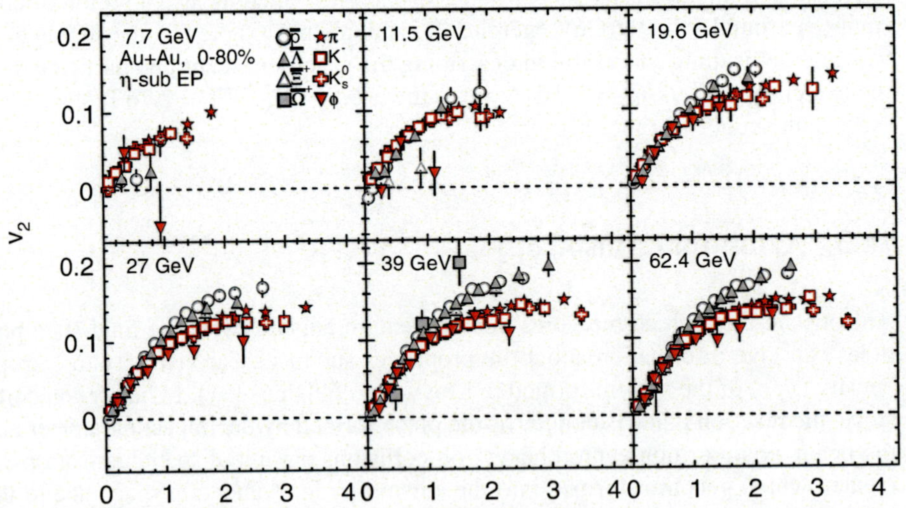

Fig. 14.8 The The energy dependence of extracted kinetic freeze-out parameters the kinetic freeze-out parameter T_{kin} and average radial flow velocity $\langle\beta\rangle$, along with the chemical freeze-out parameter for central collisions [34]

the mass of hadron. At energies from $\sqrt{s_{NN}} = 19.6$–62.4 GeV, there is a separation of baryons and mesons v_2 at intermediate p_T but for energies at $\sqrt{s_{NN}} = 7.7$ and 11.5 GeV, there is no separation. This suggests that for energies at 11.5 GeV and below, the QGP signature is missing and hence there could be no QGP formation. However, the statistics at lower energies is not very good, so the high-statistics data which has been planned to be collected in BES phase-II in STAR would be able to provide conclusive evidence.

14.4 Small Systems

Recently in ALICE, it has been possible to collect high-statistics high energy data for small systems such as pp and p-Pb [13, 35, 36]. This has allowed for more differential measurements of the observables in such a small systems. The results are very interesting in the manner that the same features are observed that have been generally considered as characteristics of heavy-ion collisions where we expect that QGP is formed. Some of the observables are discussed here.

The invariant yields of identified particles as a function of transverse momentum has been obtained for small systems in various multiplicity classes [13, 35, 36]. These spectra show similar behavior as has been observed for heavy-ion collisions. The slope of the distributions becomes flatter or harder in going from low multiplicity to higher multiplicity events [13, 35, 36]. The spectra show hardening for more massive particles [13, 35, 36]. These effects are attributed to radial flow effect in heavy-ion collisions and hence are quite interesting to observe in small systems. The same effect is quantified by fitting the blast wave model simultaneously to the π, K, and p spectra in small systems. Figure 14.9 shows the extracted fit parameters kinetic freeze-out temperature T_{kin} and transverse radial flow velocity $\langle \beta_T \rangle$ for various systems such as pp collisions at $\sqrt{s} = 7$ TeV, p-Pb collisions at $\sqrt{s_{NN}} = 5.02$ TeV,

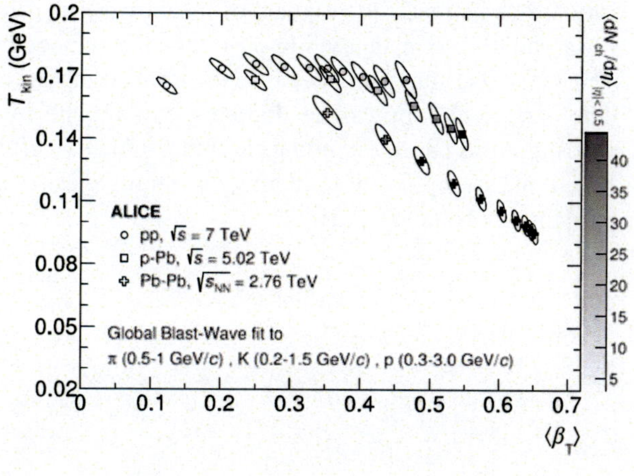

Fig. 14.9 The extracted kinetic freeze-out temperature T_{kin} and transverse radial flow velocity $\langle \beta_T \rangle$ for various systems such as pp collisions at $\sqrt{s} = 7$ TeV, p-Pb collisions at $\sqrt{s_{NN}} = 5.02$ TeV, and Pb-Pb collisions at $\sqrt{s_{NN}} = 2.76$ TeV, in different multiplicity classes [13, 15, 36–38]

and Pb-Pb collisions at $\sqrt{s_{NN}} = 2.76$ TeV, in different multiplicity classes [13, 15, 36–38]. It is quite astonishing that the blast wave model works very well even for the small systems. The extracted $T_{\rm kin}$ and $\langle \beta_T \rangle$ are similar for pp and p-Pb for similar multiplicities. However, when compared with Pb-Pb, the $\langle \beta_T \rangle$ is higher for smaller systems at similar multiplicities.

It becomes interesting to observe the behavior of various observables as a function of multiplicity with different systems plotted together. One such plot is shown in Fig. 14.10 where various particle ratios, such as $(p + \bar{p})/\phi$, $(K^{+}+ K^{-})/(\pi^{+} + \pi^{-})$, $(p + \bar{p})/(\pi^{+} + \pi^{-})$, and Λ/K^{0}_{S}, are plotted as a function of multiplicity for three different p_T regions, i.e., low-p_T (0.50–0.55 GeV/c), mid-p_T (2.50–3.00 GeV/c), and high p_T (7.00–10.00 GeV/c), for different collision systems pp, p-Pb, and Pb-Pb [13, 15, 36–38]. It is observed that all the particle ratios evolve smoothly as a function of charged particle multiplicity across different systems. This suggests that **the particle production depends only on the final-state multiplicity, irrespective of** difference in energy, and collision system.

The strangeness enhancement in heavy-ion collisions relative to pp collisions was originally proposed as a signature of QGP formation. It is also attributed to the canonical suppression in small systems. However, the microscopic understanding of strangeness enhancement is not known completely. Figure 14.11 shows the mul**tiplicity dependence of ratio of identified particle yields to the pion yield for** pp collisions at $\sqrt{s} = 7$ TeV, p-Pb collisions at $\sqrt{s_{NN}} = 5.02$ TeV, and Pb–Pb collisions at $\sqrt{s_{NN}} = 2.76$ TeV [13, 15, 36–38]. The ratios shown are the K/π, K^{*}/π, p/π, ϕ/π Λ/π, Ξ/π and Ω/π. There is a smooth evolution of all particle ratios also including strange particles as a function of charged particle multiplicity across different systems. This suggests that the strange particle production exhibits universal **final-state multiplicity dependence.**

The particle ratios involving strange particles increase with increasing multiplicity (K^{*} is an exception due to the rescattering effect). The strangeness enhancement is **observed for the first time in small systems. The enhancement is observed to increase** with increasing strangeness contents of the particles. As an example, the Ω/π ratio with total strangeness content of 3 units shows more enhancement compared to other ratios. The increase due to mass of the particles is ruled out because the p/π ratio is observed to be constant while ϕ/π shows enhancement. However, the enhancement of ϕ/π is also surprising because the total strangeness of ϕ meson ($s\bar{s}$) is zero units. The canonical suppression does not explain the increase in ϕ/π ratio [15]. The investigations based on models to understand this effect are ongoing. A recent study based on assumption of total effective strangeness of ϕ meson as 2 units (s+$\bar{s}$) could explain this behavior [25].

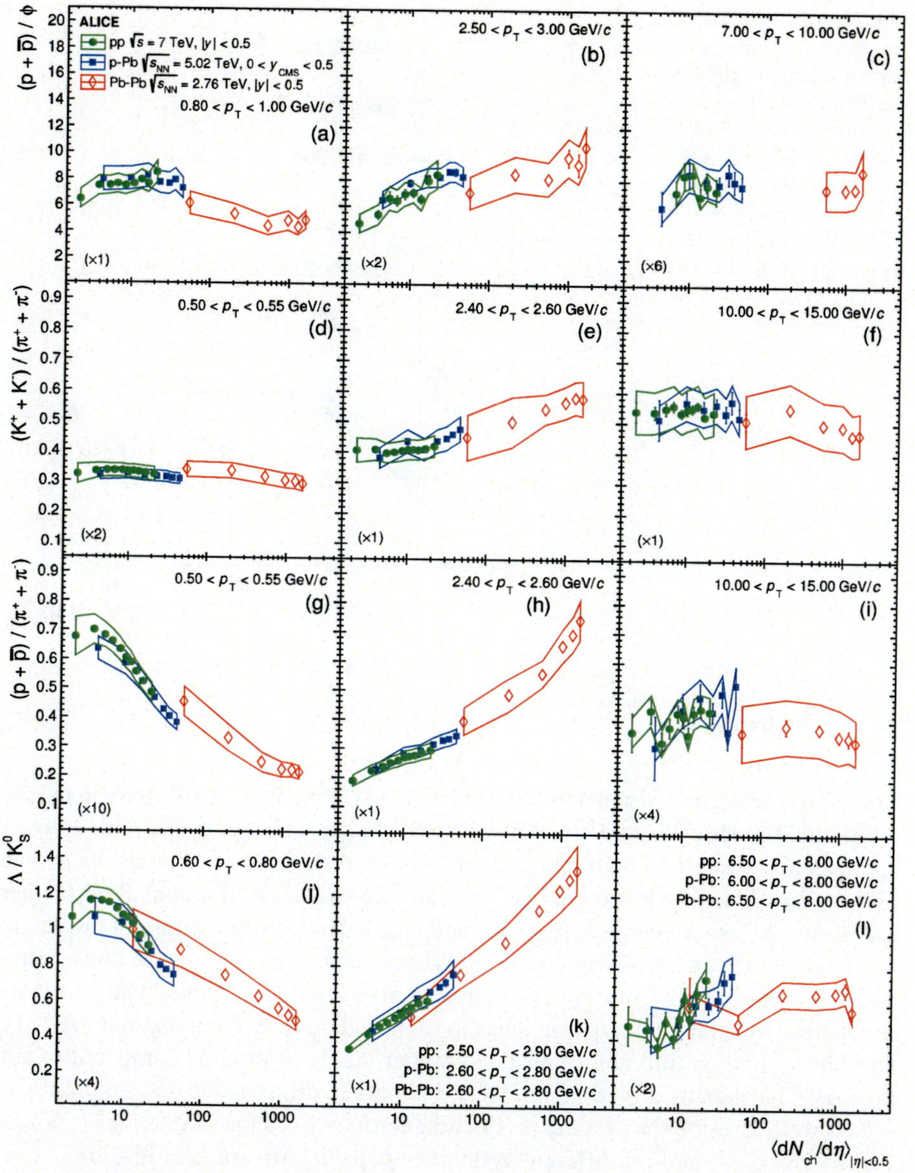

Fig. 14.10 The particles yield ratios as a function of multiplicity $\langle dN_{\rm ch}/d\eta\rangle$ for pp, p–Pb, and Pb–Pb collisions. The top panels with labels a, b, and c show $(\mathrm{p}+\bar{p})/\phi$ ratio; panels with labels d, e, and f show $(\mathrm{K}^{+}+\mathrm{K}^{-})/(\pi^{+}+\pi^{-})$ ratio, panels with labels g, h, and i show $(\mathrm{p}+\bar{p})/(\pi^{+}+\pi^{-})$ ratio; and bottom panels with labels j, k, and l show Λ/K_S^0 ratios for low $\mathrm{p_T}$ interval (left column), mid $\mathrm{p_T}$ interval (middle column), and high $\mathrm{p_T}$ interval (right column) [13, 15, 36–38]

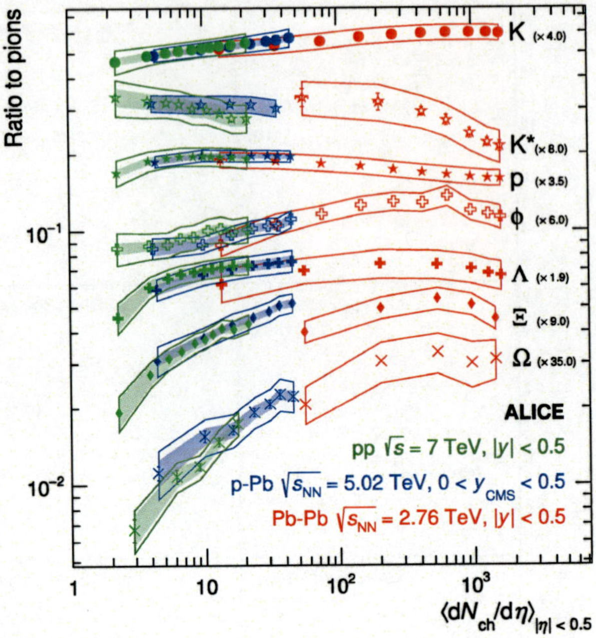

Fig. 14.11 The multiplicity dependence of ratio of identified particle yields to the pion yield for pp collisions at $\sqrt{s} = 7\,\text{TeV}$, p-Pb collisions at $\sqrt{s_{NN}} = 5.02\,\text{TeV}$, and Pb–Pb collisions at $\sqrt{s_{NN}} = 2.76\,\text{TeV}$. The ratios plotted here are K/π, K^*/π, p/π, ϕ/π Λ/π, Ξ/π and Ω/π [13, 15, 36–38]

14.5 Summary

The STAR experiment at RHIC and ALICE experiment at LHC have measured the particle production and collective phenomena in detail over the wide range of energies and systems. The transverse momentum distributions of pions, kaons, and protons in the heavy-ion collisions indicate the presence of radial flow in these collisions. At lower energies, particles and anti-particles show different production mechanism. The π^- yield is higher compared to π^+ due to isospin conservation and the contribution from decays of resonances like Δ baryons. The associated production of kaon with hyperon leads to higher yield of K^+ compared to K^-. The baryon stopping at midrapidity leads to higher yields of protons compared to anti-protons. The statistical thermal model and simple hydrodynamics-based blast wave model well describe the particle production across vast range of energies ($\sqrt{s_{NN}} =$ 7.7 to 2760 GeV) and for different systems (pp, p-Pb, Au+Au, and Pb-Pb).

The v_2 at intermediate p_T shows no separation between baryons and mesons for energies at $\sqrt{s_{NN}} = 7.7$ and 11.5 GeV. This indicates that for energies at 11.5 GeV and below, there could be no QGP formation. The multiplicity dependence measurements for small system collisions show interesting results. The results exhibit features similar to those generally associated with heavy-ion collisions. The phenomenon like collectivity is observed for small systems. The strangeness enhancement is observed for the first time in small systems. The multiplicity dependence results of small and large systems suggest that particle production depends only on the final-state multiplicity, irrespective of difference in system size and energy.

Acknowledgements The discussions with Natasha Sharma and help in preparing the manuscript is appreciated. LK acknowledges the support of the SERB Grant No. ECR/2016/000109 and would like to thank the STAR and ALICE collaborations.

References

1. J. Adam et al., Experimental and theoretical challenges in the search for the quark gluon plasma: the STAR Collaboration's critical assessment of the evidence from RHIC collisions. Nucl. Phys. A **757**, 102–183 (2005)
2. I. Arsene et al., Quark gluon plasma and color glass condensate at RHIC? The Perspective from the BRAHMS experiment. Nucl. Phys. A **757**, 1–27 (2005)
3. K. Adcox et al., QFormation of dense partonic matter in relativistic nucleus-nucleus collisions at RHIC: experimental evaluation by the PHENIX collaboration. Nucl. Phys. A **757**, 184–283 (2005)
4. B.B. Back et al., The PHOBOS perspective on discoveries at RHIC. Nucl. Phys. A **757**, 28–101 (2005)
5. C.W. Fabjan et al., ALICE: physics performance report, volume II. J. Phys. G **32**, 1295–2040 (2006)
6. M. Shifman, B. Ioffe, At the frontier of particle physics, in *Handbook of QCD*, Vol. 1–3 (World Scientific Singapore, Singapore, 2001)
7. M. Cheng et al., The QCD equation of state with almost physical quark masses. Phys. Rev. D **77**, 014511 (2008)
8. Y. Aoki, G. Endrodi, Z. Fodor, S.D. Katz, K.K. Szabo, The Order of the quantum chromodynamics transition predicted by the standard model of particle physics. Nature **443**, 675–678 (2006)
9. R.V. Gavai, S. Gupta et al., QCD at finite chemical potential with six time slices. Phys. Rev. D **78**, 114503 (2008)
10. Z. Fodor, S.D. Katz, Critical point of QCD at finite T and mu, lattice results for physical quark masses. JHEP **04**, 050 (2004)
11. B.I. Abelev et al., Identified particle production, azimuthal anisotropy, and interferometry measurements in Au+Au collisions at $\sqrt{s_{NN}}$ = 9.2- GeV. Phys. Rev. C **81**, 024911 (2010)
12. L. Kumar, Review of recent results from the RHIC beam energy scan. Mod. Phys. Lett. A **28**, 1330033 (2013)
13. B.B. Abelev et al., Multiplicity dependence of pion, kaon, proton and lambda production in p-Pb collisions at $\sqrt{s_{NN}}$ = 5.02 TeV. Phys. Lett. B **728**, 25–38 (2014)
14. J. Adam et al., Multiplicity dependence of charged pion, kaon, and (anti)proton production at large transverse momentum in p-Pb collisions at $\sqrt{s_{\rm NN}}$ = 5.02 TeV. Phys. Lett. B **760**, 720–735 (2016)
15. S. Acharya et al., Multiplicity dependence of light-flavor hadron production in pp collisions at $\sqrt{s}$ = 7 TeV. Phys. Rev. C **99**, 024906 (2009)
16. L. Adamczyk et al., Bulk properties of the medium produced in relativistic heavy-ion collisions from the beam energy scan program. Phys. Rev. C **96**, 044904 (2017)
17. J.L. Klay et al., Longitudinal flow from 2-A-GeV to 8-A-GeV Au+Au collisions at the Brookhaven AGS. Phys. Rev. Lett. **88**, 102301 (2002)
18. C. Alt et al., Pion and kaon production in central Pb + Pb collisions at 20-A and 30-A-GeV: evidence for the onset of deconfinement. Phys. Rev. C. **77**, 024903 (2007)
19. B.I. Abelev et al., Systematic measurements of Identified particle spectra in pp, d^+ Au and Au+Au collisions from STAR. Phys. Rev. C **79**, 034909 (2009)
20. B. Abelev et al., Centrality dependence of π, K, p production in Pb-Pb collisions at $\sqrt{s_{NN}}$ = 2.76 TeV. Phys. Rev. C **88**, 044910 (2013)

21. J. Cleymans et al., The Thermal model and the transition from baryonic to mesonic freeze-out. Eur. Phys. J. A **29**, 119–121 (2006)
22. W. Busza, A.S. Goldhaber, Nuclear stopping power. Phys. Lett. B **139**, 235 (2009)
23. S. Wheaton, J. Cleymans, THERMUS: a thermal model package for ROOT. Comput. Phys. Commun. **180**, 84–106 (2009)
24. A. Andronic, F. Beutler, P. Braun-Munzinger, K. Redlich, J. Stachel, Thermal description of hadron production in e+e- collisions revisited. Phys. Lett. B **675**, 312–318 (2009)
25. N. Sharma, J. Cleymans, L. Kumar, Thermal model description of p–Pb collisions at $\sqrt{s_{NN}} =$ 5.02 TeV. Eur. Phys. J. C **78**, 288 (2019)
26. J. Cleymans, K. Redlich, E. Suhonen, Canonical description of strangeness conservation and particle production. Z. Phys. C **51**, 137–141 (1991)
27. A. Andronic, P. Braun-Munzinger, K. Redlich, J. Stachel, Decoding the phase structure of QCD via particle production at high energy. Nature **561**, 321–330 (2018)
28. A. Bazavov et al., Equation of state in (2+1)-flavor QCD. Phys. Rev. D. **90**, 094503 (2014)
29. Sz Borsanyi, G. Endrodi, Z. Fodor, S.D. Katz, S. Krieg, C. Ratti, K.K. Szabo, QCD equation of state at nonzero chemical potential: continuum results with physical quark masses at order mu^2. JHEP **08**, 053 (2012)
30. E. Schnedermann, J. Sollfrank, U.W. Heinz, Thermal phenomenology of hadrons from 200-A/GeV S+S collisions. Phys. Rev. C. **48**, 2462–2475 (1993)
31. B.B. Abelev et al., Elliptic flow of identified hadrons in Pb-Pb collisions at $\sqrt{s_{NN}}$. JHEP **06**, 190 (2015)
32. J. Adams et al., Multi-strange baryon elliptic flow in Au + Au collisions at s(NN)**(1/2) = 200-GeV. Phys. Rev. Lett. **95**, 122301 (2005)
33. A. Adare et al., Scaling properties of azimuthal anisotropy in Au+Au and Cu+Cu collisions at s(NN) = 200-GeV. Phys. Rev. Lett. **98**, 162301 (2007)
34. L. Adamczyk et al., Observation of an energy-dependent difference in elliptic flow between particles and antiparticles in relativistic heavy ion collisions. Phys. Rev. Lett. **110**, 142301 (2013)
35. J. Adam, Enhanced production of multi-strange hadrons in high-multiplicity proton-proton collisions. Nature Phys. **13**, 535–539 (2017)
36. J. Adam et al., Multi-strange baryon production in p-Pb collisions at $\sqrt{s_{NN}}$=5.02 TeV. Phys. Lett. B **758**, 389–401 (2016)
37. B.B. Abelev et al., Production of charged pions, kaons and protons at large transverse momenta in pp and Pb-Pb collisions at $\sqrt{s_{\rm NN}}$ =2.76 TeV. Phys. Lett. B **736**, 196–207 (2013)
38. B.B. Abelev et al., K^0_S and Λ production in Pb-Pb collisions at $\sqrt{s_{NN}}$ = 2.76 TeV. Phys. Rev. Lett. **111**, 222301 (2013)

Chapter 15
Studies on Λ Hypernuclei and Superheavy Elements

K. P. Santhosh

Abstract The properties of various Λ hypernuclei including the binding and separation energies, decay properties, etc. are studied. An extended Bethe–Weizsäcker mass formula (BWMF) has been proposed for finding the binding energies of Λ hypernuclei. A new formula is also proposed for evaluating the hypernuclear separation energies. These two new formalisms show better agreement with the experimental results as compared to the other methods. The hypernuclear decay triggered by strong interaction has been studied next. The isotopes of hyper Po, hyper Ra, and hyper Ac are selected for the study. The alpha and cluster emissions from the hypernuclei are studied by including a Λ-nucleus potential to the well-known Coulomb and proximity potential proposed by Santhosh et al. Using the Modified Generalized Liquid Drop Model and Phenomenological Model for Production cross section, we have studied the alpha decay properties and fusion ER cross section for the SHE with Z = 121. As the nuclei $^{302-304,306}121$ shows three alpha chains, they could be synthesized and detected in a laboratory. We have studied the ER cross section for the reaction, $^{50}Ti+^{252}Es \rightarrow {}^{302}121$, $^{54}Cr+^{249}Bk \rightarrow {}^{303}121$, $^{50}Ti+^{254}Es \rightarrow {}^{304}121$ and $^{48}Ca+^{258}Md \rightarrow {}^{306}121$ and found that $^{50}Ti+^{252}Es \rightarrow {}^{302}121$ is the most probable reaction to synthesize SHE with Z = 121. We hope that our studies will be a guide line for further investigations in these fields.

15.1 Introduction

Hypernuclei are many body systems consisting of ordinary nucleons and one or more strange hyperons. Recently a large number of hypernuclei are produced experimentally and the studies on hypernuclei have received a lot of attention. Hyperons are strange baryons. One of the characteristic features of the hyperon is that it is free from Pauli's exclusion principle, which makes it easy to deeply penetrate into the nuclear interior.

K. P. Santhosh (✉)
School of Pure and Applied Physics, Kannur University, Payyanur 670327, Kerala, India
e-mail: drkpsanthosh@gmail.com

R. K. Puri et al. (eds.), *Advances in Nuclear Physics*, Springer Proceedings in Physics 257, https://doi.org/10.1007/978-981-15-9062-7_15

The observation of first hypernuclear fragment was made by Danysz and Pniewski [1] in 1952, which was a boron nucleus in which a neutron was replaced by a Λ hyperon. Various properties of hypernuclei have been studied since its first evidence [2–8]. The stability of hypernuclei can be understood by studying its binding and separation energies. Different theoretical methods are proposed for studying the binding and separation energies of hypernuclei [9–17]. The decay studies of hypernuclei suggest that within the nuclear environment because of Pauli's blocking effect, nonmesonic decay modes are dominant over the mesonic decay modes. The possibilities of decays of excited hypernuclei triggered by strong interactions [18–21] have also been a subject of study.

An extended BWMF for finding the binding energies of Λ hypernuclei and a new formula for obtaining the separation energies are presented. The alpha and cluster decays from hyper Po, hyper Ra, and hyper Ac nuclei are performed using the Coulomb and proximity potential model (CPPM) [22] with the inclusion of a Λ-nucleus potential.

Since the superheavy elements up to $Z = 118$ have been confirmed in the laboratory, the study of SHE with $Z > 118$ now becomes an important topic in the nuclear physics research. SHEs were synthesized via two methods; hot fusion reaction at JINR-FLNR, Dubna for $Z = 114-118$ [23–28] and cold-fusion reaction at GSI, Darmstadt and at RIKEN, Japan for $Z = 102-112$ [29, 30]. The element $Z = 113$ was synthesized successfully by hot fusion reaction using $^{48}Ca + ^{237}Np$ by Oganessian et al. [31] and by cold-fusion reaction using $^{70}Zn+^{209}Bi$ by Morita et al. [32]. However, it is difficult to produce SHE with $Z = 113$ in cold-fusion reactions because of the smaller production cross sections. The discovery of superheavy nuclei in the fusion reactions of $^{48}Ca + ^{238}U \rightarrow ^{249}Cf$ were reviewed by Oganessian and Utyonkov in 2015 [33]. Recently Khuyagbaatar et al. [34] predicted the ER cross section for the isotope of Ts ($Z = 117$) for which the experiment was performed at the gas-filled recoil separator TASCA and confirmed the previous findings at the Dubna Gas-Filled Recoil Separator [25, 26, 35].

The ER residue cross section depends on the projectile-target pair, center-of-mass energy which in turn depends on probability of CN formation, excitation energy, fission barrier of CN, and survival probability. Hence, the predictions of the favorable reactions and beam energy are very important for the synthesis of superheavy elements. In the present paper, we have studied the decay modes and production cross section for the SHE with $Z = 121$. The evidence for the synthesis of $Z = 121$ is not reported yet and this study may helpful for future experimental investigations.

15.2 Theory

15.2.1 Methodology to Study the Properties of Hypernuclei

The lambda–nucleon (ΛN) interaction inside a hypernucleus can be understood by studying its binding and separation energies. Different formalisms are there for studying the properties of hypernuclei. One among them is the use of semi-empirical methods for evaluating the binding and separation energies.

In a recent study [36], we have proposed a new semi-empirical formula, by extending the Bethe–Weizsäcker mass formula (BWMF), for calculating the binding energy of singe Λ hypernuclei, which is given by

$$B(N, Z, \Lambda) = a_v A_c - a_s A_c^{2/3} - a_c \frac{Z(Z-1)}{A_c^{1/3}} - a_{sym} \frac{(N-Z)^2}{A_c} + \delta - \frac{b_0}{A^{2/3}} \left[1 - \frac{b_1}{A^{2/3}}\right] + b_2. \tag{15.1}$$

Here A is the mass number of the hypernucleus, given by, A = Z + N + Λ, and A_c is the mass number of core nucleus, $A_c = Z + N$, where Z and N are the number of protons and number of neutrons. a_v, a_s, a_c and a_{sym} are the usual BWMF constants, given by, $a_v = 15.79$ MeV, $a_s = 18.34$ MeV, $a_c = 0.71$ MeV, $a_{sym} = 23.21$ MeV. δ is the pairing term, which is $12A_c^{-1/2}$ for even N even Z nuclei, $-12A_c^{-1/2}$ for odd N odd Z nuclei and 0 otherwise. b_0, b_1 and b_2 are given as, $b_0 = 119.445$ MeV, $b_1 = 1.119$ MeV, $b_2 = 33.047$ MeV.

A new formula is also suggested [36] using the least square regression to the updated experimental data of single Λ hypernuclei, given as

$$S_\Lambda = a_0 + \frac{a_1}{A^{2/3}} + \frac{a_2}{A^{4/3}}. \tag{15.2}$$

Here $a_0 = 28.442$ MeV, $a_1 = -119.445$ MeV and $a_2 = 133.651$ MeV.

These two new formulae for the binding and separation energies of Λ hypernuclei are found to be more suitable for obtaining the experimental results as well as for making theoretical predictions [36].

Another important part in the studies of hypernuclei is the decay of hypernuclei. Even though many studies have been put forward for studying the weak decay of hypernuclei, only a few numbers of studies [18–21] have been performed on the hypernuclear decay triggered by strong interaction, such as alpha and cluster emission. We have modified the Coulomb and proximity potential model (CPPM) proposed by Santhosh et al. [22] with the inclusion of a Λ-nucleus potential, for studying the alpha and cluster emissions from hypernuclei. In CPPM, the interacting potential between two nuclei is taken as the sum of Coulomb potential, proximity potential, and centrifugal potential. It is given by

$$V = \frac{Z_1 Z_2 e^2}{r} + V_p(z) + \frac{\hbar^2 l(l+1)}{2\mu r^2}, \quad for\ z > 0. \tag{15.3}$$

Here Z_1 and Z_2 are the atomic numbers of the daughter and emitted cluster, r is the distance between fragment centers, z is the distance between the near surfaces of the fragments, l represents the angular momentum and μ the reduced mass. V_p is the proximity potential given by Blocki et al. [37, 38] as

$$V_p(z) = 4\pi\gamma b\left[\frac{C_1C_2}{C_1+C_2}\right]\Phi\left(\frac{z}{b}\right) \tag{15.4}$$

with the nuclear surface tension coefficient,

$$\gamma = 0.9517\left[1 - 1.7826\frac{(N-Z)^2}{A^2}\right]MeV/fm^2. \tag{15.5}$$

Here N, Z, and A represent the neutron, proton, and mass number of the parent nuclei. Φ represents the universal proximity potential [38] and C_i is the Süsmann central radii of the fragments.

The potential for the internal part (overlap region) of the barrier is given as

$$V = a_0(L - L_0)^n, \quad for\ z < 0, \tag{15.6}$$

where $L = z + 2C_1 + C_2$ fm and $L_0 = 2C$ fm. The constants a_0 and n are determined by the smooth matching of the two potentials at the touching point.

The barrier penetrability P using the one dimensional Wentzel–Kramers–Brillouin approximation, is given as

$$P = exp\left\{-\frac{2}{\hbar}\int_a^b\sqrt{2\mu(V-Q)}dz\right\}. \tag{15.7}$$

Here the mass parameter is replaced by $\mu = \frac{mA_1A_2}{A}$, where m is the nucleon mass and A_1, A_2 are the mass numbers of daughter and emitted cluster, respectively. The turning points 'a' and 'b' are determined from the equation, $V(a) = V(b) = Q$, where Q is the energy released. The half-life time is given by

$$T_{1/2} = \left(\frac{ln2}{\lambda}\right) = \left(\frac{ln2}{\nu P}\right). \tag{15.8}$$

Here λ is the decay constant and ν is the assault frequency. The empirical vibration energy E_ν, is given as [39]

$$E_\nu = Q\left\{0.056 + 0.039exp\left[\frac{(4-A_2)}{2.5}\right]\right\},\ for\ A_2 \geq 4 \tag{15.9}$$

To incorporate the changes in potential due to Λ particle, we have included the potential, V_Λ between the non-strange normal fragment and the fragment that contains lambda particle, in the expression for the interacting potential (15.3). That is, as the alpha particle penetrates the potential produced by the Coulomb force, nuclear force, and centrifugal force, it also feels the potential generated by the Λ hyperon. The potential, V_Λ between the non-strange and strange fragments is given by

$$V_\Lambda = \int \rho_\Lambda(r_1) V_{\Lambda N}(r_1 - r)\, d^3 r_1, \tag{15.10}$$

where $\rho_\Lambda(r_1)$ is the density distribution of Λ particle. The density distribution of Λ particle is taken from the [40, 41] and has the form,

$$\rho_\Lambda(r) = \left(\pi b_\Lambda^2\right)^{-3/2} e^{-r^2/b_\Lambda^2}. \tag{15.11}$$

Here $b_\Lambda = \sqrt{\frac{4M_N + M_\Lambda}{4M_\Lambda b_\alpha}}$, where M_N and M_Λ are the mass of the nucleon and Λ particle, respectively, and $b_\alpha = 1.358\,\text{fm}$. The lambda-nucleon force is short range and the strength of lambda-nucleus potential $V_{\Lambda N}$ is smaller than the nucleon-nucleus potential. The lambda-nucleus potential, $V_{\Lambda N}$, is taken from [42] and is given by,

$$V_{\Lambda N} = \frac{V_0}{1 + exp\left[\frac{r-c}{a}\right]}. \tag{15.12}$$

Here the constants $V_0 = -27.4\,\text{MeV}$, $a = 0.6\,\text{fm}$ and $c = 1.08A^{1/3}$. By including the lambda-nucleus potential in (15.3), the half-lives for the hypernuclei can also be determined using (15.8).

15.2.2 Methodology to Find Decay Modes and Production Cross Section of SHE

The cross section of SHE production in a heavy-ion fusion reaction with subsequent emission of x neutrons is given by

$$\sigma_{ER}^{xn} = \frac{\pi}{k^2} \sum_{l=0}^{\infty} (2l+1)\, T(E,l) P_{CN}(E,l) W_{sur}^{xn}(E^*, l). \tag{15.13}$$

The probability of compound nucleus formation [43–45] is given as

$$P_{CN}(E,l) = \frac{exp\left\{-c\left(x_{eff} - x_{thr}\right)\right\}}{1 + exp\left\{\frac{E_B^* - E^*}{\Delta}\right\}}, \tag{15.14}$$

where $E^* = E_{cm} - Q - \frac{l(l+1)}{2\mu r^2}$ is the excitation energy of the compound nucleus, E_B^* denotes the excitation energy of the CN when the center-of-mass beam energy (E_{cm}) is equal to the Coulomb and proximity barrier, Δ is an adjustable parameter ($\Delta = 4\,\text{MeV}$) and x_{eff} is the effective fissility defined as

$$x_{eff} = \left[\frac{(Z^2/A)}{(Z^2/A)_{crit}}\right](1 - \alpha + \alpha f(K)). \tag{15.15}$$

With $(Z^2/A)_{crit}$, $f(K)$ and K are given by

$$(Z^2/A)_{crit} = 50.883\left[1 - 1.7286\left(\frac{(N-Z)}{A}\right)^2\right] \tag{15.16}$$

$$f(K) = \frac{4}{K^2 + K + \frac{1}{K} + \frac{1}{K^2}} \tag{15.17}$$

$$K = \left(\frac{A_1}{A_2}\right)^{1/3}, \tag{15.18}$$

where Z, N and A represent the atomic number, neutron number, and mass number, respectively. A_1 and A_2 are mass number of projectile and target, respectively. x_{thr}, c are adjustable parameters and $\alpha = 1/3$. The best fit to the cold-fusion reaction, the values of c and x_{eff} are 136.5 and 0.79, respectively. For hot fusion reaction, the best fit for $x_{eff} \leq 0.8$ is $c = 104$ and $x_{thr} = 0.69$; while $x_{eff} \geq 0.8$, the values are $c = 82$ and $x_{thr} = 0.69$. These constants are suggested by Loveland [44].

The survival probability W_{sur} is the probability for the compound nucleus to decay to the ground state of the final residual nucleus via evaporation of light particles and gamma ray for avoiding fission process. The survival probability under the evaporation of x neutrons is

$$W_{sur} = P_{xn}(E^*_{CN}) \prod_{i=1}^{i_{max}=x} \left(\frac{\Gamma_n}{\Gamma_n + \Gamma_f}\right)_{i,E^*}, \tag{15.19}$$

where the index i is equal to the number of emitted neutrons, P_{xn} is the probability of emitting exactly xn neutrons [46], E^* is the excitation energy of the compound nucleus, Γ_n and Γ_f represent the decay width of neutron evaporation and fission, respectively. To calculate Γ_n/Γ_f, Vandenbosch and Huizenga [47] have suggested a classical formalism:

$$\frac{\Gamma_n}{\Gamma_f} = \frac{4A^{2/3}a_f(E^* - B_n)}{K_0 a_n\left[2a_f^{1/2}(E^* - B_f)^{1/2} - 1\right]} exp\left[2a_n^{1/2}(E^* - B_n)^{1/2} - 2a_f^{1/2}(E^* - B_f)^{1/2}\right], \tag{15.20}$$

where A is the mass number of the nucleus considered, E^* is the excitation energy, and B_n is the neutron separation energy. The constant K_0 is taken as 10 MeV. $a_n = A/10$ and $a_f = 1.1a_n$, are the level density parameters of the daughter nucleus and the fissioning nucleus at the ground state and saddle configurations, respectively, and B_f is the fission barrier. The alpha decay half-lives are calculated using the modified generalized liquid drop model (MGLDM) [48] of our group and spontaneous fission half-lives are calculated using the formula of Santhosh et al. [49]

$$log\left(T_{1/2}/yr\right) = a\frac{Z^2}{A} + b\left(\frac{Z^2}{A}\right)^2 + c\left(\frac{N-Z}{N+Z}\right) + d\left(\frac{N-Z}{N+Z}\right)^2 + eE_{shell} + f, \tag{15.21}$$

where $a = -43.25203$, $b = 0.49192$, $c = 3674.3927$, $d = -9360.6$, $e = 0.8930$, and $f = 578.56058$. E_{shell} is the shell correction energy.

15.3 Results and Discussion

The BWMF for hypernuclei is an extension of BWMF of normal nuclei to the hypernuclear sector. The formula was developed by studying the variation of binding energy for all the experimentally identified Λ hypernuclei with $A^{-2/3}$. It was seen that there exists an asymptotic relation between the binding energy and the hypernuclear surface term, which is proportional to $A^{2/3}$. The performance of the newly proposed mass formula (15.1) has been demonstrated by evaluating the binding energies of all the thirty-five experimentally synthesized Λ hypernuclei from ${}^{4}_{\Lambda}H$ to ${}^{208}_{\Lambda}Pb$ [50–55]. The predictive power of the new formula has been revealed by evaluating the standard deviation. As compared to other theoretical formalisms, the new formula gives the minimum standard deviation. Figure 15.1 shows the variation of BE/A using (15.1) with the mass number for the experimentally synthesized Λ hypernuclei. A comparison with the experimental results is also given. The formula proposed for finding the separation energy (15.2) also gives better agreement with experimental data as

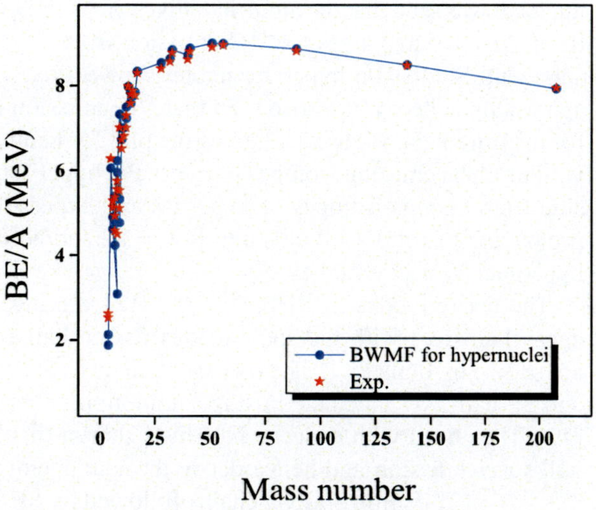

Fig. 15.1 Plot of BE/A versus mass number for all the experimentally synthesized Λ hypernuclei

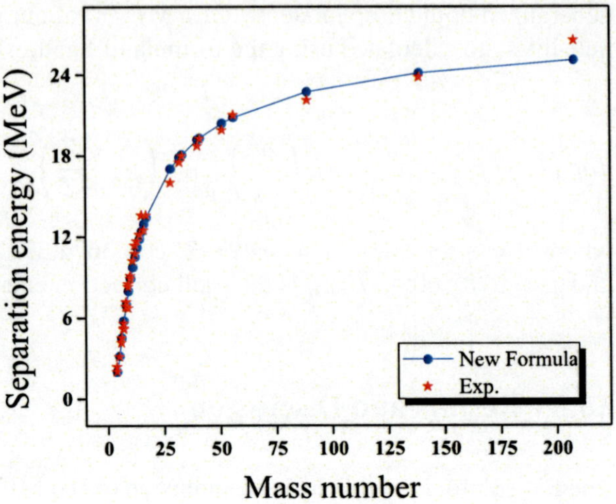

Fig. 15.2 Plot of separation energy versus mass number for all the experimentally synthesized Λ hypernuclei

compared to other theoretical formalisms. Figure 15.2 shows the plot of separation energy obtained using (15.2) versus mass number of Λ hypernuclei. The agreement between theoretical predictions and the experimental observations can be seen from the figure.

Considering to the decay of hypernuclei triggered by strong interaction, we have studied alpha emission from hyper Po nuclei, alpha, and cluster emission from hyper Ra nuclei and hyper Ac nuclei using CPPM with the inclusion of a Λ-nucleus potential. The elements Po, Ra, and Ac are well-known members of natural radioactive series. Since many isotopes of these elements are good alpha emitters, their corresponding hypernuclei may also exhibit alpha decay. The half-lives of alpha emission from hyper ${}^{187-224}{}_{\Lambda}Po$ are calculated. It was found that the isotopes ${}^{187-224}{}_{\Lambda}Po$ exhibit alpha decay. Similarly, while studying the alpha emission from ${}^{202-235}{}_{\Lambda}Ra$ nuclei, it was seen that the alpha half-lives of ${}^{202-231}{}_{\Lambda}Ra$ are within the experimental limit. Also the half-lives for ${}^{14}C$ emission from ${}^{217-229}{}_{\Lambda}Ra$ are favorable for measurement. Studies on hyper Ac nuclei showed that ${}^{207-234}{}_{\Lambda}Ac$ nuclei are unstable against alpha decay. It is also seen that ${}^{14}C$ emission from ${}^{218-229}{}_{\Lambda}Ac$ are favorable for measurement. Figure 15.3 gives the plot for neutron number versus $log_{10}T_{1/2}$ of various alpha emitting isotopes of hyper Po, hyper Ra, and hyper Ac. The ${}^{14}C$ emission from various isotopes of hyper Ra and hyper Ac is depicted in Fig. 15.4. The proton and neutron shell closures at Z = 82 and N = 126 are revealed through the hypernuclear decay studies.

The decay modes of SHE ${}^{302-304,306}121$ are investigated by comparing the α-decay half-lives with the spontaneous fission half-lives and the calculated values are shown in Table 15.1. The α-decay half-lives are calculated using MGLDM [48] and SF half-lives calculated using semi-empirical relation given by Santhosh et al. [49]. The nuclei with α-decay half-lives shorter than spontaneous fission half-lives will survive fission and hence decay through α emission. It is found that isotopes ${}^{306,304-302}121$ exhibit 3 alpha chains followed by SF.

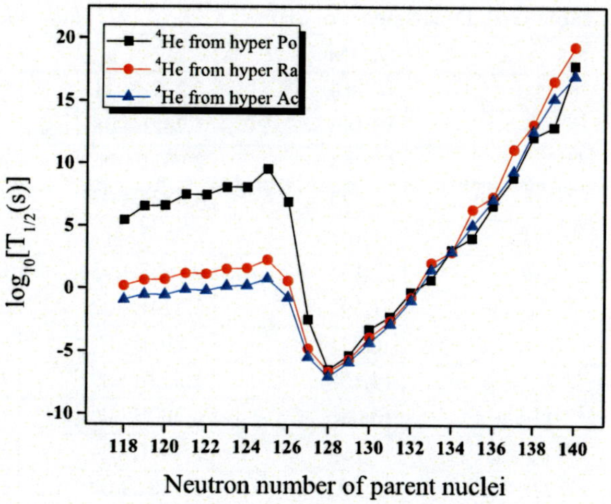

Fig. 15.3 Plot of $log_{10}T_{1/2}$ versus neutron number of parent for various alpha emitting isotopes of hyper Po, hyper Ra, and hyper Ac

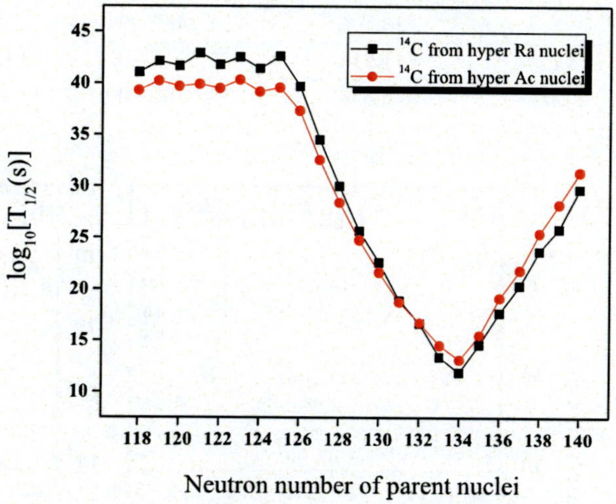

Fig. 15.4 Plot of $log_{10}T_{1/2}$ versus neutron number of parent for various ^{14}C emitting isotopes of hyper Ra and hyper Ac

The ER cross section in 2n, 3n, 4n, and 5n evaporation channel for the synthesis of isotopes $^{302-304,306}121$ using the reactions, $^{50}Ti +^{252} Es \rightarrow^{302}$ 121, $^{54}Cr + ^{249}Bk \rightarrow^{303}$ 121,$^{50}Ti +^{254} Es \rightarrow^{304}$ 121, and $^{48}Ca +^{258} Md \rightarrow^{306}$ 121 are studied. The corresponding figures representing ER cross section in each evaporation channel is presented in Fig. 15.5. The largest ER cross section obtained in each evaporation channel is listed in Table 15.2. It is found that, the 2n and 3n channel cross section is more for the reaction $^{50}Ti +^{252} Es \rightarrow^{302}$ 121, and 4n channel cross section is more for the reaction $^{50}Ti +^{254} Es \rightarrow^{304}$ 121.

Table 15.1 Decay modes of SHEs $^{302-304,306}121$. *Half − lives are in seconds*

^{Z}A	Q_α (MeV)	$T_{SF}^{(KPS)}$	$T_\alpha^{(MGLDM)}$	Decay mode
$^{302}121$	14.118	5.239E+11	1.164E−08	α
$^{298}119$	13.146	7.742E+08	1.149E−06	α
^{294}Ts	11.354	2.931E+06	7.046E−04	α
^{290}Mc	7.973	4.885E+03	8.358E+12	SF
$^{303}121$	14.168	2.354E+10	4.013E−08	α
$^{299}119$	13.135	9.793E+03	1.107E−06	α
^{295}Ts	11.593	2.814E−02	6.781E−04	α
^{291}Mc	6.821	1.297E−07	8.020E+12	SF
$^{304}121$	14.158	2.213E+09	4.031E−08	α
$^{300}119$	13.086	6.300E+06	1.333E−06	α
^{296}Ts	11.694	3.747E+04	3.785E−04	α
^{292}Mc	8.303	2.188E+02	1.370E+06	SF
$^{306}121$	14.838	6.017E+05	2.554E−09	α
$^{302}119$	13.106	1.248E+04	1.130E−06	α
^{298}Ts	11.914	1.311E+02	1.098E−04	α
^{294}Mc	8.853	1.988E+00	1.133E+04	SF

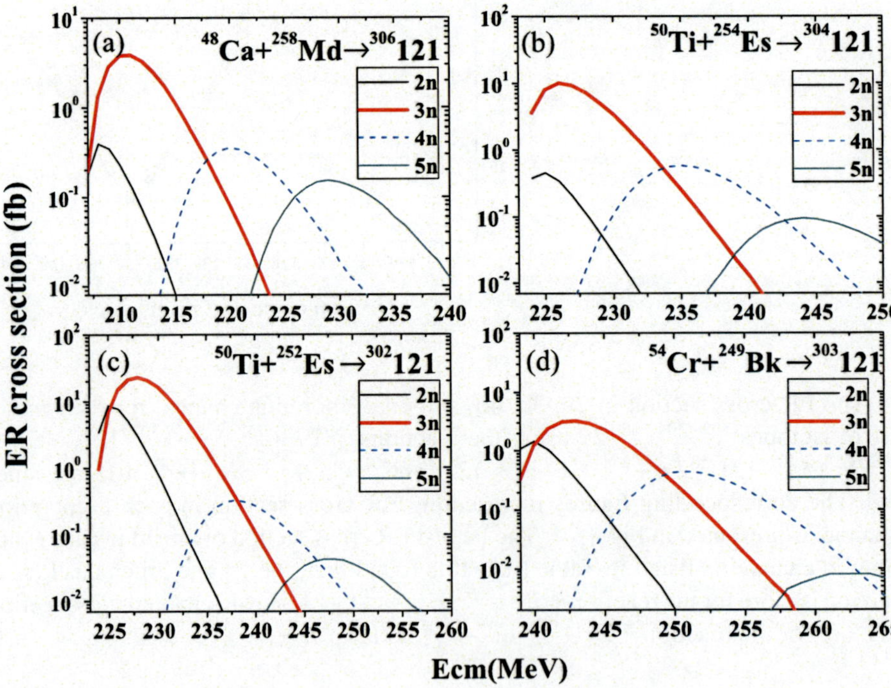

Fig. 15.5 ER cross section in 2n, 3n, 4n and 5n evaporation channel

Table 15.2 Maximum value of ER cross section obtained in each evaporation channel

Reaction	ER cross section in fb			
	2n	3n	4n	5n
$^{48}Ca +^{258} Md \rightarrow^{306}121$	0.39	3.89	0.35	0.046
$^{50}Ti +^{254} Es \rightarrow^{304}121$	0.45	10.02	0.58	0.017
$^{50}Ti +^{252} Es \rightarrow^{302}121$	8.62	23.09	0.33	0.002
$^{54}Cr +^{249} Bk \rightarrow^{303}121$	1.45	3.28	0.45	0.006

15.4 Conclusion

The binding and separation energies of Λ hypernuclei are predicted using two new formalisms. The values predicted using the new formulae are in good agreement with the experimental results. These two simple formulae provide a reliable method for finding the two important quantities of hypernuclei, that is, the binding energy and the separation energy. The hypernuclear decay triggered by strong interactions, particularly the alpha and the cluster emissions are studied using CPPM with the inclusion of a Λ-nucleus potential for the isotopes of hyper Po, hyper Ra, and hyper Ac. The study shows that the alpha and cluster decay half-lives of many of the isotopes of these hyper elements are within the experimental limit. The proton and neutron shell closure at $Z = 82$ and $N = 126$ in the hypernuclear region is also revealed from the study.

We have predicted the alpha decay half-lives and SF half-lives of the isotope SHE $^{302-304,306}121$. The isotopes $^{302-304,306}121$ shows 3 alpha chains followed by SF and hence these isotopes can be synthesized and detected in the laboratory. The studies on ER cross section for the synthesis of these isotopes using the reactions, $^{50}Ti + {}^{252}Es \rightarrow^{302} 121$, $^{54}Cr +^{249} Bk \rightarrow^{303} 121$,$^{50}Ti +^{254} Es \rightarrow^{304} 121$ and $^{48}Ca + {}^{258}Md \rightarrow^{306} 121$ are performed and it is clear that the reactions $^{50}Ti +^{254} Es \rightarrow {}^{304}121$, and $^{50}Ti +^{252} Es \rightarrow^{302} 121$ have maximum probability in 3n and 4n channel, respectively.We hope that our studies on hypernuclei and superheavy elements will be a guide line for further investigations in these fields.

References

1. M. Danysz, J. Pniewski, Delayed disintegration of a heavy nuclear fragment: I. Philos. Mag. **44**, 348–350 (1953)
2. B. Povh, Hypernuclei. Annu. Rev. Nucl. Part. Sci. **28**, 1–32 (1978)
3. H. Bandō, T. Motoba, J. Žofka, Production, Structure and Decay of Hypernuclei. Int. J. Mod. Phys. A **5**, 4021–4198 (1990)
4. T. Hasegawa, O. Hashimoto, S. Homma, T. Miyachi, T. Nagae, M. Sekimoto, T. Shibata, H. Sakaguchi, T. Takahashi, K. Aoki, H. Noumi, H. Bhang, M. Youn, Y. Gavrilov, S. Ajimura, T. Kishimoto, A. Ohkusu, K. Maeda, R. Sawafta, R.P. Redwine, Spectroscopic study of

$^{10}{}_{\Lambda}B, ^{12}{}_{\Lambda}C, ^{28}{}_{\Lambda}Si, ^{89}{}_{\Lambda}Y, ^{139}{}_{\Lambda}La$, and $^{208}{}_{\Lambda}Pb$ by the (π^+, K^+) reaction. Phys. Rev. C **53**, 1210–1220 (1996)
5. R.E. Chrien, C.B. Dover, Nuclear systems with strangeness. Annu. Rev. Nucl. Part. Sci. **39**, 113–150 (1989)
6. P.H. Pile, S. Bart, R.E. Chrien, D.J. Millener, R.J. Sutter, N. Tsoupas, J.-C. Peng, C.S. Mishra, E.V. Hungerford, T. Kishimoto, L.-G. Tang, W. von Witsch, Z. Xu, K. Maeda, D. Gill, R. McCrady, B. Quinn, J. Seydoux, J.W. Sleight, R.L. Stearns, H. Plendl, A. Rafatian, J. Reidy, Study of hypernuclei by associated production. Phys. Rev. Lett. **66**, 2585–2588 (1991)
7. E. Botta, T. Bressani, G. Garbarino, Strangeness nuclear physics: a critical review on selected topics. Eur. Phys. J. A **48**, 41 (2012)
8. A. Gal, E.V. Hungerford, D.J. Millener, Strangeness in nuclear physics. Rev. Mod. Phys. **88**, 035004 (2016)
9. J. Schaffner-Bielich, A. Gal, Properties of strange hadronic matter in bulk and in finite systems. Phys. Rev. C **62**, 034311 (2000)
10. J. Schaffner, C.B. Dover, A. Gal, C. Greiner, D.J. Millener, H. Stoecker, Multiply strange nuclear systems. Ann. Phys. (N.Y.) **235**, 35–76 (1994)
11. M.Z.R. Khan, N. Neelofer, M.A. Suhail, Semi-empirical formula for Λ-binding energies in ground states of light hypernuclei. Pramana J. Phys. **49**, 617–622 (1997)
12. N. Neelofer, M. Shoeb, M.Z.R. Khan, A one-parameter formula for estimating the Λ well depth. Pramana J. Phys. **37**, 419–424 (1991)
13. J. Schaffner, C.B. Dover, A. Gal, C. Greiner, H. Stoecker, Strange hadronic matter. Phy. Rev. Lett. **71**, 1328–1331 (1993)
14. S. Balberg, A. Gal, J. Schaffner, Application of mass formulae for multiply strange nuclei. Prog. Theor. Phys. Suppl. **117**, 325–337 (1994)
15. A.S. Botvina, J.P. Pochodzalla, Production of hypernuclei in multifragmentation of nuclear spectator matter. Phys. Rev. C **76**, 024909 (2007)
16. G. Lévai, J. Cseh, P. Van Isacker, O. Juillet, Mass formula for Λ hypernuclei based on SU(6) symmetry. Phys. Lett. B **433**, 250–256 (1998)
17. C. Samanta, P.R. Chowdhury, D.N. Basu, Generalized mass formula for non-strange and hypernuclei with SU(6) symmetry breaking. J. Phys. G: Nucl. Part. Phys. **32**, 363–375 (2006)
18. R.H. Dalitz, D.H. Davis, D.N. Tovee, Proton decay of excited hypernuclei. Nucl. Phys. A **450**, 311c–328c (1986)
19. K.P. Santhosh, Probable alpha and ^{14}C cluster emission from hyper Ac nuclei. Eur. Phys. J. A **49**, 127 (2013)
20. K.P. Santhosh, C. Nithya, Possible alpha and ^{14}C cluster emission from hyper radium nuclei in the mass region $A = 202 - 235$. J. Nucl. Phys. Mater. Sci. Radiat. Appl. **4**, 337–353 (2017)
21. K.P. Santhosh, C. Nithya, Theoretical studies on the α decay half-lives of hyper and normal isotopes of Po. Pramana J. Phys. **90**, 35 (2018)
22. K.P. Santhosh, A. Joseph, Cluster radioactivity in xenon isotopes. Pramana J. Phys. **62**, 957–965 (2004)
23. Y.T. Oganessian, V.K. Utyonkov, Y.V. Lobanov, F.S. Abdullin, A.N. Polyakov, R.N. Sagaidak, I.V. Shirokovsky, Y.S. Tsyganov, A.A. Voinov, G.G. Gulbekian, S.L. Bogomolov, B.N. Gikal, A.N. Mezentsev, S. Iliev, V.G. Subbotin, A.M. Sukhov, K. Subotic, V.I. Zagrebaev, G.K. Vostokin, M.G. Itkis, K.J. Moody, J.B. Patin, D.A. Shaughnessy, M.A. Stoyer, N.J. Stoyer, P.A. Wilk, J.M. Kenneally, J.H. Landrum, J.F. Wild, R.W. Lougheed, Synthesis of the isotopes of elements 118 and 116 in the ^{249}Cf and $^{245}Cm+^{48}Ca$ fusion reactions. Phys. Rev. C **74**, 044602 (2006)
24. Y.T. Oganessian, F.S. Abdullin, S.N. Dmitriev, J.M. Gostic, J.H. Hamilton, R.A. Henderson, M.G. Itkis, K.J. Moody, A.N. Polyakov, A.V. Ramayya, J.B. Roberto, K.P. Rykaczewski, R.N. Sagaidak, D.A. Shaughnessy, I.V. Shirokovsky, M.A. Stoyer, N.J. Stoyer, V.G. Subbotin, A.M. Sukhov, Y.S. Tsyganov, V.K. Utyonkov, A.A. Voinov, G.K. Vostokin, Investigation of the $^{243}Am+^{48}Ca$ reaction products previously observed in the experiments on elements 113, 115, and 117. Phys. Rev. C **87**, 014302 (2013)

25. Y.T. Oganessian, F.S. Abdullin, P.D. Bailey, D.E. Benker, M.E. Bennett, S.N. Dmitriev, J.G. Ezold, J.H. Hamilton, R.A. Henderson, M.G. Itkis, Y.V. Lobanov, A.N. Mezentsev, K.J. Moody, S.L. Nelson, A.N. Polyakov, C.E. Porter, A.V. Ramayya, F.D. Riley, J.B. Roberto, M.A. Ryabinin, K.P. Rykaczewski, R.N. Sagaidak, D.A. Shaughnessy, I.V. Shirokovsky, M.A. Stoyer, V.G. Subbotin, R. Sudowe, A.M. Sukhov, Y.S. Tsyganov, V.K. Utyonkov, A.A. Voinov, G.K. Vostokin, P.A. Wilk, Synthesis of a new element with atomic number Z = 117. Phys. Rev. Lett. **104**, 142502 (2010)
26. Y.T. Oganessian, F.S. Abdullin, C. Alexander, J. Binder, R.A. Boll, S.N. Dmitriev, J. Ezold, K. Felker, J.M. Gostic, R.K. Grzywacz, J.H. Hamilton, R.A. Henderson, M.G. Itkis, K. Miernik, D. Miller, K.J. Moody, A.N. Polyakov, A.V. Ramayya, J.B. Roberto, M.A. Ryabinin, K.P. Rykaczewski, R.N. Sagaidak, D.A. Shaughnessy, I.V. Shirokovsky, M.V. Shumeiko, M.A. Stoyer, N.J. Stoyer, V.G. Subbotin, A.M. Sukhov, Y.S. Tsyganov, V.K. Utyonkov, A.A. Voinov, G.K. Vostokin, Experimental studies of the ^{249}Bk+^{48}Ca reaction including decay properties and excitation function for isotopes of element 117, and discovery of the new isotope ^{277}Mt. Phys. Rev. C **87**, 054621 (2013)
27. Y.T. Oganessian, F.S. Abdullin, C. Alexander, J. Binder, R.A. Boll, S.N. Dmitriev, J. Ezold, K. Felker, J.M. Gostic, R.K. Grzywacz, J.H. Hamilton, R.A. Henderson, M.G. Itkis, K. Miernik, D. Miller, K.J. Moody, A.N. Polyakov, A.V. Ramayya, J.B. Roberto, M.A. Ryabinin, K.P. Rykaczewski, R.N. Sagaidak, D.A. Shaughnessy, I.V. Shirokovsky, M.V. Shumeiko, M.A. Stoyer, N.J. Stoyer, V.G. Subbotin, A.M. Sukhov, Y.S. Tsyganov, V.K. Utyonkov, A.A. Voinov, G.K. Vostokin, Production and decay of the heaviest nuclei ^{293,294}117 and ^{294}118. Phys. Rev. Lett. **109**, 162501 (2012)
28. Y.T. Oganessian, Synthesis and Properties of Even-even Isotopes with Z = 110–116 in ^{48}Ca Induced Reactions. J. Nucl. Radiochem. Sci. **3**, 5–8 (2002)
29. K. Morita, K. Morimoto, D. Kaji, T. Akiyama, S. Goto, H. Haba, E. Ideguchi, K. Katori, H. Koura, H. Kudo, T. Ohnishi, A. Ozawa, T. Suda, K. Sueki, F. Tokanai, T. Yamaguchi, A. Yoneda, A. Yoshida, Experiment on synthesis of an isotope ^{277}112 by ^{208}Pb+^{70}Zn reaction. J. Phys. Soc. Jpn. **76**, 043201 (2007)
30. J.H. Hamilton, S. Hofmann, Y.T. Oganessian, Search for superheavy nuclei. Annu. Rev. Nucl. Part. Sci. **63**, 383–405 (2013)
31. Y.T. Oganessian, V.K. Utyonkov, Y.V. Lobanov, F.S. Abdullin, A.N. Polyakov, R.N. Sagaidak, I.V. Shirokovsky, Y.S. Tsyganov, A.A. Voinov, G.G. Gulbekian, S.L. Bogomolov, B.N. Gikal, A.N. Mezentsev, V.G. Subbotin, A.M. Sukhov, K.V. Subotic, I. Zagrebaev, G.K. Vostokin, M.G. Itkis, R.A. Henderson, J.M. Kenneally, J.H. Landrum, K.J. Moody, D.A. Shaughnessy, M.A. Stoyer, N.J. Stoyer, P.A. Wilk, Synthesis of the isotope ^{282}113 in the ^{237}NP+^{48}Ca fusion reaction. Phys. Rev. C **76**, 011601(R) (2007)
32. K. Morita, K. Morimoto, D. Kaji, T. Akiyama, S. Goto, H. Haba, E. Ideguchi, R. Kanungo, K. Katori, H. Koura, H. Kudo, T. Ohnishi, A. Ozawa, T. Suda, K. Sueki, H. Xu, T. Yamaguchi, A. Yoneda, A. Yoshida, Y. Zhao, Experiment on the Synthesis of Element 113 in the Reaction ^{209}Bi(^{70}Zn,n)^{278}113. J. Phys. Soc. Jpn. **73**, 2593–2596 (2004)
33. Y.T. Oganessian, V.K. Utyonkov, Super-heavy element research. Rep. Prog. Phys. **78**, 036301–036323 (2015)
34. J. Khuyagbaatar, A. Yakushev, C.E. Düllmann, D. Ackermann, L.-L. Andersson, M. Asai, M. Block, R.A. Boll, H. Brand, D.M. Cox, M. Dasgupta, X. Derkx, A. Di Nitto, K. Eberhardt, J. Even, C. Fahlander, M. Evers, U. Forsberg, J.M. Gates, N. Gharibyan, P. Golubev, K.E. Gregorich, J.H. Hamilton, W. Hartmann, R.-D. Herzberg, F.P. Heßberger, D.J. Hinde, J. Hoffmann, R. Hollinger, A. Hübner, E. Jäger, B. Kindler, J.V. Kratz, J. Krier, N. Kurz, M. Laatiaoui, S. Lahiri, R. Lang, B. Lommel, M. Maiti, S. Minami, A. Mistry, C. Mokry, H. Nitsche, J.P. Omtvedt, G.K. Pang, P. Papadakis, D. Renisch, J. Roberto, D. Rudolph, J. Runke, K.P. Rykaczewski, L.G. Sarmiento, M. Schädel, B. Schausten, A. Semchenkov, D.A. Shaughnessy, P. Steinegger, J. Steiner, E.E. Tereshatov, P. Thörle-Pospiech, K. Tinschert, T. Torres De Heidenreich, N. Trautmann, A. Türler, J. Uusitalo, D.E. Ward, M. Wegrzecki, N. Wiehl, S.M. Van Cleve, V. Yakusheva, Fusion reaction ^{48}Ca + ^{249} Bk leading to formation of the element Ts (Z = 117). Phys. Rev. C **99**, 054306 (2019)

35. Y.T. Oganessian, F.S. Abdullin, C. Alexander, J. Binder, R.A. Boll, S.N. Dmitriev, J. Ezold, K. Felker, J.M. Gostic, R.K. Grzywacz, J.H. Hamilton, R.A. Henderson, M.G. Itkis, K. Miernik, D. Miller, K.J. Moody, A.N. Polyakov, A.V. Ramayya, J.B. Roberto, M.A. Ryabinin, K.P. Rykaczewski, R.N. Sagaidak, D.A. Shaughnessy, I.V. Shirokovsky, M.V. Shumeiko, M.A. Stoyer, N.J. Stoyer, V.G. Subbotin, A.M. Sukhov, Y.S. Tsyganov, V.K. Utyonkov, A.A. Voinov, G.K. Vostokin, Study of the properties of the superheavy nuclei Z = 117 produced in the $^{249}Bk + ^{48}Ca$ reaction, in *EPJ Web of Conferences*, vol. 66 (2014), p. 02073
36. K.P. Santhosh, C. Nithya, New semi-empirical formulae for the binding and separation energies of single Λ hypernuclei. Eur. Phys. J. Plus **133**, 343 (2018)
37. J. Blocki, J. Randrup, W.J. Swiatecki, C.F. Tsang, Proximity Forces. Ann. Phys. (NY) **105**, 427–462 (1977)
38. J. Blocki, W.J. Swiatecki, A generalization of the proximity force theorem. Ann. Phys. (NY) **132**, 53–65 (1981)
39. D.N. Poenaru, M. Ivascu, A. Sandulescu, W. Greiner, Atomic nuclei decay modes by spontaneous emission of heavy ions. Phys. Rev. C **32**, 572–581 (1985)
40. T. Motoba, H. Bandō, K. Ikeda, Light p-shell Λ-hypernuclei by the microscopic three-cluster model. Prog. Theor. Phys. **70**, 189–221 (1983)
41. K. Hagino, T. Koike, Relation between shrinkage effect and compression of rotational spectrum in $^{7}_{\Lambda}Li$ hypernucleus. Phys. Rev. C **84**, 064325 (2011)
42. P.B. Siegel, M. Farrow Reid, A square-well potential model to describe lambda-hypernuclei. Am. J. Phys. **58**, 1016–1017 (1990)
43. V. Zagrebaev, W. Greiner, Synthesis of superheavy nuclei: a search for new production reactions. Phys. Rev. C **78**, 034610 (2008)
44. W. Loveland, Synthesis of transactinide nuclei using radioactive beams. Phys. Rev. C **76**, 014612 (2007)
45. J. Zhang, C. Wang, Z. Ren, Calculation of evaporation residue cross sections for the synthesis of superheavy nuclei in hot fusion reactions. Nucl. Phys. A **909**, 36–49 (2013)
46. J.D. Jackson, A schematic model for (p, xn) cross sections in heavy elements. Can. J. Phys. **34**, 767–779 (1956)
47. R. Vandenbosch, J.R. Huizenga, *Nuclear fission* (Academic, New York, 1973), p. 422
48. K.P. Santhosh, C. Nithya, H. Hassanabadi, D.T. Akrawy, α-decay half-lives of superheavy nuclei from a modified generalized liquid-drop model. Phys. Rev. C **98**, 024625 (2018)
49. K.P. Santhosh, C. Nithya, α-decay chains of superheavy nuclei with $Z = 125$. Phys. Rev. C **97**, 044615 (2018)
50. E. Botta, T. Bressani, A. Feliciello, On the binding energy and the charge symmetry breaking in $A \leq 16$ Λ-hypernuclei. Nucl. Phys. A **960**, 165–179 (2017)
51. F. Schulz, P. Achenbach, S. Aulenbacher, J. Beriči, S. Bleser, R. Böhm, D. Bosnar, L. Correa, M.O. Distler, A. Esser, H. Fonvieille, I. Friščić, Y. Fujii, M. Fujita, T. Gogami, H. Kanda, M. Kaneta, S. Kegel, Y. Kohl, W. Kusaka, A. Margaryan, H. Merkel, J. Pochodzalla, M. Mihovilovič, U. Müller, S. Nagao, S.N. Nakamura, J.A. Sanchez Lorente, B.S. Schlimme, M. Schoth, C. Sfienti, S. Širca, M. Steinen, Y. Takahashi, L. Tang, M. Thiel, K. Tsukada, A. Tyukin, A. Weber, Ground-state binding energy of $^{4}_{\Lambda}H$ from high-resolutiondecay-pionspectroscopy. Nucl. Phys. A **954**, 149–160 (2016)
52. H. Bandō, T. Motoba, J. Žofka, Production, structure and decay of hyper nuclei. Int. J. Mod. Phys. A **5**, 4021–4198 (1990)
53. P.H. Pile, S. Bart, R.E. Chrien, D.J. Millener, R.J. Sutter, N. Tsoupas, J.-C. Peng, C.S. Mishra, E.V. Hungerford, T. Kishimoto, L.-G. Tang, W. von Witsch, Z. Xu, K. Maeda, D. Gill, R. McCrady, B. Quinn, J. Seydoux, J.W. Sleight, R.L. Stearns, H. Plendl, A. Rafatian, J. Reidy, Study of hypernuclei by associated production. Phys. Rev. Lett. **66**, 2585 (1991)
54. T. Gogami, C. Chen, D. Kawama, P. Achenbach, A. Ahmidouch, I. Albayrak, D. Androic, A. Asaturyan, R. Asaturyan, O. Ates, P. Baturin, R. Badui, W. Boeglin, J. Bono, E. Brash, P. Carter, A. Chiba, E. Christy, S. Danagoulian, R. De Leo, D. Doi, M. Elaasar, R. Ent, Y. Fujii, M. Fujita, M. Furic, M. Gabrielyan, L. Gan, F. Garibaldi, D. Gaskell, A. Gasparian, Y. Han, O. Hashimoto, T. Horn, B. Hu, E.V. Hungerford, M. Jones, H. Kanda, M. Kaneta, S. Kato, M.

Kawai, H. Khanal, M. Kohl, A. Liyanage, W. Luo, K. Maeda, A. Margaryan, P. Markowitz, T. Maruta, A. Matsumura, V. Maxwell, A. Mkrtchyan, H. Mkrtchyan, S. Nagao, S.N. Nakamura, A. Narayan, C. Neville, G. Niculescu, M.I. Niculescu, A. Nunez, Nuruzzaman, Y. Okayasu, T. Petkovic, J. Pochodzalla, X. Qiu, J. Reinhold, V.M. Rodriguez, C. Samanta, B. Sawatzky, T. Seva, A. Shichijo, V. Tadevosyan, L. Tang, N. Taniya, K. Tsukada, M. Veilleux, W. Vulcan, F.R. Wesselmann, S.A. Wood, T. Yamamoto, L. Ya, Z. Ye, K. Yokota, L. Yuan, S. Zhamkochyan, L. Zhu, High resolution spectroscopic study of $^{10}_{\Lambda}Be$. Phys. Rev. C **93**, 034314 (2016)

55. G.M. Urciuoli, F. Cusanno, S. Marrone, A. Acha, P. Ambrozewicz, K.A. Aniol, P. Baturin, P.Y. Bertin, H. Benaoum, K.I. Blomqvist, W.U. Boeglin, H. Breuer, P. Brindza, P. Bydzovsky, A. Camsonne, C.C. Chang, J.-P. Chen, S. Choi, E.A. Chudakov, E. Cisbani, S. Colilli, L. Coman, B.J. Craver, G. De Cataldo, C.W. de Jager, R. De Leo, A.P. Deur, C. Ferdi, R.J. Feuerbach, E.R. Fratoni, S. Frullani, F. Garibaldi, O. Gayou, F. Giuliani, J. Gomez, M. Gricia, J.O. Flansen, D. Hayes, D.W. Higinbotham, T.K. Higinbotham, C.E. Hyde, H.F. Ibrahim, M. lodice, X. Jiang, L.J. Kaufman, K. Kino, B. Kross, L. Lagamba, J.J. LeRose, R.A. Lindgren, M. Lindgren, D.J. Margaziotis, P.Z. Markowitz, E. Meziani, K. McCormick, R.W. Michaels, D.J. Millener, T. Miyoshi, B. Moffit, P.A. Monaghan, M. Moteabbed, C. Munoz Camacho, S. Nanda, E. Nappi, V.V. Nelyubin, B. Norum, E.Y. Okasyasu, K.D. Paschke, C.F. Perdrisat, E. Piasetzky, V.A. Punjabi, Y. Qiang, P.E. Reimer, J. Reinhold, B. Reitz, R.E. Roche, V.M. Rodriguez, A. Saha, F. Santavenere, A.J. Sarty, J. Segal, A. Shahinyan, J. Singh, S. Sirca, R. Snyder, P. Solvignon, H.M. Sotona, R. Subedi, V.A. Sulkosky, T. Suzuki, H. Ueno, P.E. Ulmer, P. Veneroni, E. Voutier, B.B. Wojtsekhowski, X. Zheng, C. Zorn, Spectroscopy of $^{9}_{\Lambda}Li$ by electroproduction. Phys. Rev. C **91**, 034308 (2015)

Chapter 16
Systematic Study of Po Compound Nuclei Using Evaporation Residue, Fission Cross-Section, and Neutron Multiplicity as a Probe

Ruchi Mahajan

Abstract Statistical model calculations for neutron multiplicity, evaporation residue, and fission cross-section have been performed for $^{18}O + ^{192}Os$ populating ^{210}Po compound nucleus in the excitation energy range 52.43–83.51 MeV. Experimental fusion cross-section has been fitted using CCFULL code. Evaporation residue and fission cross-section are then fitted using Bohr–Wheeler's formalism including shell effects in the level density and fission barrier by using scaling factor in the range 1.0–0.75. The results of the calculations are in good agreement with the experimental data. In this chapter, we are presenting the results of these calculations.

16.1 Introduction

The understanding of the fission dynamics of a nucleus, particularly in the mass region ≈200, continues to be challenging task for the nuclear physicists [1]. In order to have a better insight of the fission dynamics in this mass region, many statistical as well as dynamical approaches have been used. The key ingredients of these approaches include the spin distribution of the compound nucleus (CN), nuclear level density parameters, shell effects in fission barrier, and the potential energy surface. In general, the fission barrier used in these models has a macroscopic (liquid drop) part and a microscopic (shell correction) part. The nuclear level density parameter is also sensitive to the shell correction and deformation. Even though significant progress has been made in the understanding of the fission process, there are ambiguities in choosing the parameters of the theoretical models [2–5]. At low energies, CN decays predominantly by emission of particles and fission. Experimental observations clearly show that fusion cross-section is significantly reduced in medium mass region even for very asymmetric systems due to onset of non-compound nuclear processes like quasi fission (QF) [6, 7]. Evaporation residues (ERs) are the pure signatures of CN formation and can become a useful probe to study the statistical as well as dynamical aspects of the fusion–fission reactions.

R. Mahajan (✉)
Department of Physics, Panjab University, Chandigarh 160014, India
e-mail: ruchimahajan4@gmail.com

R. K. Puri et al. (eds.), *Advances in Nuclear Physics*, Springer Proceedings in Physics 257, https://doi.org/10.1007/978-981-15-9062-7_16

Sagaidak et al. [7] have analyzed the ER and fission excitation functions data obtained in complete fusion reactions leading to Po compound nuclei in the framework of the standard statistical model. They have observed a drop in the Po fission barriers with the decrease in a neutron number, which is supported by the presence of entrance channel effects and collective excitations in the CN decay. Mahata et al. [8] have also suggested the lowering the fission barrier (saddle point) from the statistical model (SM) analysis of the prefission neutron multiplicity data for ^{210}Po in excitation energy range of ≈40–60 MeV.

It is now well established that the fission process is somewhat delayed with respect to the statistical picture of CN decay due to the presence of dissipation. It affects many experimental observables like pre-scission particle multiplicities, fission probability, and mass-energy distribution of fission fragments. Neutron emission that acts as a clock to measure the fission time scale has proved to be very useful in investigating the mechanism of nuclear fission [9]. This is because neutrons are emitted in succession from a hot CN till it fissions and thus the pre-scission neutron multiplicity becomes a measure of the time scale of fission. Therefore, in order to address these problems in the actinide region, a detailed study of the decay products of the CN, such as ERs, CN fission fragments are necessary. In this regard, ER, fission cross-section and neutron multiplicity measurements can be a useful probe to understand the fusion–fission dynamics. In view of this, we have extended our theoretical investigation to understand the fission dynamics in actinide region by considering different isotopes of Po [10, 11].

With these motivations, we have performed SM calculations for ^{210}Po CN populated through ^{16}O +^{192}Os in the excitation energy range 52.43–83.51 MeV for which experimental data on ER, fission cross-section, and neutron multiplicity is already available in the literature [9]. In this chapter, results of the SM calculations have been presented.

This first section of this chapter deals with various ingredients used in the SM and the results of the analysis obtained for ER and fission cross-section. The results of neutron multiplicity for ^{210}Po CN are given in Sect. 16.2. The last section contains the interpretation of the results obtained from the SM analysis.

16.2 Statistical Model Analysis of Evaporation Residue and Fission Cross-Section

Sagaidak et al. [7] have also performed the similar systematic study of Po compound nuclei utilizing the standard approach, in which CN production (fusion) and de-excitations are considered independently. The barrier-passing (BP) model takes care of the fusion part, whereas the CN de-excitation is treated within the standard statistical approach (SSM). Both of them are implemented within the HIVAP code [12]. For the present work, theoretical calculations were performed using Bohr–Wheeler formalism including shell-corrections in the level density and fission barrier in order to fit the experimental data for ER and fission cross-sections. The experimental fusion cross-section has been reproduced in good agreement by using coupled channel calculations (CCFULL) [13]. Then, in order to fit the experimental data, different scaling factors (K_f) for the finite-range liquid drop model fission barrier in the range 1.0–0.75 were used.

In the framework of SM, the possible decay channels of a CN are the emission of neutrons, protons, α, and giant dipole resonance (GDR) γ-rays [14]. The main assumption of the SM is that the system forms a fully equilibrated CN after capture of projectile and contribution from non-compound nuclear processes such as QF and fast-fission are negligible. The Bohr–Wheeler fission width used in the present calculations is given by [15]:

$$\Gamma_{BW} = \frac{1}{2\pi\rho(E^*)} \int_0^{E^*-V_B} d\epsilon \rho^*(E^* - V_B - \epsilon) \tag{16.1}$$

Here, V_B is the fission barrier and the nuclear potential is obtained from the finite-range liquid drop model (FRLDM). The level density parameter used in the present work is taken from the work of Ignatyuk et al. [16], which takes into account the nuclear shell structure at low excitation energies and goes over to its asymptotic form at high excitation energies as given below

$$a(E^*) = a\left(1 + \frac{f(E^*)}{E^*}\delta M\right) \tag{16.2}$$

where

$$f(E^*) = 1 - \exp^{\frac{-E^*}{E_D}} \tag{16.3}$$

Here, a is the asymptotic level density parameter and E_D determines the rate at which the shell effects disappear at high excitation energy and δM is the shell correction in the LDM masses, i.e.,

$$\delta M = M_{experimental} - M_{LDM} \tag{16.4}$$

A value of 18.5 MeV was used for E_D, which was obtained from an analysis of s-wave neutron resonances [17]. The shell-corrected temperature-dependent fission barrier is given by

$$V_B(T) = K_f V_{LDM} - \delta M \exp^{\frac{-E^*}{E_D}} \tag{16.5}$$

where K_f is the scaling factor [7], V_{LDM} is the fission barrier from the finite-range rotating LDM potential, and E^* is the CN excitation energy. In our analysis, ER and fission cross-sections are fitted with the adjustment of scaling factor K_f in the fission barrier. In this work, shell correction is applied only to the ground state mass, and it is assumed that the shell correction at the saddle deformation can be neglected [18–20]. The above assumption of neglecting the shell correction at the saddle deformation follows from the work of Myers and Swiatecki [18]. A particular decay channel is selected by performing Monte Carlo sampling between all the particles and γ emission widths.

16.2.1 Spin Distribution from CCFULL

The spin distribution of CN is an important component of the SM and for the present work and it is obtained from the coupled channel calculations code CCFULL [21]. Fusion of two separate nuclei to form a composite system, for low incident energy (near or sub-barrier) and light system, is predominantly governed by quantum tunneling through the coulomb barrier. Extensive experimental and theoretical studies have revealed that fusion reactions at energies near and below the coulomb barrier are strongly influenced by the couplings to the relative motion of the colliding nuclei to several nuclear intrinsic motions [13]. For heavy-ion fusion reactions, to a good approximation, the angular momentum of the relative motion in each channel can be replaced by the total angular momentum J [22, 23]. The program CCFULL solves the coupled-channels equations to compute the fusion cross-sections and mean angular momenta of CN, taking into account the couplings to all orders. The coupled channel equations then can be given as

$$\left[\frac{\hbar}{2\mu}\frac{d^2}{dr^2} + \frac{J(J+1)\hbar^2}{2\mu^2} + V_N^{(o)}(r) + \frac{Z_P Z_T e^2}{r} + \epsilon_n - E\right]\phi_n(r) + \sum_m V_{nm}(r)\phi_m(r) = 0 \tag{16.6}$$

where r is the radial component of the coordinate of the relative motion and μ is the reduced mass, E is the bombarding energy in center of mass frame and ϵ_n is the excitation energy of the nth channel. V_{nm} are the matrix elements of the coupling Hamiltonian, which in the collective model consists of coulomb and nuclear components. V_N^o is the nuclear potential in the entrance channel. The coupled channel equations are solved exactly by imposing the boundary conditions. The solution of the coupled channel equations with the proper boundary conditions is given by linear combination of χ_{nm} as

$$\phi_m(r) = \sum_n T_n \chi_{nm}(r) \tag{16.7}$$

$$\chi_{nm}(r) = C_{nm} H_J^{(-)}(k_m r) + D_{mn} H_J^{(+)}(k_m r),\ r \rightarrow r_{max} \tag{16.8}$$

where C_{nm} and D_{nm} are determined either by matching the logarithmic derivatives at r_{max}. The fusion cross-sections and mean angular momentum of CN are given by

$$\sigma_{fus}(E) = \sum_l \sigma_l(E) = \frac{\pi}{k_o^2} \sum_l (2l+1) P_l(E) \tag{16.9}$$

$$< l >= \sum_l l\sigma_l(E) / \sum_l \sigma_l(E) \tag{16.10}$$

where P_l(E) is penetrability.

It works on the ingoing-wave boundary condition inside the Coulomb barrier to account for fusion, along with the isocentrifugal approximation, which works well for heavy ions. The nuclear potential in the entrance channel is defined by parameters V_0, r_0, and a_0; where V_0 is the depth parameter of the Woods–Saxon potential, r_0 is the radius parameter, and a_0 is the surface diffuseness parameter.

Depending upon the value of E(4^+)/E (2^+), nuclei can be classified as vibrator or rotor. If this ratio is 3.3, the nucleus is treated as rotor and vibrator if this value is 2. In case of $^{18}O + ^{192}Os$ system, the projectile ^{18}O is treated as a vibrator and target ^{192}Os is treated as a rotor. The deformation parameters along with the value of E(4^+)/E (2^+) are given in Table 16.1.

The potential parameter used in the present Coupled Channel calculations was chosen by fitting the experimental capture cross-section and is shown in Fig. 16.1. The fitted values are V_0, r_0, and a_0 are given in Table 16.2.

From Fig. 16.1, it is clear that the energy points above the Coulomb barrier are fitted without including the coupling effects, whereas energy points well below the

Table 16.1 Values of β_2, β_4, and E(4^+)/E (2^+)

Nucleus	β_2	β_4	E(4^+)/E (2^+)
^{18}O	0.355	–	1.8
^{192}Os	0.167	−0.081	2.82

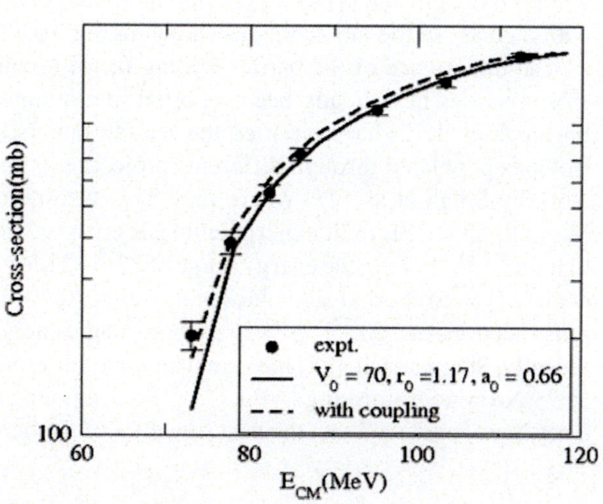

Fig. 16.1 Experimental capture cross-section (full dots) for $^{18}O + ^{192}Os$ as a function of E_{CM} (center of mass energy). Dashed line shows coupling and solid line shows no coupling

Table 16.2 Fitting parameters from CCFULL code

Nucleus	V_0	r_0	a_0
^{210}Po	70 MeV	1.17 fm	0.66 fm

coulomb barrier are fitted taking into consideration the coupling effects. This is because coupling among the intrinsic degrees becomes more dominant at energies near and close to the coulomb barrier.

After fitting, CCFULL gives the spin distribution (for capture cross-sections) as an output file and this file has been used as an input for the spin distribution of compound nuclei for SM code to fit the experimental ER and fission cross-section data. Then, in order to fit the experimental data for ER and fission cross-section, final theoretical calculations were performed using Bohr–Wheeler formalism including shell correction in the level density and fission barrier. For reproducing the data, different scaling factors (K_f) in the range 1.0–0.75 has been used. The ER and fission cross-sections calculated from SM using Bohr–Wheeler formalism are enumerated in Table 16.3. The fitted fission and ER cross-sections are shown in Fig. 16.2.

From Fig. 16.2, it becomes clear that scaling factor has to be reduced from $K_f =$ 1.0 to 0.75 to describe the excitation function in the whole range of CN excitation energy. Since scaling factor is directly related to fission barrier so, decreasing the scaling factor means fission barrier has to be reduced. It was observed that the SM results using Bohr–Wheeler approach over predicts the ER cross-section especially at high excitation energies and under predicts the fission cross-section throughout the entire energy range under study. Also, at lower energies, fission cross-section is a very small fraction of total fusion cross-section. Hence, to fit the ER cross-section in the desired range, we have to increase the fission cross-section, and this is done by reducing the fission barrier. In other words, we can say that for reducing the fission barrier we have to reduce the scaling factor (K_f) to fit the ER and fission cross-section data. For the present system, we have found that a scaling factor increases with increase in the lab energy as shown in Fig. 16.3.

The importance of the barrier scaling factor required to fit the experimental ER cross-sections has already been reported in a number of earlier works [7, 24–27]. Sagaidak et al. [7] have reported the scaling factors less than 1 for a number of Po isotopes populated through different (projectile + target) combinations. In another work by Singh et al. [25] where they have also reported the scaling factor values ($K_f = 0.65 - 1.0$) to fit experimental ER cross-sections for ^{19}F + ^{194,196,198}Pt populating ^{213,215,217}Fr in the energy range of 82–122 MeV. In a similar work, by Mohanto et al. [24] also reported K_f values in the range (0.70–1.10) to fit the experimental ER cross-sections of ^{30}Si,^{31}P + ^{170}Er systems in the energy range of 110–150 MeV. Very recently, Sharma et al. [28] have analyzed the ER cross-section data for ^{48}Ti+^{144}Sm, ^{142,150}Nd systems forming ^{192}Po, ^{190,198}Pb compound nuclei in the framework of SM. They have concluded that the best fit values of the barrier scaling factor K_f obtained

Table 16.3 ER and fission cross-sections calculated using SM for $^{18}O + ^{192}Os$ as a function of E_{Lab} (MeV). σ_{ER} (mb) and $\sigma_{fission}$ (mb) are the ER and fission cross-sections calculated using Bohr–Wheeler fission width, respectively

			$K_f = 1.0$		$K_f = 0.95$		$K_f = 0.90$		$K_f = 0.80$		$K_f = 0.75$	
E_{Lab} (MeV)	σ_{ER} (exp.)	$\sigma_{fission}$ (exp.)	σ_{ER}	$\sigma_{fission}$	σ_{ER}	$\sigma_{fission}$	σ_{ER}	$\sigma_{fission}$	σ_{ER}	$\sigma_{fission}$	σ_{ER}	$\sigma_{fission}$
80	195±20	6.3±0.6	220.2	1.4	219.5	2.1	218.5	3.1	215.0	6.6	212.2	9.3
85	364±30	24.4±1.2	404.4	5.8	401.4	8.8	398.7	11.5	388.8	21.4	381.3	28.9
90	487±30	66±3	550.5	18.0	545.5	23.1	539.5	29.0	518.5	50.1	501.1	67.4
95	594±30	131±8	686.6	39.9	676.0	50.5	660.2	66.4	624.0	102.6	543.6	132.9
104	660±30	335±8	799.6	194.9	767.3	227.3	734.8	259.8	635.1	359.4	578.6	415.9
113	630±30	582±17	729.9	502.4	690.2	542.1	644.9	587.4	535.6	696.8	474.8	757.5
123	570±30	885±25	606.4	833.3	560.2	879.5	523.2	916.5	419.1	1020.6	364.4	1075.3

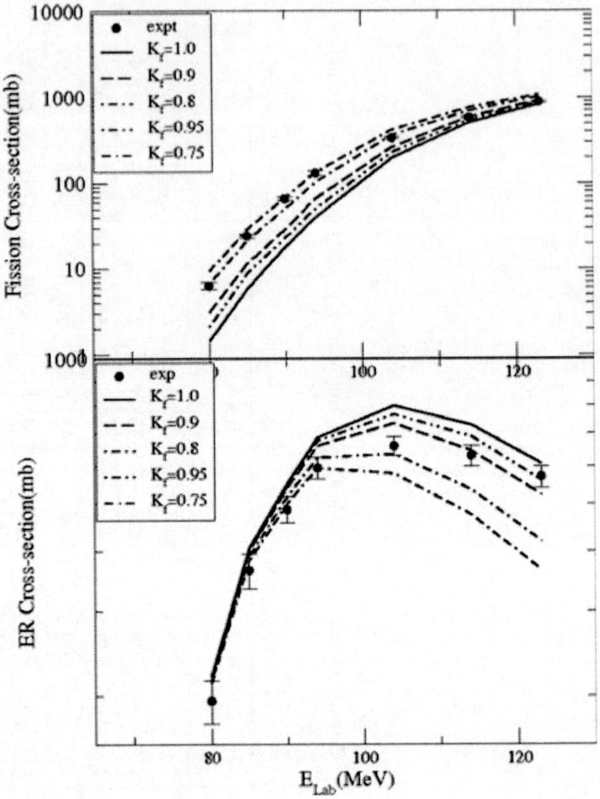

Fig. 16.2 Solid circles are the experimental data and different lines are the theoretical calculations for different scaling factor, K_f (as given inside the diagram): **a** For Fission cross-section **b** For ER cross-section

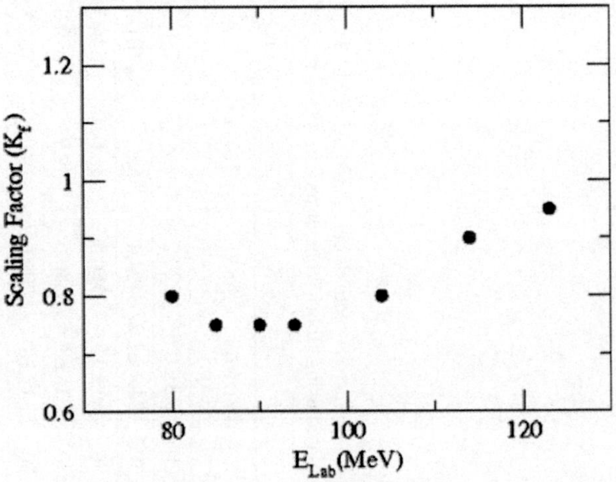

Fig. 16.3 Variation of scaling factor (K_f) with E_{Lab} (MeV)

from the SM calculations are in the range (0.75–0.85) for the CN ^{190}Pb, whereas the K_f value for ^{198}Pb and ^{192}Po are (0.65–0.70) and (1.0–1.25), respectively.

The results discussed in this section strongly indicate the significance of using excitation energy-dependent scaling of the finite-range rotating liquid drop model fission barrier in order to fit the experimental data for ER and fission cross-section.

16.3 Statistical Model Analysis of Neutron Multiplicity

The dissipation effect in fusion–fission process is well established. The excess in multiplicities with respect to SM predictions indicates the presence of a dynamic hindrance to fission. Recently, Singh et al. [29] have measured the pre-scission neutron multiplicities for ^{19}F + ^{194,196,198}Pt populating ^{213,215,217}Fr in the excitation energy range of 46.6–91.8 MeV and discussed the effect of shell effect on nuclear dissipation. In another set of measurements by Sandal et al. [30] also support the fission hindrance due to nuclear dissipation.

In the present case, we have performed SM calculations for ^{210}Po populated by ^{18}O +^{192}Os in the excitation energy range of 65.23–91.74 MeV. Experimental data on pre-scission neutron multiplicity is available in the literature for ^{210}Po [9]. In this section, we are comparing the results obtained from SM calculations with the experimental results to see the isotopic and excitation energy dependence of the dissipation strength. Here, SM calculations have been performed using Bohr–Wheeler and Kramer's formalism for ^{210}Po by including and excluding shell effects in level density parameter and the fission barrier.

The fission width where effect of dissipation that is included is given by Kramers [31]. The other ingredients used in the present calculations like Bohr–Wheeler fission width, level density parameter, and fission barrier have been discussed in Sect. 16.2 of this chapter.

Upper panel of Fig. 16.4 shows the results of calculations for pre-scission neutron multiplicity with no shell effects in level density parameter and barrier height for ^{210}Po and lower panel shows the results with inclusion of shell effects.

The comparison of calculated pre-scission neutron multiplicities with the experimental values clearly shows that the predictions using Bohr–Wheeler fission width considerably underestimate the pre-scission neutron multiplicity at all the energies. It is also observed that the experimental values at all the energies cannot be reproduced by a single value of β. However, the calculations which take into consideration Kramers' fission width are in well agreement with the experimental data.

Figure 16.5 shows the lab energy dependence of best fit β value for ^{210}Po where the shaded areas represent the uncertainty in β with the experimental error in M_{pre}. We first note in this figure that the inclusion of shell effects in the calculation results in higher values for best fit β.

Inclusion of shell effects affects both the particle/γ decay widths (due to change in level density parameter) and also the fission width (due to change in both fission barrier and level density parameter). Consequently, the relative strength of the fission

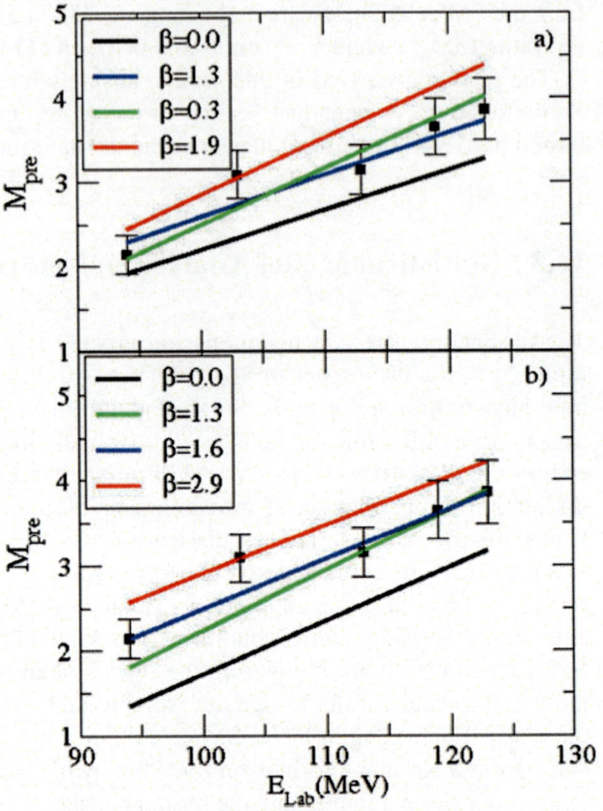

Fig. 16.4 Experimental pre-scission neutron multiplicities (filled squares) along with SM calculation results for ^{210}Po with **a** exclusion of shell effects **b** inclusion of shell effects

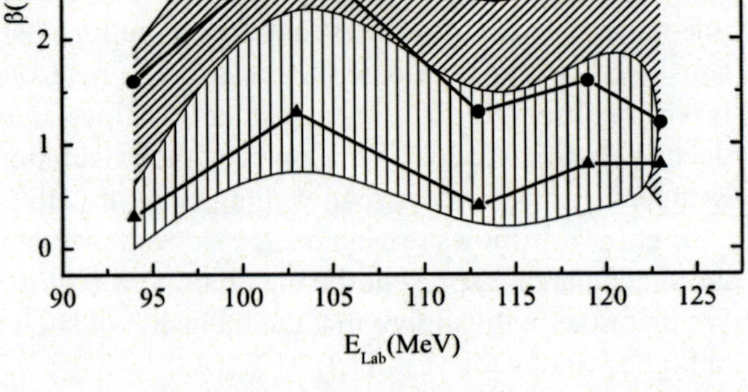

Fig. 16.5 The best fit value of β obtained for ^{210}Po

width with respect to the other decay channels changes which results in a higher value for β to fit the experimental data. It is further interesting to note in this figure that the β values for the above two cases tend to merge to a common value at higher excitation energies which is a consequence of reduced shell effects at higher excitations.

The comparison of the above results with ^{48}Ti+^{144,154}Sm system [10] clearly indicates that the strength of nuclear dissipation required to fit the neutron multiplicity data is very high as compared to results for ^{18}O+^{192}Os. These calculations clearly suggest the role of entrance channel dynamics for the systems considered here.

Lastly, we observe that the best fit β values at high excitation energies for ^{210}Po are very close for both the cases. This indicates that shell effects on β, if any, are small at high excitation energies.

16.4 Conclusion

SM calculations have been performed for ^{210}Po CN populated through ^{16}O +^{192}Os in the excitation energy range 52.43–83.51 MeV for which experimental data on ER, fission cross-section, and neutron multiplicity is already available in the literature [9]. These calculations have been performed by using Bohr–Wheeler and Kramers' formalism by including and excluding shell effects in the level density and the fission barrier. The results of the ER and fission cross-section obtained from SM calculations strongly indicate the need of using excitation energy-dependent scaling factor of the finite-range rotating liquid drop model fission barrier in order to fit the experimental data. The comparison of the pre-scission neutron multiplicities obtained from the SM calculations with the experimental values clearly shows that the Bohr–Wheeler fission width considerably underestimates the pre-scission neutron multiplicity at all the energies. However, the calculations which take into consideration Kramers' fission width are in well agreement with the experimental data. The comparison of the SM results presented in this chapter with the results obtained for ^{48}Ti+^{144,154}Sm system clearly indicate the role of entrance channel dynamics for the systems [10] considered in the present work.

References

1. K. Mahata, S. Kailas, Reexamination of fission in the A - 200 mass region with excitation energy near 50 MeV. Phys. Rev. C **95**, 054616 (2017)
2. C. Schmitt, K. Mazurek, P. Nadtochy, Description of isotopic fission-fragment distributions within the Langevin approach. Phys. Rev. C **91**, 041603(R) (2015)
3. K. Mahata, S. Kailas, S.S. Kapoor, Shell corrections at the saddle point for mass region 200. Phys. Rev. C **74**, 041301(R) (2006)
4. K. Mahata, S. Kailas, A. Shrivastava, A. Chatterjee, A. Navin, P. Singh, S. Santra, B. Tomar, Fusion of ^{19}F with ^{188,192}Os. Nucl. Phys. A **720**, 209 (2003)

5. S.E. Vigdor, D.G. Kovar, P. Sperr, J. Mahoney, A. Menchaca-Rocha, C. Olmer, M.S. Zisman, Energy dependence of the fusion and elastic scattering of $^{16}O + ^{40}Ca$. Phys. Rev. C **20**, 2147 (1979)
6. A.C. Berriman, D.J. Hinde, M. Dasgupta, C.R. Morton, R.D. Butt, J.O. Newton, Unexpected inhibition of fusion in nucleus-nucleus collisions. Nature (London) **413**, 144 (2001)
7. R.N. Sagaidak, A.N. Andreyev, Fission barriers for Po nuclei produced in complete fusion reactions with heavy ions. Phys. Rev. C **79**, 054613 (2009)
8. K. Mahata, S. Kailas, S.S. Kapoor, Fission barrier, damping of shell correction, and neutron emission in the fission of A 200. Phys. Rev. C **92**, 034602 (2015)
9. D.J. Hinde, R.J. Charity, G.S. Foote, J.R. Leigh, J.O. Newton, S. Ogazac, A. Chatterjee, Neutron multiplicities in heavy-ion induced fission: timescale of fusion-fission. Nucl. Phys. A **452**, 550–572 (1986)
10. R. Mahajan, B.R. Behera, M. Thakur, G. Kaur, P. Sharma, K. Kapoor, A. Kumar, P. Sugathan, A. Jhingan, A. Chatterjee, N. Saneesh, A. Yadav, R. Dubey, N. Kumar, H. Singh, A. Saxena, S. Pal, Systematic study of $^{192,202.206,210}Po$ compound nuclei using neutron multiplicity as a probe. Phys. Rev. C **98**, 034601 (2018)
11. G.G. Chubaryan, M.G. Itkis, S.M. Luk'yanov, V.N. Okolovich, Y. E. Penionzhkevich, A.Y. Rusanov, V.S. Salamatin, G.N. Smirenkin, The fission of heated nuclei in reactions involving heavy ions: static and dynamical aspects. Yad. Fiz. **56**, 3; [Phys. At. Nucl. **56**, 286] (1993)
12. W. Reisdorf, M. Schdel, How well do we understand the synthesis of heavy elements by heavy-ion induced fusion. Z. Phys. A **343**, 47 (1992)
13. K. Hagino, N. Rowley, A.T. Kruppa, A program for coupled-channel calculations with all order couplings for heavy-ion fusion reactions. Comput. Phys. Commun. **123**, 143 (1999)
14. P. Frobrich, I.I. Gontchar, Langevin description of fusion, deep-inelastic collisions and heavy-ion-induced fission. Phys. Rep. **292**, 132 (1998)
15. N. Bohr, J.A. Wheeler, The mechanism of nuclear fission. Phys. Rev. **56**, 426 (1939)
16. A.V. Ignatyuk, G.M. Smirenkin, A. Tishin, Shell effects in the symmetric-modal fission of pre-actinide nuclei. Sov. J. Nucl. Phys. **21**, 255 (1975)
17. W. Reisdorf, Analysis of fissionability data at high excitation energies. Z. Phys. A **300**, 227 (1986)
18. D. Myers, W.J. Swiatecki, A model of nuclear rotation. Nucl. Phys. A **601**, 141 (1996)
19. A.V. Ignatyuk, M.G. Itkis, S.I. Mulgin, Modeling fission in the cascade-exciton model. Fiz. Elem.Chastits At. Yadra **16**, 709. [Sov. J. Part Nucleus **16**, 307] (1985)
20. V.Y. Denisov, S. Hofmann, Formation of superheavy elements in cold fusion reactions. Phys. Rev. C **61**, 034606 (2000)
21. P. Grange, S. Hassani, H.A. Weidenmuller, A. Gavron, J.R. Nix, A.J. Sierk, Effect of nuclear dissipation on neutron emission prior to fission. Phys. Rev. C **34**, 209 (1986)
22. K. Hagino, N. Takigawa, A.B. Balantekin, J.R. Bennett, Path integral approach to no-Coriolis approximation in heavy-ion collisions. Phys. Rev. C **52**, 286 (1985)
23. R. Lindsay, N. Rowley, Noncollective excitations in low-energy heavy-ion reactions. J. Phys. G **10**, 805 (1994)
24. G. Mohanto, N. Madhavan, S. Nath, J. Gehlot, I. Mukul, A. Jhingan, T. Varughese, A. Roy, R.K. Bhowmik, I. Mazumdar, D.A. Gothe, P.B. Chavan, J. Sadhukhan, S. Pal, M. Kaur, V. Singh, A.K. Sinha, V.S. Ramamurthy, Evaporation residue excitation function and spin distribution for $^{31}P + ^{170}Er$. Phys. Rev. C **88**, 034606 (2013)
25. V. Singh, B.R. Behera, M. Kaur, A. Kumar, K.P. Singh, N. Madhavan, S. Nath, J. Gehlot, G. Mohanto, A. Jhingan, I. Mukul, T. Varughese, J. Sadhukhan, S. Pal, S. Goyal, A. Saxena, S. Santra, S. Kailas, Measurement of evaporation residue excitation functions for the $^{19}F + ^{194,196,198}Pt$ reactions. Phys. Rev. C **89**, 024609 (2014)
26. R. du Rietz, E. Williams, D.J. Hinde, M. Dasgupta, M. Evers, C.J. Lin, D.H. Luong, C. Simenel, A. Wakhle, Mapping quasifission characteristics and timescales in heavy element formation reactions. Phys. Rev. C **88**, 054618 (2013)
27. R. Sandal, B.R. Behera, V. Singh, M. Kaur, A. Kumar, G. Kaur, P. Sharma, N. Madhavan, S. Nath, J. Gehlot, A. Jhingan, K.S. Golda, H. Singh, S. Mandal, S. Verma, E. Prasad, K.M.

Varier, A.M. Vinod kumar, A. Saxena, J. Sadhukhan, S. Pal, Probing nuclear dissipation via evaporation residue excitation functions for the $^{16,18}O + ^{194,198}Pt$ reactions. Phys. Rev. C **91**, 044621 (2015)

28. P. Sharma, B.R. Behera, R. Mahajan, M. Thakur, G. Kaur, K. Kapoor, K. Rani, N. Madhavan, S. Nath, J. Gehlot, R. Dubey, I. Mazumdar, S.M. Patel, M. Dhibar, M.M. Hosamani, Khushboo, N. Kumar, A. Shamlath, G. Mohanto, S. Pal, Evaporation residue cross-section measurements for ^{48}Ti-induced reactions. Phys. Rev. C **96**, 034613 (2017)
29. V. Singh, B.R. Behera, M. Kaur, A. Kumar, P. Sugathan, K.S. Golda, A. Jhingan, M.B. Chatterjee, R.K. Bhowmik, D. Siwal, S. Goyal, J. Sadhukhan, S. Pal, A. Saxena, S. Santra, S. Kailas, Neutron multiplicity measurements for $^{19}F + ^{194,196,198}Pt$ systems to investigate the effect of shell closure on nuclear dissipation. Phys. Rev. C **87**, 064601 (2013)
30. R. Sandal, B.R. Behera, V. Singh, M. Kaur, A. Kumar, G. Singh, K.P. Singh, P. Sugathan, A. Jhingan, K.S. Golda, M.B. Chatterjee, R.K. Bhowmik, S. Kalkal, D. Siwal, S. Goyal, S. Mandal, E. Prasad, K. Mahata, A. Saxena, J. Sadhukhan, S. Pal, Effect of N/Z in pre-scission neutron multiplicity for $^{16,18}O + ^{194,198}Pt$ systems. Phys. Rev. C **87**, 014604 (2013)
31. H.A. Kramers, Brownian motion in a field of force and the diffusion model of chemical reactions. Physica (Amsterdam) **7**, 284 (1940)

Chapter 17
Momentum and Density Dependence of the Nuclear Mean Field Using Finite Range Simple Effective Interaction: A Tool for Heavy-Ion Collision Dynamics

T. R. Routray, X. Viñas, and B. Behera

Abstract The finite range Simple Effective Interaction (SEI) has been constructed with minimum number of parameters that can simulate the correct trend of momentum dependence of nuclear mean field, as ascertained from the heavy-ion collision experiments at intermediate energy range. Nine parameter combinations of the total eleven interaction parameters required for the complete study of symmetric and isospin asymmetric nuclear matter have been carefully determined, which gives momentum dependence of the nucleonic mean field and equation of state similar in trend to the predictions of microscopic calculations. In the determination of the parameters, one needs to assume the standard values of only three nuclear matter saturation properties, namely, the density, energy per particle, and symmetry energy at saturation in normal nuclear matter. The present formulation has the advantage of varying the density dependence of nuclear symmetry energy where the momentum dependence of the nuclear mean field remains unchanged, and the vice-versa. This can provide quality inputs to the transport model equations solved in the analysis of data of heavy-ion collision experiments in the intermediate energy range. The SEI has also the ability of predicting the finite nucleus properties similar in quality as that of any of the existing effective interaction in the field.

T. R. Routray (✉) · B. Behera
School of Physics, Sambalpur University, Jyoti-Vihar, Burla 768019, Odisha, India
e-mail: trr1@rediffmail.com

X. Viñas
Department de Física Quàntica i Astrofísica and Institut de Ciències del Cosmos (ICCUB), Facultat de Física, Universitat de Barcelona, Martí i Franquès 1, 08028 Barcelona, Spain

R. K. Puri et al. (eds.), *Advances in Nuclear Physics*, Springer Proceedings in Physics 257, https://doi.org/10.1007/978-981-15-9062-7_17

17.1 Introduction

In the heavy-ion (HI) collision at intermediate and high energy, the flow and particle production are modeled by transport model equations using hydrodynamics. The Boltzmann–Uehling–Uhlenbeck (BUU) equation is a popular transport equation [1] developed that includes the stochastic collisions between individual nucleons, particle production, effects of the Pauli principle, and propagation between collisions controlled by a mean field. Here, the mean field otherwise termed as the single-particle potential is the key ingredient and in the early stage of development of the subject the calculations were done using the simple version of the mean field,

$$U(\rho) = a\frac{\rho}{\rho_0} + b\left(\frac{\rho}{\rho_0}\right)^{\gamma}. \tag{17.1}$$

The experimental data on pion production [2] and collective flow [3] could be reproduced with this form of the mean field in (17.1) for the value of γ that predicts a high value of nuclear matter (NM) incompressibility, $K(\rho_0) \sim 380\,\text{MeV}$, is in disagreement with the microscopic calculations based on realistic interactions that reproduces the nucleon–nucleon (NN) scattering data [4]. In the attempt to resolve this anomaly, different groups [5, 6] have examined the impact of considering momentum dependence in the mean field. It was found that the momentum dependent mean field could reproduce the flow data of HI collision for which the equation of state (EOS) is relatively softer having $K(\rho_0) \sim 215\,\text{MeV}$, in contrast to the requirement of a very stiff EOS in case of momentum independent mean field of type given in (17.1). This finding demonstrated the crucial importance of momentum dependent aspect of the nuclear mean field that was usually not given emphasis in earlier nuclear calculations.

The momentum **k** (we shall use the notations **k** P in the unit of $\hbar$) dependent mean field used in the analysis of transverse momentum flow data is proposed by Gale, Bertsch and Das Gupta (GBD) [5] has the form,

$$U_{GBD}(\rho, \vec{k}) = a\left(\frac{\rho}{\rho_0}\right) + b\left(\frac{\rho}{\rho_0}\right)^{\gamma} + \frac{c}{\rho_0}\int d^3k \frac{f(\vec{r}, \vec{k})}{1 + \left(\frac{\vec{k} - <\vec{k}>}{\Lambda}\right)^2}$$
$$+ \frac{c}{\rho_0}\frac{\rho}{1 + \left(\frac{\vec{k} - <\vec{k}>}{\Lambda}\right)^2}, \tag{17.2}$$

where $\vec{k}$ is the momentum of the particle, $< k >$ is the local momentum average, $f(\vec{r}, \vec{k})$ is the phase space occupation density, Λ is the momentum scale, and a, b, c and γ are parameters. Soon after the success of reproducing the flow data with this momentum dependent mean field (17.2), an improved version was proposed with the nomenclature momentum dependent Yukawa interaction (MDYI) [7] given as

$$U_{MDYI}(\rho, \vec{k}) = a\left(\frac{\rho}{\rho_0}\right) + b\left(\frac{\rho}{\rho_0}\right)^{\gamma} + 2\frac{c}{\rho_0}\int d^3k' \frac{f(\vec{r}, \vec{k'})}{1 + (\frac{\vec{k}-\vec{k'}}{\Lambda})^2}. \quad (17.3)$$

The origin of the nomenclature comes from the fact that the exchange part of the mean field for a Yukawa form of interaction would result into the momentum dependent expression of MDYI.

This momentum dependent mean field was used for transport model calculations for a pretty long duration [5, 7–13] with the focus on the stiffness of symmetric nuclear matter (SNM), i.e., nuclear matter incompressibility $K(\rho_0)$. Apart from the dependence on $K(\rho_0)$ the flow data also exhibited sensitiveness to the N/Z ratio of the two colliding nuclei. It was realized that the plasma matter formed at the instant of collision and its subsequent dynamical evolution should be governed by the nucleonic mean fields in isospin asymmetric matter (ANM). The mean field experienced by a neutron in isospin asymmetric medium is different from that of a proton, and modifications in the BUU transport theory [14] were initiated to take the isospin effect into account side-by-side with the formulations for nucleonic mean fields [15–17] to be used in the modified BUU theory. Here, now, arises two questions to be answered, (a) the density dependence aspect and (b) the momentum dependent aspect of the difference in the neutron and proton mean fields in isospin ANM. The former, as we shall see in the next section, is connected to the nuclear symmetry energy (NSE), whereas the latter to the neutron-proton effective mass splitting.

The iso-scaling study [18–21] where one scales nuclei of different masses formed during the expansion of nucleonic plasma created at the instant of collision between two nuclear system at intermediate/high energy reveals the crucial relevance of nuclear symmetry, $E_s(\rho)$. The energy per particle in ANM, $e(\rho, \beta)$ is popularly approximated as

$$e(\rho, \beta) = e(\rho) + \beta^2 E_s(\rho), \quad (17.4)$$

$$\text{where } \beta = \frac{\rho_n - \rho_p}{\rho_n + \rho_p} = (1 - 2Y_p),$$

$Y_p = \frac{\rho_p}{\rho}$ being the proton fraction and $e(\rho)$ is the energy per particle in SNM. The density dependence of $e(\rho)$ determines the EOS in SNM. This together with the density dependence of $E_s(\rho)$ determines the EOS of ANM. The density dependence of the NSE, $E_s(\rho)$, has remained elusive in the supra saturation region although some constraints are proposed [22–24] to pin down the behavior of $E_s(\rho)$ in the sub-saturation density region.

There have also been attempts from the analysis of the observables in HI collision experiments [25–27] to ascertain the influence of the difference in the k-dependence of neutron (n) and proton (p) mean fields in the isospin asymmetric medium. Although the signature of the influence has been found, quantitative measurements of the influence could not be met with due to several other factors. From the analysis of N-nucleus scattering study, the data for the possible differences in the k-dependence of the n- and p-mean fields at normal density $\rho = \rho_0$ are available up to energy

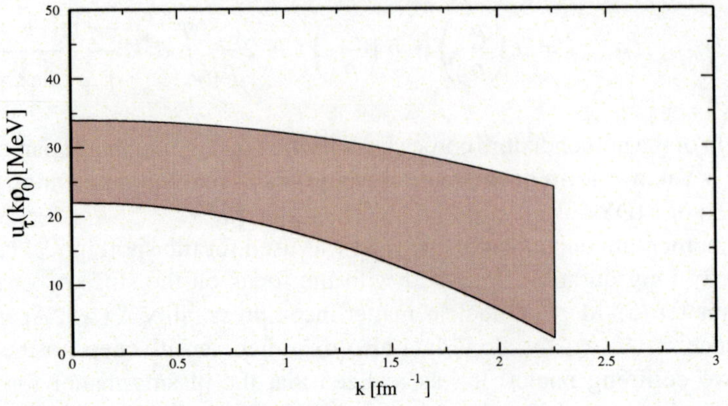

Fig. 17.1 Experimentally extracted [29] results of the iso-vector part of the mean field, $u_\tau(k, \rho_0)$, as a function of momentum (k) in normal nuclear matter

100 MeV and that too with large uncertainty [28–31]. This is shown as the shaded region in Fig. 17.1 and is known as Lane potential. The Lane potential is defined in terms of the iso-vector part of the mean field which is the difference of the n- and p-mean fields averaged over the isospin asymmetry, β.

The quantity that characterizes the momentum dependence of the nucleonic mean field is the effective mass defined as

$$\left(\frac{m*}{m}\right)_{n(p)} = \left[1 + \frac{m}{\hbar^2 k}\frac{\partial u_{n(p)}(k, \rho, Y_p)}{\partial k}\right]^{-1}, \qquad (17.5)$$

where u_n and u_p are the momentum dependent n- and p-mean fields, respectively, in isospin asymmetric nuclear medium. In the limit of SNM, both n- and p-effective masses are the same under the isospin invariance of NN force.

So it is pertinent that in explaining the flow data of HI collision, the correct momentum and density dependence of the nucleonic mean field are the fundamental requirement. These two characteristic dependences of the nuclear mean field can be studied independent of each other, as will be shown in the next section, where the density dependence of the mean field will have direct connection with the nuclear matter EOS. Properties of the mean field for each of these two areas can be divided into two parts, (i) in SNM and (ii) in ANM. Information in SNM in both the areas of the nuclear mean field have been achieved to some reasonable extents. But our knowledge in ANM for both these areas is still poor. The overall state of affair is summarized in the following chart.

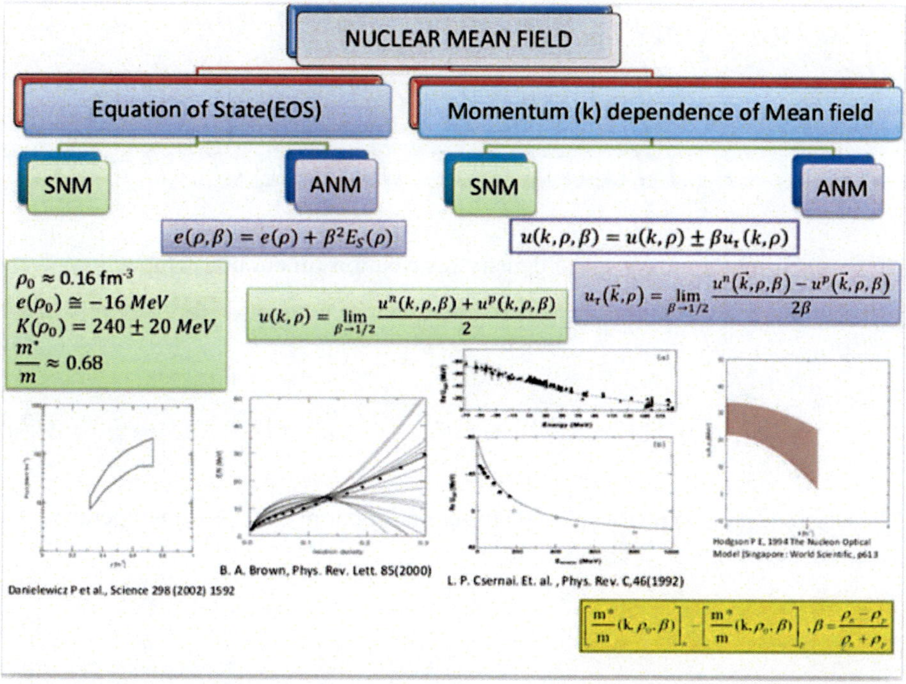

In the following section, we shall show that a finite range part in the NN-effective force is an essential requirement to simulate the k-dependence in the mean field and the k- and ρ-dependence of the nuclear mean field, can be considered independent of each other. This we shall work out for the general case of NN-effective force in the framework of non-relativistic mean field theory.

17.2 Finite Range Effective Interaction and Momentum Dependence of the Mean Field

One can consider four different effective interactions, namely, the direct and exchange interaction between pairs of like (l) and unlike (ul) nucleons: $v_d^{(l)}(r)$, $v_d^{(ul)}(r)$, $v_{ex}^{(l)}(r)$ and $v_{ex}^{(ul)}(r)$. These effective interactions are averaged over angles, spins, and isospins of the two interacting nucleons and depend on the relative separation, r, between them as well as on the total nucleon density $\rho = \rho_n + \rho_p$ of the medium in which they are interacting. The energy density functional $H_T(\rho_n, \rho_p)$ in ANM obtained from these effective interactions can be expressed as

$$H_T(\rho_n, \rho_p) = \frac{\hbar^2}{2m} \int [f_T^n(\vec{k}) + f_T^p(\vec{k})] k^2 d^3k + V^T(\rho_n, \rho_p), \quad (17.6)$$

where the first term is the kinetic energy part and second one is the interaction contribution given by

$$v^T(\rho_n,\rho_p)=\frac{1}{2}(\rho_n^2+\rho_p^2)\int v_d^l(r)d^3r+\rho_n\rho_p\int v_d^{(ul)}(r)d^3r$$
$$+\frac{1}{2}\int\int[f_T^n(\vec{k})f_T^n(\vec{k}')+f_T^p(\vec{k})f_T^p(\vec{k}')]g_{ex}^l(|\vec{k}-\vec{k}'|)d^3kd^3k'$$
$$+\frac{1}{2}\int\int[f_T^n(\vec{k})f_T^p(\vec{k}')+f_T^p(\vec{k})f_T^n(\vec{k}')]g_{ex}^{ul}(|\vec{k}-\vec{k}'|)d^3kd^3k',$$

$f_T^{n(p)}(\vec{k})$ being the phase space density distribution function at temperature T and $g_{ex}^{l(ul)}(|\vec{k}-\vec{k}'|)$ is Fourier transform of the exchange interaction for like (unlike) pair $v_{ex}^{l(ul)}(r)$,

$$g_{ex}^{l(ul)}(|\vec{k}-\vec{k}'|)=\int e^{i(\vec{k}-\vec{k}').\vec{r}}v_{ex}^{l(ul)}(r)d^3r. \quad (17.7)$$

The neutron (proton) phase space distribution function $f_T^{n(p)}(\vec{k})$ at temperature T is given by Fermi–Dirac (FD) distribution

$$f_T^{n(p)}(\vec{k})=\frac{\xi}{(2\pi)^3}\eta_T(k)=\frac{\xi}{(2\pi)^3}\frac{1}{1+e^{(\epsilon_T^{n(p)}(k,\rho,Y_p)-\mu_T^{n(p)})/T}}, \quad (17.8)$$

where ξ is the spin-isospin degeneracy factor, $\epsilon_T^{n(p)}$ and $\mu_T^{n(p)}$ are neutron (proton) single particle energy and chemical potential and T is in MeV; and the integration over the momentum space gives neutron (proton) density,

$$\int f_T^{n(p)}(\vec{k})d^3k=\rho_{n(p)}. \quad (17.9)$$

The neutron and proton single particle energies can be obtained from the $H_T(\rho_n,\rho_p)$ in (17.6) by taking the respective functional derivatives,

$$\epsilon_T^{n(p)}(k,\rho,Y_p)=\frac{\hbar^2k^2}{2m}+u_T^{n(p)}(k,\rho,Y_p), \quad (17.10)$$

where the 1st term in the right-hand side (RHS) results from the kinetic energy part of H_T and the 2nd one $u_T^{n,p}$ are the respective single particle potentials or mean fields resulting from the interaction part V^T, which are explicitly given by

$$u_T^n(k,\rho,Y_p)=\rho_n\int v_d^l(r)d^3r+\rho_p\int v_d^{ul}(r)d^3r$$
$$+\int f_T^n(\vec{k}')g_{ex}^l(|\vec{k}-\vec{k}'|)d^3k'$$
$$+\int f_T^p(\vec{k}')g_{ex}^{ul}(|\vec{k}-\vec{k}'|)d^3k', \quad (17.11)$$

and

$$u_T^p(k, \rho, Y_p) = \rho_p \int v_d^l(r) d^3r + \rho_n \int v_d^{ul}(r) d^3r$$
$$+ \int f_T^p(\vec{k'}) g_{ex}^l(|\vec{k} - \vec{k'}|) d^3k'$$
$$+ \int f_T^n(\vec{k'}) g_{ex}^{ul}(|\vec{k} - \vec{k'}|) d^3k'. \tag{17.12}$$

It may be seen that by interchanging the indices n and p in u^n, one recovers u^p and vice-verse. We shall now consider the nuclear matter in the zero-temperature limit.

17.2.1 Nuclear Matter at T = 0 MeV

At T = 0 MeV, the nuclear matter is considered to be in ground state where all the states are occupied by n and p from the bottom up to a maximum referred to as Fermi levels for neutron and proton. Thus the phase space density distribution functions $f_{T=0}^n$ and $f_{T=0}^p$ take the form of step-functions and the neutron and proton densities given in (17.9) are expressed in terms of n-, p-Fermi momenta k_n and k_p as

$$\rho_n = \frac{k_n^3}{3\pi^2} \; and \; \rho_p = \frac{k_p^3}{3\pi^2}. \tag{17.13}$$

The finite temperature expression of energy density H_T given in (17.6) together with (17.7) takes the following form for T = 0 MeV,

$$H_(\rho_n, \rho_p) = \frac{3\hbar^2}{10m}(k_n^2\rho_n + k_p^2\rho_p) + V(\rho_n, \rho_p), \tag{17.14}$$

where

$$V(\rho_n, \rho_p) = \frac{1}{2}(\rho_n^2 + \rho_p^2) \int v_d^l(r) d^3r + \rho_n\rho_p \int v_d^{ul}(r) d^3r$$
$$+ \frac{\rho_n^2}{2} \int \left[\frac{3j_1(k_n r)}{k_n r}\right]^2 v_{ex}^l(r) d^3r$$
$$+ \frac{\rho_p^2}{2} \int \left[\frac{3j_1(k_p r)}{k_p r}\right]^2 v_{ex}^l(r) d^3r$$
$$+ \rho_n\rho_p \int \frac{3j_1(k_n r)}{k_n r} \frac{3j_1(k_p r)}{k_p r} v_{ex}^{ul}(r) d^3r, \tag{17.15}$$

with j_1 being spherical Bessel function of orders 1. The n- and p-single particle potentials at finite T in (17.11) and (17.12) take the following forms for T = 0 case,

$$u^n(k,\rho,Y_p) = \rho_n \int v_d^l(r) d^3r + \rho_p \int v_d^{ul}(r) d^3r + \rho_n \int \frac{3j_1(k_n r)}{k_n r} v_{ex}^l(r) j_0(kr) d^3r + \rho_p \int \frac{3j_1(k_p r)}{k_p r} v_{ex}^{ul}(r) j_0(kr) d^3r + u_R(\rho_n, \rho_p), \tag{17.16}$$

and

$$u^p(k,\rho,Y_p) = \rho_p \int v_d^l(r) d^3r + \rho_n \int v_d^{ul}(r) d^3r + \rho_p \int \frac{3j_1(k_p r)}{k_p r} v_{ex}^l(r) j_0(kr) d^3r + \rho_n \int \frac{3j_1(k_n r)}{k_n r} v_{ex}^{ul}(r) j_0(kr) d^3r + u_R(\rho_n, \rho_p), \tag{17.17}$$

where $j_0(kr)$ is the zeroth order Bessel function as a function of the momentum k of the traversing nucleon and $u_R(\rho_n, \rho_p)$ is the rearrangement energy arising out of any explicit dependence of the interactions on the total density, $\rho = \rho_n + \rho_p$,

$$u_R(\rho_n, \rho_p) = \frac{\rho_n^2 + \rho_p^2}{2\rho} \int \rho \frac{\partial v_d^l(r)}{\partial \rho} d^3r + \frac{\rho_n \rho_p}{\rho} \int \rho \frac{\partial v_d^{ul}(r)}{\partial \rho} d^3r + \frac{1}{2\rho} \int \left[\left(\rho_n \frac{3j_1(k_n r)}{k_n r} \right)^2 + \left(\rho_p \frac{3j_1(k_p r)}{k_p r} \right)^2 \right] \rho \frac{\partial v_{ex}^l(r)}{\partial \rho} d^3r + \frac{\rho_n \rho_p}{\rho} \int \frac{3j_1(k_n r)}{k_n r} \frac{3j_1(k_p r)}{k_p r} \rho \frac{\partial v_{ex}^{ul}(r)}{\partial \rho} d^3r. \tag{17.18}$$

In the limit of symmetric nuclear matter, $\rho_n = \rho_p = \frac{\rho}{2}$ and $k_n = k_p = k_f$, where $k_f = (\frac{3\pi^2 \rho}{2})^{1/3}$ is the Fermi momentum in SNM. The zero-temperature expression for energy density in ANM given in (17.14) together with (17.15) in the limit of SNM becomes

$$H(\rho) = \frac{3\hbar^2}{10m} k_f^2 \rho + \frac{\rho^2}{4} \int (v_d^l(r) + v_d^{ul}(r)) d^3r + \frac{\rho^2}{4} \int \left(\frac{3j_1(k_f r)}{k_f r} \right)^2 (v_{ex}^l(r) + v_{ex}^{ul}(r)) d^3r. \tag{17.19}$$

The single particle potentials, the n and p feels in ANM as given in (17.16) and (17.17), are now the same in the limit of SNM and given as

$$u(k,\rho) = \frac{\rho}{2}\int (v_d^l(r) + v_d^{ul}(r))d^3r$$
$$+ \frac{\rho}{2}\int \left(\frac{3j_1(k_f r)}{k_f r}\right)(v_{ex}^l(r) + v_{ex}^{ul}(r))j_0(kr)d^3r + u_R(\rho), \tag{17.20}$$

where the rearrangement term $u_R(\rho)$ is

$$u_R(\rho) = \frac{\rho}{4}\int \rho\left(\frac{\partial v_d^l}{\partial\rho} + \frac{\partial v_d^{ul}}{\partial\rho}\right)d^3r + \frac{\rho}{4}\int \left(\frac{3j_1(k_f r)}{k_f r}\right)^2 \rho\left(\frac{\partial v_{ex}^l}{\partial\rho} + \frac{\partial v_{ex}^{ul}}{\partial\rho}\right)d^3r. \tag{17.21}$$

It is evident from (17.20) (as well as from (17.16) and (17.17)) that the k-dependence of the mean field is simulated through the exchange part of the interaction. From the expression of single particle energy in SNM, $\epsilon(k,\rho) = \frac{\hbar^2 k^2}{2m} + u(k,\rho)$, where $u(k,\rho)$ is given in (17.20), it can be verified that

$$\epsilon(k = k_f, \rho) = e(\rho) + \rho\frac{de(\rho)}{d\rho}, \tag{17.22}$$

where $e(\rho) = \frac{H(\rho)}{\rho}$ is the energy per particle in SNM that can be obtained from (17.19). This is as per requirement of the Hugenholtz–Van hove (HV) theorem at Fermi surface. In course of verification of HV theorem from (17.19) and (17.20), one requires the following relations for the spherical Bessel function,

$$\frac{d}{dx}[x^{-l}j_l(x)] = -x^{-l}j_{l+1}(x),$$

and

$$j_2(x) = \frac{3j_1(x)}{x} - j_0(x). \tag{17.23}$$

17.2.2 Iso-scalar and Iso-vector Parts of the Mean Field

On expanding the n- and p-mean fields, given in (17.16) and (17.17), in a Taylor series around $\beta = 0$, we can write

$$u^{n/p}(k,\rho,\beta) \cong u(k,\rho,\beta=0) \pm \beta u_\tau(k,\rho), \tag{17.24}$$

where +(−) sign is for neutron (proton) and $u_\tau(k,\rho)$ is identified with $\frac{\partial u^{n/p}(k,\rho,\beta)}{\partial\beta}|_{\beta=0}$ and referred to as the iso-vector part of the mean field in ANM. The first term in the RHS of (17.24) is the iso-scalar part of the mean field identified as

$$u(k,\rho,\beta=0)=\lim_{\beta\to 0}\frac{u^n(k,\rho,\beta)+u^p(k,\rho,\beta)}{2}, \tag{17.25}$$

and is identical to the mean field $u(k,\rho)$ in SNM given in (17.20). In the second term of the RHS, the iso-vector part of the mean field in ANM, $u_\tau(k,\rho)$ can be defined as

$$u_\tau(k,\rho)=\lim_{\beta\to 0}\frac{u^n(k,\rho,\beta)-u^p(k,\rho,\beta)}{2\beta}. \tag{17.26}$$

The explicit expression for $u_\tau(k,\rho)$ can be obtain by differentiating u^n and u^p in (17.16) and (17.17), respectively, with respect to β and evaluate (17.26) in the limit $\beta\to 0$, that gives

$$\begin{aligned}u_\tau(k,\rho)=&\frac{\rho}{2}\int\left(v_d^l(r)-v_d^{ul}(r)\right)d^3r\\&+\frac{\rho}{2}\int j_0(kr)j_0(k_f r)\left(v_{ex}^l(r)-v_{ex}^{ul}(r)\right)d^3r.\end{aligned} \tag{17.27}$$

In arriving at this expression, the formulae for spherical Bessel functions in (17.23) are used. The iso-vector part of the mean field gives an account of how the n- and p-mean fields evolve with the evolution of isospin asymmetry in the medium. The k-dependence of $u_\tau(k,\rho)$ is also simulated through the exchange part of the interaction, a feature similar to the iso-scalar mean field $u(k,\rho)$, with the difference that in $u(k,\rho)$ the exchange interactions between like and unlike pairs of nucleons appear as sum, i.e., $(v_{ex}^l+v_{ex}^{ul})$, whereas their difference $(v_{ex}^l-v_{ex}^{ul})$ appears in the iso-vector part of the mean field $u_\tau(k,\rho)$.

In both, $u(k,\rho)$ and $u_\tau(k,\rho)$, their respective k- and ρ-dependence are involved in a complicated way. The separation of the k-dependence in each case can be done by taking out the respective saturation property and one can write $u(k,\rho)$ and $u_\tau(k,\rho)$ as

$$u(k,\rho)=u(k=k_f,\rho)+[u(k,\rho)-u(k=k_f,\rho)], \tag{17.28}$$

and

$$u_\tau(k,\rho)=u_\tau(k=k_f,\rho)+[u_\tau(k,\rho)-u_\tau(k=k_f,\rho)]. \tag{17.29}$$

The iso-scalar part of the mean field $u(k,\rho)$ in (17.28) can be readily expressed in terms of the EOS of SNM by using the HV theorem as given by

$$u(k,\rho) = e(\rho) + \rho\frac{de(\rho)}{d\rho} - \frac{\hbar^2 k_f^2}{2m} + u^{ex}(k,\rho), \tag{17.30}$$

where $u^{ex}(k,\rho)$ is the functional that contains the k-dependence of the mean field in SNM and is given as

$$u^{ex}(k,\rho) = \frac{\rho}{2}\int [j_0(kr) - j_0(k_f r)]\frac{3j_1(k_f r)}{k_f r}\left(v_{ex}^l(r) + v_{ex}^{ul}(r)\right) d^3r. \tag{17.31}$$

The saturation property of the iso-vector part of the mean field $u_\tau(k = k_f, \rho)$ is related to the NSE, $E_s(\rho)$. Under the Taylor series expansion of energy density in ANM $H(\rho,\beta)$ around $\beta = 0$,

$$H(\rho,\beta) \cong H(\rho) + \frac{1}{2}\beta^2\frac{\partial^2 H(\rho,\beta)}{\partial\beta^2}\,|_{\beta=0}\,. \tag{17.32}$$

The NSE is defined as

$$E_s(\rho) = \frac{1}{2\rho}\frac{\partial^2 H(\rho,\beta)}{\partial\beta^2}\,|_{\beta=0}\,. \tag{17.33}$$

Using (17.14) and (17.15) for $H(\rho,\beta)$ in (17.33), the expression for $E_s(\rho)$ becomes

$$\begin{aligned} E_s(\rho) = {} & \frac{\hbar^2 k_f^2}{6m} + \frac{\rho}{4}\int \left(v_d^l - v_d^{ul}\right) d^3r + \frac{\rho}{4}\int j_0^2(k_f r)\left(v_{ex}^l - v_{ex}^{ul}\right) d^3r \\ & - \frac{\rho}{4}\int j_1^2(k_f r)\left(v_{ex}^l + v_{ex}^{ul}\right) d^3r. \end{aligned} \tag{17.34}$$

Further, on evaluating the effective mass in SNM in the limit $k \to k_f$, $\frac{m*}{m} = \left[1 + \frac{m}{\hbar^2 k}\frac{\partial u}{\partial k}\right]^{-1}_{k=k_f}$, by using the mean field given in (17.20), one obtains

$$\frac{\hbar^2 k_f^2}{6m}\left(\frac{m}{m*}(k = k_f,\rho) - 1\right) = -\frac{\rho}{4}\int j_1^2(k_f r)(v_{ex}^l + v_{ex}^{ul})d^3r. \tag{17.35}$$

The iso-vector part of the mean field given in (17.27) at $k = k_f$ can be expressed, in terms of NSE, $E_s(\rho)$ and effective mass in SNM by using (17.34) and (17.35). Thus the expression for the iso-vector part $u_\tau(k,\rho)$ in (17.29), where the k-dependence has been separated out by subtracting the saturation contribution, becomes

$$u_\tau(k,\rho) = 2E_s(\rho) - \frac{\hbar^2 k_f^2}{3m^*(k = k_f,\rho)} + u_\tau^{ex}(k,\rho), \tag{17.36}$$

where $u_\tau^{ex}(k,\rho)$ is the functional that simulates the k-dependence for $u_\tau(k,\rho)$ and is given by

$$u_\tau^{ex}(k,\rho) = \frac{\rho}{2}\int [j_0(kr) - j_0(k_f r)] j_0(k_f r) \left(v_{ex}^{l}(r) - v_{ex}^{ul}(r)\right) d^3r. \tag{17.37}$$

It can be seen from (17.31) and (17.37) that for momentum k at Fermi momentum, $k = k_f$, the functionals $u^{ex}(k,\rho)$ and $u_\tau^{ex}(k,\rho)$ vanish. The iso-scalar and iso-vector parts of the mean field, $u(k,\rho)$ and $u_\tau(k,\rho)$ in (17.30) and (17.36) at $k = k_f$ are determined from the density dependence of $e(\rho)$ in SNM and $E_s(\rho)$ in ANM, respectively. Hence, the nuclear mean field at Fermi momentum decides the EOS of nuclear matter. Further, for δ-function interaction these k-dependent functionals u^{ex} and u_τ^{ex} also vanish. Thus, pure contact interactions cannot simulate k-dependence in the mean field which is an indispensable aspect as has been proved from the HI collision study. In order to simulate the k-dependence in the single particle potential, the NN-effective interaction must contain a finite range part and the simplest way to achieve this objective is by taking a single finite range term of conventional form. Moreover, the k-dependence of the mean field is not influenced by the δ-function interaction. Therefore, a NN-effective interaction can be constructed in the simplest form containing a single finite range term and a zero-range part,

$$v_{eff}(\vec{r}) = t_0(1 + x_0 P_\sigma)\delta(\vec{r}) + \frac{t_3}{6}(1 + x_3 P_\sigma)\left(\frac{\rho(\vec{R})}{1 + b\rho(\vec{R})}\right)^{\gamma}\delta(\vec{r})$$
$$+ (W + BP_\sigma - HP_\tau - MP_\sigma P_\tau) f(r), \tag{17.38}$$

where $f(r)$ is the functional form of the finite range interaction of any conventional form, Yukawa/Gaussian/exponential, containing the range of the interaction, α. The zero-range part has been taken analogous to the Skyrme force, containing a density independent t_0-term and density dependent t_3-term. The ρ^γ-density dependence of Skyrme forces has been modified by including the denominator $(1 + b\rho)^\gamma$ in order to ensure that the SNM should not have supra-luminous behavior, i.e., velocity of sound in SNM should not exceed the velocity of light, c. The constrain on the parameter b on account of this is found in [32] that reads

$$b \geq \frac{1}{\rho_0}\left(\left[\frac{mc^2}{T_{f_0}/5} - e(\rho_0)\right]^{\frac{1}{\gamma+1}} - 1\right)^{-1}, \tag{17.39}$$

where $T_{f_0} = \hbar^2 k_{f_0}^2/2m$ is the Fermi kinetic energy at normal nuclear matter saturation density ρ_0, with $k_{f_0} = (\frac{3\pi^2\rho_0}{2})^{1/2}$ being the Fermi momentum, $e(\rho_0)$ is the energy per particle at saturation and mc^2 is the nucleonic mass. γ is the parameter in the exponent of the density dependent term of the interaction in (17.38). The interaction proposed in (17.38), hereafter shall be referred to as simple effective interaction (SEI), contains altogether 11-number of parameters, namely, $\alpha, b, \gamma, t_0, x_0, t_3, x_3, W, B, H\, and\, M$.

17.3 Energy Density and Single Particle Potentials in ANM and SNM Using SEI

The energy density in ANM at T = 0 given in (17.14) together with (17.15) for the SEI takes the form,

$$
\begin{aligned}
H(\rho_n, \rho_p) = &\frac{3\hbar^2}{10m}(k_n^2\rho_n + k_p^2\rho_p) + \frac{\epsilon_0^l}{2\rho_0}(\rho_n^2 + \rho_p^2) \\
&+ \frac{\epsilon_0^{ul}}{\rho_0}\rho_n\rho_p + \left[\frac{\epsilon_\gamma^l}{2\rho_0^{\gamma+1}}(\rho_n^2 + \rho_p^2) + \frac{\epsilon_\gamma^{ul}}{\rho_0^{\gamma+1}}\rho_n\rho_p\right]\left(\frac{\rho(\vec{R})}{1 + b\rho(\vec{R})}\right)^\gamma \\
&+ \frac{\epsilon_{ex}^l}{2\rho_0}(\rho_n^2 J(k_n) + \rho_p^2 J(k_p)) + \frac{\epsilon_{ex}^{ul}}{\rho_0}\rho_n\rho_p J(k_n, k_p),
\end{aligned}
\tag{17.40}
$$

where

$$
J(k_{n(p)}) = \frac{\int \frac{9j_1^2(k_{n(p)}r)}{(k_{n(p)}r)^2} f(r)d^3r}{\int f(r)d^3r},
\tag{17.41}
$$

and

$$
J(k_n, k_p) = \frac{\int \frac{3j_1(k_n r)}{k_n r}\frac{3j_1(k_p r)}{k_p r} f(r)d^3r}{\int f(r)d^3r},
\tag{17.42}
$$

$f(r)$ being the form factor of the finite range interaction and the superscript indices “l” and “ul” prescribe the contributions coming from the interactions between like-pair and unlike-pair of nucleons. The energy density in (17.40) now contains 09-parameters, namely, $\alpha, b, \gamma, \epsilon_0^l, \epsilon_0^{ul}, \epsilon_\gamma^l, \epsilon_\gamma^{ul}, \epsilon_{ex}^l$, and ϵ_{ex}^{ul}. The new parameters are connected to the interaction parameters through the relations given as follows:

$$\epsilon_0^l = \rho_0 \left[\frac{t_0}{2}(1-x_0) + \left(W + \frac{B}{2} - H - \frac{M}{2} \right) \int f(r) d^3 r \right], \quad (17.43a)$$

$$\epsilon_0^{ul} = \rho_0 \left[\frac{t_0}{2}(2+x_0) + \left(W + \frac{B}{2} \right) \int f(r) d^3 r \right], \quad (17.43b)$$

$$\epsilon_\gamma^l = \frac{t_3}{12} \rho_0^{\gamma+1} (1-x_3), \quad (17.43c)$$

$$\epsilon_\gamma^{ul} = \frac{t_3}{12} \rho_0^{\gamma+1} (2+x_3), \quad (17.43d)$$

$$\epsilon_{ex}^l = \rho_0 \left(M + \frac{H}{2} - B - \frac{W}{2} \right) \int f(r) d^3 r, \quad (17.43e)$$

$$\epsilon_{ex}^{ul} = \rho_0 \left(M + \frac{H}{2} \right) \int f(r) d^3 r. \quad (17.43f)$$

The neutron (proton) single particle potential in (17.16), (17.17) now becomes

$$\begin{aligned} u^{n(p)}(k, \rho, Y_P) &= \epsilon_0^l \frac{\rho_{n(p)}}{\rho_0} + \epsilon_0^{ul} \frac{\rho_{p(n)}}{\rho_0} \\ &+ \left[\frac{\epsilon_\gamma^l}{\rho_0^{\gamma+1}} \rho_{n(p)} + \frac{\epsilon_\gamma^{ul}}{\rho_0^{\gamma+1}} \rho_{p(n)} \right] \left(\frac{\rho}{1+b\rho} \right)^\gamma \\ &+ \epsilon_{ex}^l \frac{\rho_{n(p)}}{\rho_0} I(k, k_{n(p)}) + \epsilon_{ex}^{ul} \frac{\rho_{p(n)}}{\rho_0} I(k, k_{p(n)}) \\ &+ \left[\frac{\epsilon_\gamma^l (\rho_n^2 + \rho_p^2)}{2\rho_0^{\gamma+1}} + \frac{\epsilon_\gamma^{ul} \rho_n \rho_p}{\rho_0^{\gamma+1}} \right] \frac{\gamma \rho^{\gamma-1}}{(1+b\rho)^{\gamma+1}}, \end{aligned} \quad (17.44)$$

where

$$I(k, k_{n(p)}) = \frac{\int j_0(kr) \frac{3 j_1(k_{n(p)} r)}{k_{n(p)} r} f(r) d^3 r}{\int f(r) d^3 r}. \quad (17.45)$$

In SNM, the energy density $H(\rho)$ and single particle potential $u(k, \rho)$ can be obtained from (17.40) and (17.44), respectively, by substituting $\rho_n = \rho_p = \frac{\rho}{2}$ and $k_n = k_p = k_f$, and the resulting expressions are

$$H(\rho) = \frac{3\hbar^2 k_f^2 \rho}{10m} + \frac{\epsilon_0}{2} \frac{\rho^2}{\rho_0} + \frac{\epsilon_\gamma}{2\rho_0^{\gamma+1}} \rho^2 \left(\frac{\rho}{1+b\rho} \right)^\gamma + \frac{\epsilon_{ex}}{2\rho_0} \rho^2 J(k_f), \quad (17.46)$$

and

$$u(k, \rho) = \frac{\epsilon_0}{\rho_0} \rho + \frac{\epsilon_\gamma}{\rho_0^{\gamma+1}} \left(1 + b\rho + \frac{\gamma}{2} \right) \left(\frac{\rho}{1+b\rho} \right)^{(\gamma+1)} + \frac{\epsilon_{ex}}{\rho_0} \rho I(k, k_f), \quad (17.47)$$

where the strength parameters ϵ_0, ϵ_γ and ϵ_{ex} of SNM are related to the corresponding "l" and "ul" strengths of ANM as

$$\epsilon_0 = \frac{\epsilon_0^l + \epsilon_0^{ul}}{2}, \quad \epsilon_\gamma = \frac{\epsilon_\gamma^l + \epsilon_\gamma^{ul}}{2} \; and \; \epsilon_{ex} = \frac{\epsilon_{ex}^l + \epsilon_{ex}^{ul}}{2} \tag{17.48}$$

and their expressions in terms of interaction parameters can be obtained from using (17.43a)–(17.43f). The functions $J(k_f)$ and $I(k, k_f)$ can be written from (17.41) and (17.45), respectively, by replacing $k_{n(p)}$ by k_f. The explicit expressions of the functions $J(k_f)$ and $I(k, k_f)$ can be written provided the form factor $f(r)$ of the finite range interaction is specified. The two conventional forms, Yukawa and Gaussian, are used widely in the nuclear calculations. For Yukawa form, $f(r) = \frac{e^{-r/\alpha}}{(r/\alpha)}$, the analytical expressions for $J(k_f)$ and $I(k, k_f)$ are given as

$$J_Y(k_f) = \left[\left(\frac{3\Lambda^6}{32k_f^6} + \frac{9\Lambda^4}{8k_f^4}\right)\ln\left(1 + \frac{4k_f^2}{\Lambda^2}\right) - \frac{3\Lambda^4}{8k_f^4} + \frac{9\Lambda^2}{4k_f^2} - \frac{3\Lambda^3}{k_f^3} tan^{-1}\left(\frac{2k_f}{\Lambda}\right)\right], \tag{17.49}$$

$$I_Y(k, k_f) = \frac{3\Lambda^2(\Lambda^2 + k_f^2 - k^2)}{8kk_f^3}\ln\left(\frac{\Lambda^2 + (k + k_f)^2}{\Lambda^2 + (k - k_f)^2}\right) + \frac{3\Lambda^2}{2k_f^2} - \frac{3\Lambda^3}{2k_f^3}\left[tan^{-1}\left(\frac{k + k_f}{\Lambda}\right) - tan^{-1}\left(\frac{k - k_f}{\Lambda}\right)\right]. \tag{17.50}$$

For Gaussian form, $f(r) = e^{-r^2/\alpha^2}$, the corresponding expressions are

$$J_G(k_f) = \left[\frac{3\Lambda^6}{16k_f^6} - \frac{9\Lambda^4}{8k_f^4} + \left(\frac{3\Lambda^4}{8k_f^4} - \frac{3\Lambda^6}{16k_f^6}\right)e^{\frac{-4k_f^2}{\Lambda^2}} + \frac{3\Lambda^3}{2k_f^3}\int_0^{\frac{2k_f}{\Lambda}} e^{-t^2}dt\right], \tag{17.51}$$

$$I_G(k, k_f) = \frac{3\Lambda^4}{8kk_f^3}\left[exp\left[-\left(\frac{k + k_f}{\Lambda}\right)^2\right] - exp\left[-\left(\frac{k - k_f}{\Lambda}\right)^2\right]\right] + \frac{3\Lambda^3}{4k_f^3}\int_{\frac{(k-k_f)}{\Lambda}}^{\frac{(k+k_f)}{\Lambda}} e^{-t^2}dt. \tag{17.52}$$

The momentum scale Λ in these (17.49)–(17.52) is given by $\Lambda = \frac{1}{\alpha}$ for the Yukawa and $\Lambda = \frac{2}{\alpha}$ for the Gaussian form of the interactions, where α is the corresponding range of the interactions. It may be noted that at a given density ρ, the functions $I_Y(k, k_f)$ and $I_G(k, k_f)$ vanish in the limit of high momentum k. Similarly, the functions $J_Y(k_f)$ and $J_G(k_f)$ vanish in the limit of very high density ρ.

17.3.1 Parameter Determination in SNM

Out of the six parameters in SNM, $\alpha, \gamma, b, \epsilon_{ex}, \epsilon_0, \epsilon_\gamma$, the range α and the exchange strength ϵ_{ex} are associated with the exchange contribution of the mean field expression given in (17.37) that is responsible for simulating the k-dependence. Thus, these two parameters need to be determined so as to give a proper account of $u^{ex}(k, \rho)$ over a wide range of density and momentum. An important feature of N-nucleus potential obtained from the analysis of transverse momentum flow data at intermediate energies reveals that $u(k, \rho_0)$ turns out to be repulsive for kinetic energy of nucleon $\frac{\hbar^2 k^2}{2m} > 300\,\text{MeV}$ [7–10]. By using (17.28) and the HV theorem, $u(k, \rho_0)$ can be expressed as

$$u(k, \rho_0) = [e(\rho_0) - T_{f0}] + \epsilon_{ex}[I(k, \rho_0) - I(k = k_{f0}, \rho_0)], \qquad (17.53)$$

where $T_{f0} = \frac{\hbar^2 k_{f0}^2}{2m}$ is the Fermi energy at saturation density ρ_0. By exploiting the condition $u(k = k_{300}, \rho_0) = 0$ where k_{300} corresponds to the momentum for kinetic energy 300 MeV of the nucleon, one can write from (17.53),

$$[I(k_{300}, \rho_0) - I(k = k_{f0}, \rho_0)] = \frac{T_{f0} - e(\rho_0)}{\epsilon_{ex}}, \qquad (17.54)$$

where the left-hand side (LHS) is a function of momentum scale Λ. Expressing Λ in the unit of k_{f0}, $\Lambda = \lambda k_{f0}$, the LHS $S(\lambda) = [I(k_{300}, \rho_0) - I(k = k_{f0}, \rho_0)]$ is shown as a function of λ in Fig. 17.2 for both Yukawa and Gaussian forms of SEI. The range α and exchange strength ϵ_{ex} are calculated for the respective minimum value of λ in Yukawa and Gaussian cases from (17.54), where one requires the value of $e(\rho_0)$ and T_{f0} only. Using the standard values of $e(\rho_0) = -16\,\text{MeV}$ and $T_{f0} = 37\,\text{MeV}$ (that corresponds to $\rho_0 = 0.16102\,\text{fm}^{-3}$) the values of these two

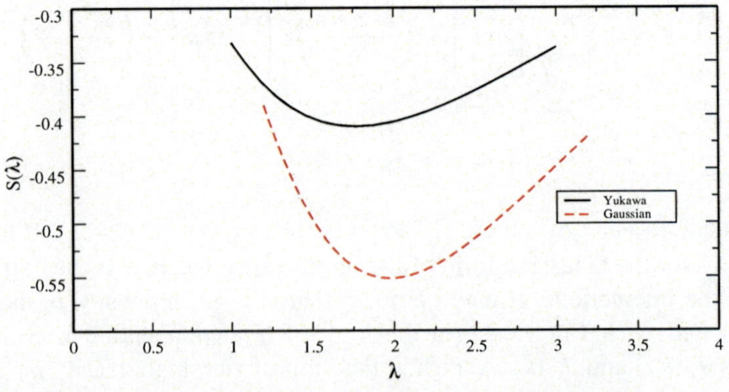

Fig. 17.2 $S(\lambda)$ as a function of λ for both Yukawa and Gaussian forms of SEI

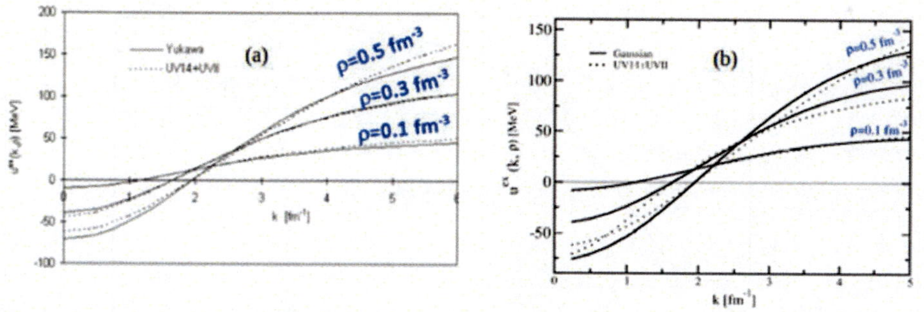

Fig. 17.3 $u^{ex}(k, \rho)$ in (17.31) as a function of k for three densities $\rho = 0.1$, 0.3 and $0.5\,\text{fm}^{-3}$ for Yukawa form (panel (**a**)) and Gaussian form (panel (**b**)) compared with the predictions of the UV14+UVII [34]

parameters are calculated to be, $\epsilon_{ex} = -129.2\,\text{MeV}$, $\alpha = 0.4231\,\text{fm}$ (for SEI Yukawa form) and $\epsilon_{ex} = -96.24\,\text{MeV}$, $\alpha = 0.7597\,\text{fm}$ (for SEI Gaussian form). Fixation of the range and exchange strength parameter using the minimization procedure adopted here is more fundamental than adjusting them either from the optical potential fit or from the finite nuclei study. By knowing these two parameters, α and ϵ_{ex}, one can evaluate $u^{ex}(k, \rho)$ in (17.31) and study the k-dependence of the mean field in SNM at given density ρ. The results for $u^{ex}(k, \rho)$ at three values of density $\rho = 0.1, 0.3$ *and* $0.5\,\text{fm}^{-3}$ are shown in Fig. 17.3a, b as a function of momentum k for Yukawa and Gaussian form factors, respectively [33]. The results of the realistic interaction UV14 + UVII [34] are also shown in the same figures for comparison and it can be seen that in both cases the results compare well over a wide range of momentum and density. It is to be noted that the fixation of α and ϵ_{ex} determines completely the k-dependence of the mean field in SNM.

The effective mass defined in (17.5) can now be calculated for SNM by evaluating the expression,

$$\frac{m*}{m}(k, \rho) = \left[1 - \frac{3m}{\hbar^2}\frac{\rho}{\rho_0}\epsilon_{ex}\int \frac{\frac{j_1(kr)}{kr}\frac{j_1(k_f r)}{k_f r} f(r) r^2 d^3r}{\int f(r) d^3r}\right]^{-1}, \tag{17.55}$$

for the Yukawa and Gaussian form factors of $f(r)$. In the limit of large k, the second term in the square bracket vanishes resulting into $\frac{m*}{m}(k \to \infty) = 1$. On the other hand, in the limit $k \to 0$,

$$\frac{m*}{m}(k = 0, \rho) = \left[1 - \frac{m}{\hbar^2}\frac{\rho}{\rho_0}\int \frac{\frac{j_1(k_f r)}{k_f r} f(r) d^3r}{\int f(r) d^3r}\right]^{-1}. \tag{17.56}$$

The variation of the effective mass m*/m as a function of k in normal nuclear matter, $\rho = \rho_0$, is shown in the Fig. 17.4 for both the forms of SEI. The pre-

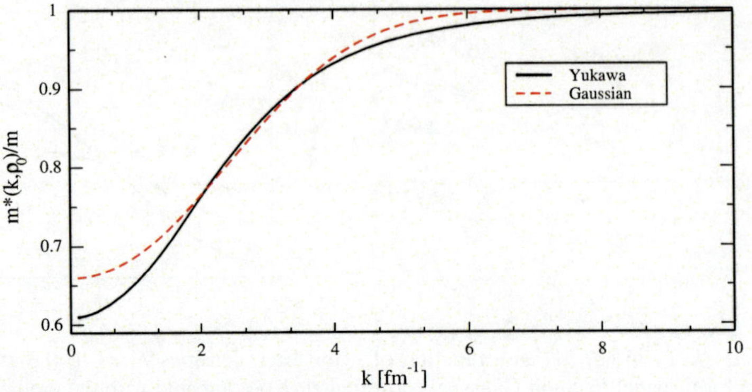

Fig. 17.4 Effective mass in SNM as a function of momentum k at normal nuclear matter density ($\rho = \rho_0$) for Yukawa and Gaussian form of SEI

Table 17.1 Properties of SNM calculated with the knowledge of the two parameters ϵ_{ex} and α. See the text for details

Properties	Yukawa	Gaussian
ϵ_{ex} (MeV)	–129.2	–96.24
λ	1.769	1.976
$u_R(\rho_0)$	15.93	16.91
$\frac{m*}{m}(k = k_{F0}, \rho_0)$	0.686	0.709
$\frac{m*}{m}(k = 0, \rho_0)$	0.609	0.660
$u(k = \infty, \rho_0)$ (MeV)	35.89	12.68
$u(k = 0, \rho_0)$ (MeV)	–73.11	–70.05

dictions of effective mass at Fermi momentum k_{f0} in normal nuclear matter are $\frac{m*}{m}(k = k_{f0}, \rho_0) = 0.686$ and 0.709, in case of Yukawa and Gaussian forms of SEI, respectively. The results of m*/m in the asymptotic limit k = 0 in normal nuclear matter calculated from (17.56) for both the Yukawa and Gaussian forms of SEI are given in Table 17.1 along with the results of the other properties those can be calculated with the knowledge of the two parameters, α and ϵ_{ex} only. These properties of SNM at normal nuclear matter density are $u(k = 0, \rho_0)$, $u(k = k_{f_0}, \rho_0)$, $u(k = \infty, \rho_0)$ and the rearrangement energy $u_R(\rho_0)$. The results obtained for the standard values of the saturation properties, $e(\rho_0) = -16\,\text{MeV}$ and $T_{f0} = 37\,\text{MeV}$, are given in Table 17.1. The expression for the rearrangement energy, $u_R(\rho)$, used for its calculation is obtained by isolating the saturation properties of the effective interaction from the rearrangement part of the single particle potential in (17.20):

$$u(k=k_f,\rho)-2e(\rho)=-\frac{6\hbar^2k_f^2}{10m}+\rho\int\left[j_0(kr)-\frac{3j_1(k_fr)}{k_fr}\right]\frac{3j_1(k_fr)}{k_fr}v_{ex}(r)d^3r+u_R(\rho). \tag{17.57}$$

By using the relation $u(k_f,\rho)=e(\rho)+\rho\frac{de(\rho)}{d\rho}-\frac{\hbar^2k_f^2}{2m}$, the expression for $u_R(\rho)$ results into

$$u_R(\rho)=\frac{\hbar^2k_f^2}{10m}-e(\rho)+\rho\frac{de(\rho)}{d\rho}-\frac{\epsilon_{ex}\rho}{\rho_0\int f(r)d^3r}\int\left[j_0(kr)-\frac{3j_1(k_fr)}{k_fr}\right]\frac{3j_1(k_fr)}{k_fr}f(r)d^3r, \tag{17.58}$$

which can be calculated at saturation density ρ_0 with the knowledge of only α, ϵ_{ex}, $e(\rho_0)$, ρ_0 and by using (17.49)–(17.52) for both forms of SEI. The values of $u_R(\rho_0)$ given in Table 17.1 for both the forms of SEI are calculated from (17.58).

On the other hand, by using the explicit expressions of $u(k=k_f,\rho)$ and $e(\rho)$ for the SEI from (17.47) and (17.46), respectively, one obtains

$$u(k=k_f,\rho)-2e(\rho)=-\frac{6\hbar^2k_f^2}{10m}+\frac{\gamma}{2}\frac{\epsilon_\gamma}{(1+b\rho_0)^{\gamma+1}}+\frac{\epsilon_{ex}\rho}{\rho_0\int f(r)d^3r}\int\left[j_0(k_fr)-\frac{3j_1(k_fr)}{k_fr}\right]\frac{3j_1(k_fr)}{(k_fr)}f(r)d^3r. \tag{17.59}$$

Comparing (17.59) with (17.57), the alternative definition for rearrangement energy is

$$u_R(\rho)=\frac{\gamma}{2}\frac{\epsilon_\gamma}{(1+b\rho_0)^{\gamma+1}}, \tag{17.60}$$

which is determined independently from the density dependent part of the interaction that has no connection to the momentum dependence of the mean field. Both the expressions (17.60) and (17.58) give the same result for $u_R(\rho)$.

With the knowledge of the parameters α and ϵ_{ex}, one can also study the momentum dependence of the single particle potential in normal nuclear matter, $u(k,\rho_0)$, i.e., the optical potential. Using (17.25), $u(k,\rho_0)$ can be written as

$$u(k,\rho_0)=e(\rho_0)-\frac{\hbar^2k_{f0}^2}{2m}-\frac{\rho\epsilon_{ex}}{\rho_0\int f(r)d^3r}\int[j_0(kr)-j_0(k_{f0}r)]\frac{3j_1(k_{f0}r)}{k_{f0}r}f(r)d^3r. \tag{17.61}$$

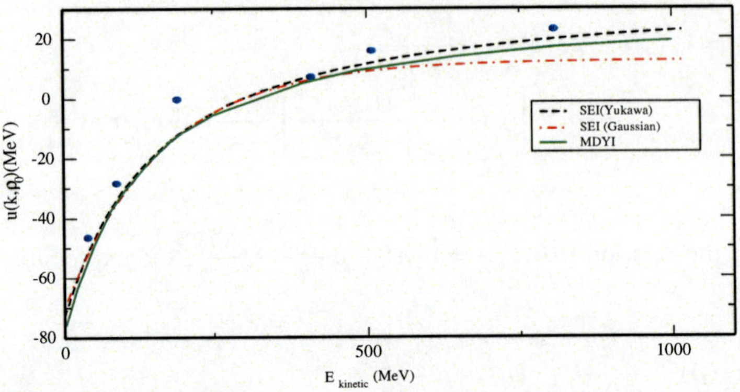

Fig. 17.5 Real part of optical potential $u(k, \rho_0)$ as a function of kinetic energy of the nucleons for both Yukawa and Gaussian form of SEI. The results of MDYI interaction [7] along with the data extracted from the optical model analysis of scattering study [35] (filled circle) are also shown for comparison

This is shown as a function of k in the Fig. 17.5 up to kinetic energy 1 GeV of the nucleon for both the forms of SEI. The results of the MDYI interaction [7] along with the results extracted from the optical model analysis of scattering data [35] are also shown in the figure. It can be seen that the prediction of k-dependence of $u(k, \rho_0)$ for Yukawa form agrees to the MDYI results and optical model data quite well over the whole range of energy. However, in case of the Gaussian form of SEI, it approaches a lower asymptotic value much faster and does not produce well data at energy $E \geq 400$ MeV.

The complete behavior of the single particle potential in SNM $u(k, \rho)$ (both ρ and k-dependence) can be studied only after fixing all the six parameters, out of which we have ascertained upto now only two namely, α and ϵ_{ex}. The parameters b can be calculated from (17.39) after fixing γ. Out of the rest three parameters $\epsilon_0, \epsilon_\gamma$, and γ, the two parameters ϵ_0 and ϵ_γ are determined from the saturation properties, $\frac{de(\rho)}{d\rho}\,|_{\rho=\rho_0}= 0$ and $e(\rho_0)$, and γ has been kept as a free parameter. The admissible values of γ are the ones for which the pressure-density curve ($P \sim \rho$) passes through the regions shown in the Fig. 17.6 those represent the results extracted from the analysis of HI-collisions data [36] and sub-threshold k^+ production data [37].

In both the cases of Yukawa and Gaussian forms of SEI, the values of γ can be taken upto $\gamma = 1$. The value of γ in the exponent of the density dependent term of the SEI determines the nuclear matter incompressibility $K(\rho)$, defined as

$$K(\rho) = 9\rho\frac{d\epsilon(k=k_f,\rho)}{d\rho} = 9\rho^2\frac{d^2e(\rho)}{d\rho^2} + 18\rho\frac{de(\rho)}{d\rho}. \qquad (17.62)$$

At saturation density $\rho = \rho_0$, the $K(\rho_0) = 9\rho^2\frac{d^2e(\rho)}{d\rho^2}\,|_{\rho=\rho_0}$ and the values of $K(\rho_0)$ for $\gamma = \frac{1}{3}, \frac{1}{2}, \frac{2}{3}$, and 1 are 219.93, 237.47, 253.64, and 282.24 MeV for Yukawa and

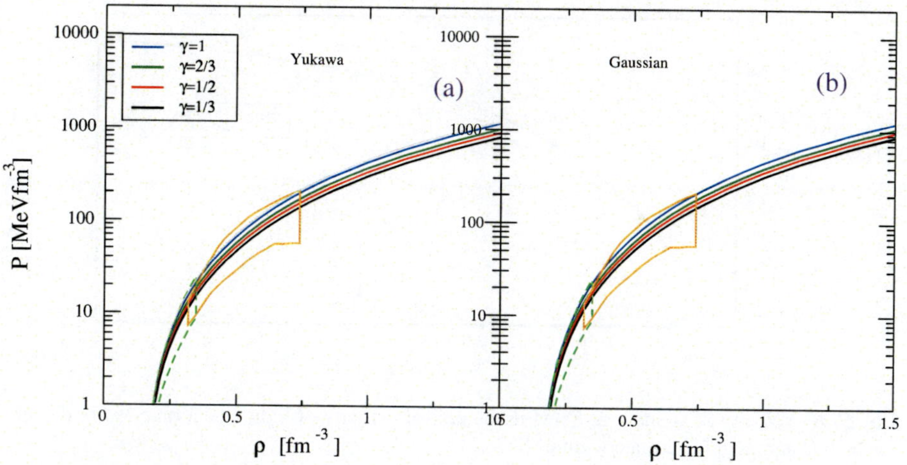

Fig. 17.6 Pressure in SNM as a function of density ρ for different sets corresponding to $\gamma = \frac{1}{3}, \frac{1}{2}, \frac{2}{3}, 1$ of the SEI for Yukawa form in panel (**a**) and Gaussian form in panel (**b**). The results extracted from the heavy-ion collision experiments are shown by the region enclosed by the curve (orange) [36]. Similarly, results obtained from the sub-threshold k^+ production study are the area enclosed by the dashed curve (green) [37]

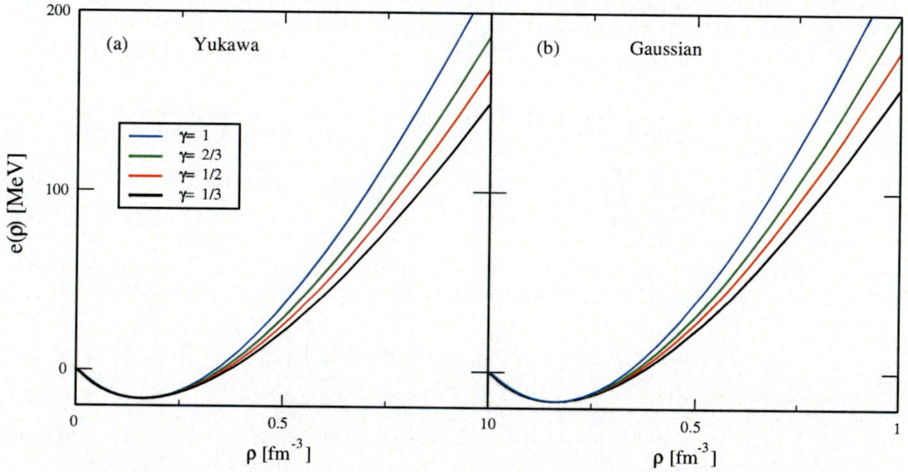

Fig. 17.7 Energy per particle $e(\rho)$ as a function of density for the 04-sets of EOS corresponding to $\gamma = \frac{1}{3}, \frac{1}{2}, \frac{2}{3}$, and 1 for both Yukawa and Gaussian form of SEI in panels (**a**) and (**b**), respectively

227.57, 246.19, 263.36, and 293.73 MeV for Gaussian form of SEI. The results for energy per particle $e(\rho)$ as a function of density for the four sets of EOSs corresponding to $\gamma = \frac{1}{3}, \frac{1}{2}, \frac{2}{3}$, and 1 are shown in Fig. 17.7 for both Yukawa and Gaussian form of SEI in panels (a) and (b), respectively. The velocity of sound, v, in SNM is shown in the unit of velocity of light c as a function of density ρ in Fig. 17.8 for both Yukawa and Gaussian form of SEI.

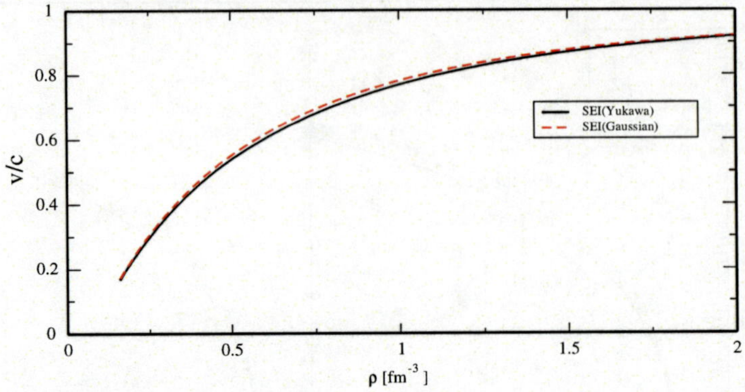

Fig. 17.8 Velocity of sound v in SNM in the unit of velocity of light c as a function of density ρ for both Yukawa and Gaussian form of SEI

17.3.2 Symmetric Nuclear Matter at Finite Temperature, $T \neq 0$

The energy density $H_T(\rho)$ and the single particle potential $u_T(k, \rho)$ in SNM at finite T for the SEI can be written from (17.6) and (17.11) as

$$H_T(\rho) = \frac{\hbar^2}{2m}\int f_T(\vec{k})k^2 d^3k + \frac{\epsilon_0}{2}\frac{\rho^2}{\rho} + \frac{\epsilon_\gamma}{2\rho_0^{\gamma+1}}\rho^2\left(\frac{\rho}{1+b\rho}\right)^\gamma + \frac{\epsilon_{ex}}{2\rho_0}\iint f_T(\vec{k})f_T(\vec{k'})g_{ex}(|\vec{k}-\vec{k'}|)d^3k d^3k', \quad (17.63)$$

and

$$u_T(k,\rho) = \epsilon_0\frac{\rho}{\rho_0} + \frac{\epsilon_\gamma}{\rho_0^{\gamma+1}}\left(1+b\rho+\frac{\gamma}{2}\right)\left(\frac{\rho}{1+b\rho}\right)^{\gamma+1} + \frac{\epsilon_{ex}}{\rho_0}\int f_T(\vec{k'})g_{ex}(|\vec{k}-\vec{k'}|)d^3k'. \quad (17.64)$$

The effect of T, as can be seen from (17.63) and (17.64), is simulated through the momentum space distribution function $f_T(k)$ appearing in the kinetic and finite range exchange terms. Evaluation of $f_T(k)$, given in terms of FD distribution function in (17.8), requires the single particle energy, $\epsilon_T(k,\rho) = \frac{\hbar^2 k^2}{2m} + u_T(k,\rho)$, which in turn contains $f_T(k)$ and therefore imply a self-consistent calculation. A method of successive iteration, starting with zero-temperature $\epsilon(k,\rho)$ as initial input is implemented for the self-consistent evaluation of $f_T(\vec{k})$ at a given density ρ and temperature T. With the initial input $\epsilon^{(0)}(k,\rho)$, $\rho = \int f_T(\vec{k})d^3k$, gives the $\mu_T^{(0)}$ for given ρ and T. With $\epsilon^{(0)}$ and $\mu^{(0)}$, the zeroth order $f_T^{(0)}(\vec{k})$ is defined that is used to evaluate $\epsilon_T^{(1)}$.

Using $\epsilon_T^{(1)}$, the $\mu_T^{(1)}$ is obtained from the evaluation of the density, $\rho = \int f_T(\vec{k}) d^3k$. The process is repeated till the desired accuracy is achieved. This procedure simultaneously gives the distribution function $f_T(\vec{k})$, the single particle energy $\epsilon_T(k, \rho)$ and chemical potential $\mu_T(\rho)$ [38]. The $f_T(\vec{k})$ thus determined is used to evaluate $H_T(\rho)$. The pressure $P_T(\rho)$ in SNM can be obtained as

$$P_T(\rho) = \mu_T(\rho)\rho - F_T(\rho) = \mu_T(\rho)\rho - H_T(\rho) + TS, \tag{17.65}$$

where $F_T(\rho)$ is the free energy density and S is the entropy density at density ρ and temperature T. The entropy density S is expressed in terms of the occupancy distribution function $\eta_T(\vec{k})$ as

$$S = -\frac{\xi}{(2\pi)^3} \int \Big[\eta_T(\vec{k}) \ln \eta_T(\vec{k}) + [1 - \eta_T(\vec{k})] ln[1 - \eta_T(\vec{k})]\Big] d^3k. \tag{17.66}$$

17.3.3 Parameters for ANM

In order to study the properties of ANM, one needs to know the individual parameters for the like and unlike contributions, $\epsilon_0^l, \epsilon_0^{ul}, \epsilon_\gamma^l, \epsilon_\gamma^{ul}, \epsilon_{ex}^l$ *and* ϵ_{ex}^{ul}, which are subject to the constraint given in (17.48). Here, it is assumed implicitly that the ranges for interaction between like and unlike pairs of nucleons are the same.

The momentum and temperature dependence of the mean field in nuclear matter are solely simulated through the finite range exchange part. Therefore, the splitting of the exchange strength ϵ_{ex} into ϵ_{ex}^l and ϵ_{ex}^{ul} should be decided by utilizing constraint following from the study in the related area. Here, the splitting is decided from finite temperature calculation of nuclear matter. The splitting of ϵ_{ex} into ϵ_{ex}^l and ϵ_{ex}^{ul} is subject to the constraint $\epsilon_{ex}^l + \epsilon_{ex}^{ul} = 2\epsilon_{ex}$. Further, if ϵ_{ex}^l is known, then the occupancy distribution function in pure neutron matter (PNM) can be calculated for given ρ and T. Thus, the entropy density in PNM can be obtained from (17.66). The entropy density in PNM is shown in Fig. 17.9 in panels (a) and (b) as a function of density for arbitrary splittings of ϵ_{ex} into ϵ_{ex}^l and ϵ_{ex}^{ul}, at T = 40 and 60 MeV, respectively. The results of SNM at same ρ and T is also shown in the figures as solid curves. The findings can be summarized as, if $0 \leqslant \epsilon_{ex}^l < \frac{2}{3}\epsilon_{ex}$, then the PNM result for entropy density will cross-over the SNM one at some density which corresponds to a larger value for higher ϵ_{ex}^l value in the range. For $\frac{2}{3}\epsilon_{ex} < \epsilon_{ex}^l \leqslant 2\epsilon_{ex}$, the results for entropy density in PNM lie below that of SNM at all densities. For $\epsilon_{ex}^l = \frac{2}{3}\epsilon_{ex}$, it is found that the PNM results for entropy density approaches that of SNM result asymptotically in the region of large density. These findings are true at all temperature T. Once the splitting for ϵ_{ex} into ϵ_{ex}^l and ϵ_{ex}^{ul} is known, then the n- and p-effective masses in ANM can be studied. The n- and p-effective masses calculated from (17.5) as a function of Y_p for $k = k_{f0}$ and $\rho = \rho_0$ predict the n-effective mass

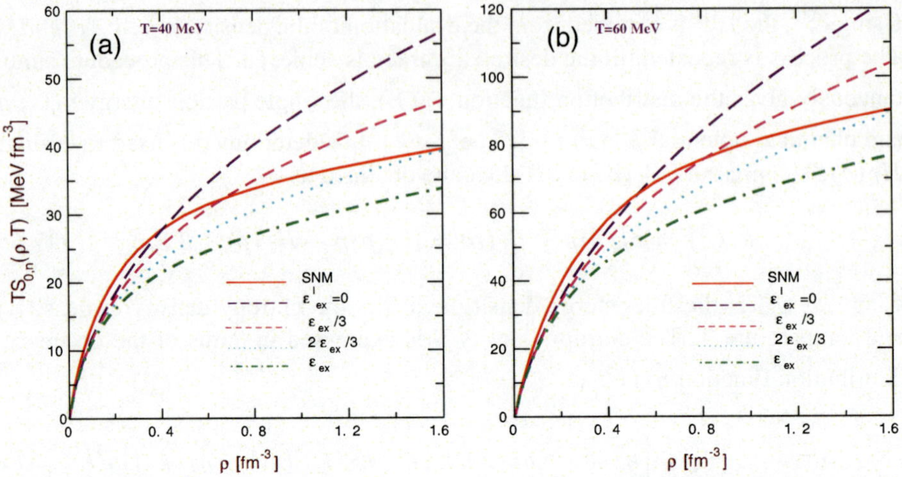

Fig. 17.9 Entropy density in SNM and PNM as a function of density for Yukawa form of SEI at temperatures T = 40 MeV (panel (**a**)) and T = 60 MeV (panel (**b**)). The curves of PNM in both the figures correspond to different choices of finite range exchange strength parameter, ϵ^{l}_{ex}

$(\frac{m*}{m})_n$ above the p-effective mass $(\frac{m*}{m})_p$ if $0 \leq \epsilon^{l}_{ex} < \epsilon_{ex}$. The splitting between them, i.e., $(\frac{m*}{m})_n - (\frac{m*}{m})_p$, is maximum for $\epsilon^{l}_{ex} = 0$ and it decreases as ϵ^{l}_{ex} approaches ϵ_{ex}. At $\epsilon^{l}_{ex} = \epsilon^{ul}_{ex} = \epsilon_{ex}$, the splitting is zero and the n- and p-effective masses are the same. On the other hand, for $\epsilon_{ex} < \epsilon^{l}_{ex} \leq 2\epsilon_{ex}$, the p-effective mass lies above the n-effective mass. The majority model calculations, including the DBHF one [39], predict that the n-effective mass lies above the p-effective mass. So by going with the majority view, the value of ϵ^{l}_{ex} shall lie in the range $0 \leqslant \epsilon^{l}_{ex} < \epsilon_{ex}$. Moreover, under the isospin consideration, the PNM being an one-component system, whereas the SNM is a two-component one. From the first principle of thermodynamics it is known that under the same condition of density and temperature, a one-component system is less disorder than a multi-component system. With the above considerations, the splitting ϵ_{ex} is further restricted to the range $\frac{2}{3}\epsilon_{ex} \leq \epsilon^{l}_{ex} < \epsilon_{ex}$. In absence of any other constraint available to further decide on the high density behavior of entropy density in PNM, we have considered the splitting to be $\epsilon^{l}_{ex} = \frac{2}{3}\epsilon_{ex}$, for which the entropy in PNM approaches that of SNM asymptotically but does not exceed it. For this value of splitting, the results for n- and p-effective mass splitting, $[(\frac{m*}{m})_n - (\frac{m*}{m})_p]$ in normal nuclear matter, compares well with the DBHF prediction [39] over the whole range of asymmetry β, as can be seen in Fig. 17.10.

Now in order to decide the splitting of the rest two parameters ϵ_0 and ϵ_γ, one requires two constraints. One of them is the value of NSE at ρ_0, $E_s(\rho_0)$ and the other one is its first derivative, $E'_s(\rho_0) = \rho_0 \frac{dE_s(\rho)}{d\rho}\mid_{\rho=\rho_0}$. A standard value of $E_s(\rho_0)$ is taken within its accepted range, here $E_s(\rho_0) = 30$ MeV is used. Next, by assigning an arbitrary value to $E'_s(\rho_0)$, one can obtain the splittings of ϵ_0 and ϵ_γ, and thereby all the nine parameters required for the study of ANM are known. One can vary $E'_s(\rho_0)$ for the assumed value of $E_s(\rho_0)$ and for each resulting EOS of ANM, one can study

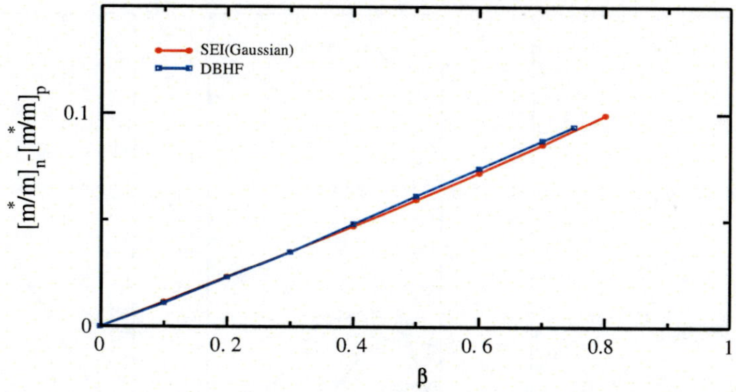

Fig. 17.10 n-p effective mass splitting in ANM at normal nuclear matter density as a function of isospin asymmetry β for $\epsilon_{ex}^{l} = \frac{2}{3}\epsilon_{ex}$. The corresponding DBHF results with Bonn B potential [39] is also shown

the density dependence of NSE, $E_s(\rho)$. In panel (a) of Fig. 17.11, this is shown for the variation of $E_s'(\rho_0)$ with in the range 17–24 MeV for the standard value of $E_s(\rho_0) = 30\,\text{MeV}$. We want to determine the optimal value of $E_s'(\rho_0)$ and to this end we consider the neutron star (NS) scenario. The core of the neutron star is mostly composed of neutrons (n), proton (p), electron (e), and muons (μ) in beta-stable charge neutral condition. The beta-stability condition is expressed as

$$\mu_n - \mu_p = \mu_e = \mu_\mu, \tag{17.67}$$

and the charge neutrality as

$$Y_p = Y_e + Y_\mu, \tag{17.68}$$

where $\mu_i, i = n, p, e, \mu$ are the respective chemical potentials and $Y_i = \frac{\rho_i}{\rho}$ with i = n, p, e, μ are the corresponding particle fractions, ρ being the total density of the neutron star matter (NSM). By solving the (17.67) and (17.68) simultaneously for a given EOS of ANM, the equilibrium proton fraction Y_p for the NSM at a given density ρ can be obtained. The equilibrium Y_p as a function of density ρ are shown in panel (b) of Fig. 17.11 for the respective EOSs of ANM of panel (a). Once the particle fractions are known as a function of ρ of NSM, then the total energy density of the NSM can be written as the sum of the nucleonic and the leptonic parts,

$$H^{NSM} = H_N(\rho, Y_p) + H_e(\rho_e, Y_e) + H_\mu(\rho, Y_\mu). \tag{17.69}$$

The asymmetric contribution of the energy density is crucial in deciding the particle fractions in NSM. So the asymmetric part of the nucleonic contribution in NSM, $H_N(\rho, Y_p)$ is calculated by taking out the symmetric contribution $H(\rho, Y_p = \frac{1}{2})$,

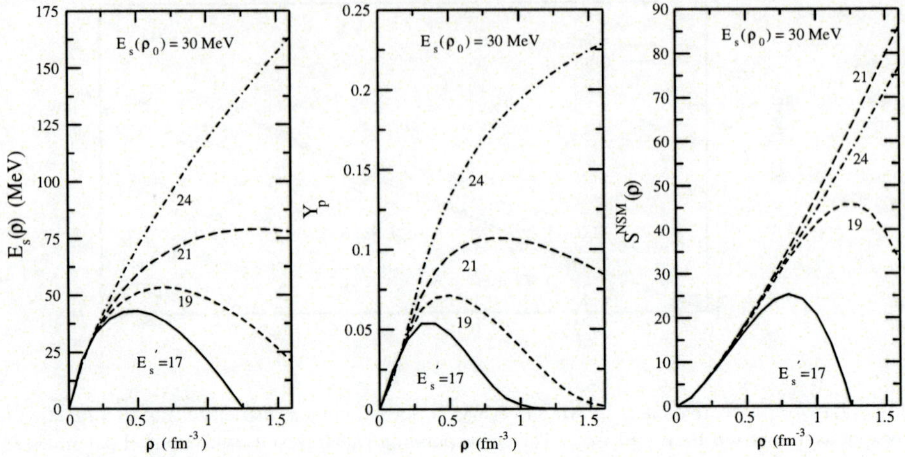

Fig. 17.11 Symmetry energy $E_s(\rho)$ in panel (**a**), equilibrium proton fraction Y_p in NSM in panel (**b**) and the asymmetric nucleonic contribution in NSM, $S^{NSM}(\rho)$, in panel (**c**) as a function of density, ρ, for $E_s(\rho_0) = 30\,\text{MeV}$ and different choices of $E_s'(\rho_0)$ for Yukawa form of SEI

$$S^{NSM}(\rho) = H_N(\rho, Y_p) - H\left(\rho, Y_p = \frac{1}{2}\right). \tag{17.70}$$

The results of $S^{NSM}(\rho)$ as a function of ρ for the EOSs of ANM are shown in panel (c) of Fig. 17.11. One can find that there is a characteristic $E_s'(\rho_0)$ (for the assumed $E_s(\rho)$ value) for which the asymmetric contribution to the nucleonic part of NSM is maximum. In this case, the characteristic $E_s'(\rho_0)$ is found to be 21.4 MeV for Yukawa and 22.2 MeV for Gaussian form of SEI for the assumed value of $E_s(\rho_0) = 30\,\text{MeV}$. In absence of any strong constraint to decide the density dependence of NSE $E_s(\rho)$, the one that corresponds to the maximum $S^{NSM}(\rho)$ can be considered to decide the splitting of ϵ_0 and ϵ_γ. $L(\rho_0) = 3E_s'(\rho_0)$ is referred to as the slope parameter. For this characteristic $E_s'(\rho_0)$, the predicted density dependence of NSE, $E_s(\rho)$ is neither stiff nor very soft and does not predict Direct URCA occurrence in the NSs.

It is to be noted that while studying the density dependence of NSE by varying $E_s'(\rho_0)$ value, that corresponds to different curves in Fig. 17.11a, the momentum dependence of the mean field in SNM remains invariant and for all the EOSs corresponding to these curves, the n- and p-effective mass splitting is the same. On the other hand, one can vary the n-, p-effective mass splitting by considering different splittings of ϵ_{ex} into ϵ_{ex}^{l} and ϵ_{ex}^{ul}, where the density dependence of NSE, $E_s(\rho)$ will remain invariant. In deciding the nine parameters required for the complete study of ANM, SNM, and PNM, one requires to assume only three standard values of nuclear matter saturation properties $e(\rho_0)$, ρ_0 and $E_s(\rho_0)$. For the standard values $e(\rho_0) = -16\,\text{MeV}$, $\rho_0 = 0.161024\,\text{fm}^{-3}$ (corresponding to $T_{f0} = 37\,\text{MeV}$) and $E_s(\rho_0) = 30\,\text{MeV}$, the values of the nine parameters along with the saturation properties for the Yukawa and Gaussian forms of SEI are given in Table 17.2 for

Table 17.2 Values of the nine parameters of ANM for Yukawa (upper line) and Gaussian (lower line) forms of SEI for $\gamma = 1/2$ together with their nuclear matter saturation properties (see text for details)

γ	b (fm)	α (fm)	ϵ_{ex} (MeV)	ϵ_{ex}^{l} (MeV)	ϵ_0 (MeV)	ϵ_0^{l} (MeV)	ϵ_γ (MeV)	ϵ_γ^{l} (MeV)
$\frac{1}{2}$	0.5792	0.4232	−129.2	−86.16	−49.69	−45.23	72.83	62.21
$\frac{1}{2}$	0.5792	0.7577	−96.24	−64.16	−78.22	−64.52	77.35	66.23

Nuclear matter properties at saturation density

γ	ρ_0 (fm^{-3})	$e(\rho_0)$ (MeV)	$K(\rho_0)$ (MeV)	$\frac{m^*}{m}(\rho_0, k_{f0})$	$E_s(\rho_0)$ (MeV)	$L(\rho_0)$ (MeV)
$\frac{1}{2}$	0.1610	−16.0	237.47	0.686	30.0	63.2
$\frac{1}{2}$	0.1610	−16.0	246.19	0.709	30.0	66.6

the EOS corresponding to $\gamma = 1/2$. It is pertinent to note that the different EOSs of ANM resulting either from the variation of $E_s'(\rho_0)$ or from the various splittings of ϵ_{ex} into ϵ_{ex}^{l} and ϵ_{ex}^{ul}, shall have no influence on the predictions of SNM and all will correspond to the same EOS of SNM.

17.4 Summary and Conclusion

The simple effective interaction proposed in (17.38) in the framework of non-relativistic mean field theory provides a fair account of the observed momentum and density dependence of the mean field in symmetric nuclear matter. In isospin asymmetric nuclear matter, our understanding on these k- and ρ-dependence of the mean field is still poor. A continuous analysis of the observables produced in HI collision experiments, such as, n/p ratio, π^+/π^- ratio, and sub-threshold k^+ meson production using the transport model calculations along with the constraints coming from other areas, namely, neutron star phenomenology and finite nuclei studies can be a viable way to narrow down the existing uncertainties, in particular, the density dependence of NSE and n-p effective mass splitting. In the analysis of HI collision data using transport model equations, the variation of ρ-dependence of NSE keeping the k-dependence of the mean field unchanged, and the vice-versa can be performed in a systematic manner without affecting the predictions in SNM by the the use of the SEI, as discussed in the last section. Thus the analysis of the ρ- and k-dependence of the mean field properties of ANM can be performed in a consistent and independent manner by use of SEI with the similar simplicity as provided by Skyrme interaction. The SEI is also used in the studies of neutron star properties as well as finite nuclei. In the studies of neutron stars, apart from mass-radius calculations, the SEI has been successfully used to analyze the crust-core transition, r-mode oscillations and emission of gravitational waves, tidal polarizability, etc. [40, 41]. The SEI has also been used in the calculations of finite nuclei properties in the framework of Density Functional theory (DFT) [42] as well as in more microscopic Hartree–Fock–Bogoliubov

(HFB) formalism [43]. The results of binding energies of 620 and charge radii of 313 even-even deformed nuclei have been obtained with root mean square deviations as low as 1.5 MeV and 0.025 fm, respectively, [43] which is comparable to the predictions of any other effective interaction. The SEI has been also used in the study of fission dynamics [43]. It is to be noted that the finite nuclei calculation with SEI interaction is performed by keeping only one parameter of the SEI open, rest 10-parameters (of the total 11-parameters of SEI) are fixed from the nuclear matter study [44]. Therefore, the predictions of the finite nuclei properties of the SEI can be considered in a sense, as ab-initio.

References

1. G.F. Bertsch, S.D. Gupta, A guide to microscopic models for intermediate energy heavy ion collisions. Phys. Rev. **160**, 189 (1988)
2. R. Stock et al., Compression effects in relativistic nucleus-nucleus collisions. Phys. Rev. Lett. **49**, 1236 (1982); J.W. Haris et al., Pion production in high-energy nucleus-nucleus collisions. Phys. Rev. Lett. **58**, 463 (1987)
3. H.A. Gustafsson et al., Collective flow observed in relativistic nuclear collisions. Phys. Rev. Lett. **52**, 1590 (1984); G. Buchwald et al., Kinetic energy flow in Nb (400 AMeV) + Nb: evidence for hydrodynamic compression of nuclear matter. Phys. Rev. Lett. **52**, 1594 (1984)
4. B. Friedman, V.R. Pandharipande, Hot and cold, nuclear and neutron matter. Nucl. Phys. A **361**, 502 (1981)
5. C. Gale, G. Bertsch, S.D. Gupta, Heavy-ion collision theory with momentum-dependent interactions. Phys. Rev. C **35**, 1666 (1987)
6. J. Aichelin, A. Rosenhauer, G. Peilert, H. Stoecker, W. Greiner, Importance of momentum-dependent interactions for the extraction of the nuclear equation of state from high-energy heavy-ion collisions. Phys. Rev. Lett. **58**, 1926 (1987)
7. L.P. Csernai, G. Fai, C. Gale, E. Osnes, Nuclear equation of state with momentum-dependent interactions. Phys. Rev. C **46**, 736 (1992)
8. M. Prakash, T.T.S. Kuo, S.D. Gupta, Momentum dependence, Boltzmann-Uehling-Uhlenbeck calculations, and transverse momenta. Phys. Rev. C **37**, 2253 (1988)
9. G.M. Welke, M. Prakash, T.T.S. Kuo, S.D. Gupta, C. Gale, Azimuthal distributions in heavy ion collisions and the nuclear equation of state. Phys. Rev. C **38**, 2101 (1988)
10. C. Gale, G.M. Welke, M. Prakash, S.J. Lee, S.D. Gupta, Transverse momenta, nuclear equation of state, and momentum-dependent interactions in heavy-ion collisions. Phys. Rev. C **41**, 1545 (1990)
11. Q. Pan, P. Danielewicz, From sideward flow to nuclear compressibility. Phys. Rev. Lett. **70**, 2062 (1993)
12. J. Zhang, S.D. Gupta, C. Gale, Momentum-dependent nuclear mean fields and collective flow in heavy-ion collisions. Phys. Rev. C **50**, 1617 (1994)
13. F. Haddad, F. Sebille, M. Farine, V. de la Mota, P. Schuck, B. Jouault, Effects of Gogny-type interactions on the nuclear flow. Phys. Rev. C **52**, 2013 (1995)
14. B.A. Li, C.M. Ko, Z.Z. Ren, Equation of state of asymmetric nuclear matter and collisions of neutron-rich nuclei. Phys. Rev. Lett. **78**, 1644 (1997); B.A. Li, Neutron-proton differential flow as a probe of isospin-dependence of the nuclear equation of state. Phys. Rev. Lett. **85**, 4221 (2000); B.A. Li, Probing the high density behavior of the nuclear symmetry energy with high energy heavy-ion collisions. Phys. Rev. Lett. **88**, 192701 (2002)
15. I. Bombaci, *Isospin Physics in Heavy-Ion Collisions at Intermediate Energies*, ed. by B.A. Li, W.U. Schroder (Nova Science, New York, 2001), p. 35

16. M. Prakash, I. Bombaci, M. Prakash, P.J. Ellis, R. Knorren, J.M. Lattimer, Composition and structure of protoneutron stars. Phys. Rep. **280**, 1 (1997)
17. C.B. Das, S.D. Gupta, C. Gale, B.A. Li, Momentum dependence of symmetry potential in asymmetric nuclear matter for transport model calculations. Phys. Rev. C **67**, 034611 (2003)
18. A. Ono, P. Danielewicz, W.A. Friedman, W.G. Lynch, M.B. Tsang, Symmetry energy for fragmentation in dynamical nuclear collisions. Phys. Rev. C **70**, 041604 (R) (2004)
19. G.A. Souliotis, D.V. Shetty, A. Keksis, E. Bell, M. Jandel, M. Veselsky, S.J. Yennello, Heavy-residue isoscaling as a probe of the symmetry energy of hot fragments. Phys. Rev. C **73**, 024606 (2006)
20. D.V. Shetty, S.J. Yennello, G.A. Souliotis, Density dependence of the symmetry energy and the nuclear equation of state: a dynamical and statistical model perspective. Phys. Rev. C **76**, 024606 (2007)
21. J. Xu, L.W. Chen, B.A. Li, H.R. Ma, Temperature effects on the nuclear symmetry energy and symmetry free energy with an isospin and momentum dependent interaction. Phys. Rev. C **75**, 014607 (2007)
22. M.B. Tsang, Y. Zhang, P. Danielewicz, M. Famiano, Z. Li, W.G. Lynch, A.W. Steiner, Constraints on the density dependence of the symmetry energy. Phys. Rev. Lett. **102**, 122701 (2009)
23. A. Klimkiewicz et al., Nuclear symmetry energy and neutron skins derived from pygmy dipole resonances. Phys. Rev. C **76**, 051603 (2007)
24. L. Trippa, G. Colò, E. Vigezzi, Giant dipole resonance as a quantitative constraint on the symmetry energy. Phys. Rev. C **77**, 061304 (2008)
25. J. Rizzo, M. Colonna, M. Di Toro, V. Greco, Transport properties of isospin effective mass splitting. Nucl. Phys. A **732**, 202 (2004)
26. B.A. Li, C.B. Das, S.D. Gupta, C. Gale, Effects of momentum-dependent symmetry potential on heavy-ion collisions induced by neutron-rich nuclei. Nucl. Phys. A **735**, 563 (2004)
27. T. Lesinski, K. Bennaceur, T. Duguet, J. Mayer, Isovector splitting of nucleon effective masses, ab initio benchmarks and extended stability criteria for Skyrme energy functionals. Phys. Rev. C **74**, 044315 (2006)
28. A.M. Lane, Isobaric spin dependence of the optical potential and quasi-elastic (p, n) reactions. Nucl. Phys. **35**, 676 (1962)
29. P.E. Hodgson, *The Nucleon Optical Model* (World Scientific, Singapore, 1994), p. 613
30. G.W. Hoffmann, W.R. Coker, Coupled-channel calculations of the energy dependence of the (p, n) charge-exchange reaction. Phys. Rev. Lett. **29**, 227 (1972)
31. A.J. Koning, J.P. Delaroche, Local and global nucleon optical models from 1 keV to 200 MeV. Nucl. Phys. A **713**, 231 (2003)
32. B. Behera, T.R. Routray, R.K. Satpathy, Causal violation of the speed of sound and the equation of state of nuclear matter. J. Phys. G: Nucl. Part. Phys. **23**, 445 (1997)
33. B. Behera, T.R. Routray, R.K. Satpathy, Momentum and density dependence of the mean field in nuclear matter. J. Phys. G: Nucl. Part. Phys. **24**, 2073 (1998)
34. R.B. Wiringa, Single-particle potential in dense nuclear matter. Phys. Rev. C **38**, 2967 (1988)
35. L.G. Arnold et al., Energy dependence of the p - ^{40}Ca optical potential: a Dirac equation perspective. Phys. Rev. C **25**, 936 (1982)
36. P. Danielewicz, R. Lacey, W.G. Lynch, Determination of the equation of state of dense matter. Science **298**, 1592 (2002)
37. W.G. Lynch, M.B. Tsang, Y. Zhang, P. Danielewicz, M. Famiano, Z. Li, A.W. Steiner, Probing the symmetry energy with heavy ions. Prog. Part. Nucl. Phys. **62**, 427 (2009)
38. B. Behera, T.R. Routray, S.K. Tripathy, Temperature dependence of the nuclear symmetry energy and equation of state of charge neutral n+ p+ e+ μ matter in beta equilibrium. J. Phys. G: Nucl. Part. Phys. **36**, 125105 (2009)
39. F. Sammarruca, The microscopic approach to nuclear matter and neutron star matter. Int. J. Mod. Phys. E **19**, 1259 (2010)
40. T.R. Routray, X. Viñas, D.N. Basu, S.P. Pattnaik, M. Centelles, L.M. Robledo, B. Behera, Exact versus Taylor-expanded energy density in the study of the neutron star crustâcore transition. J. Phys. G: Nucl. Part. Phys. **43**, 105101 (2016)

41. S.P. Pattnaik, T.R. Routray, X. Viñas, D.N. Basu, M. Centelles, K. Madhuri, B. Behera, Influence of the nuclear matter equation of state on the r-mode instability using the finite-range simple effective interaction. J. Phys. G: Nucl. Part. Phys. **45**, 055202 (2018)
42. B. Behera, X. Viñas, M. Bhuyan, T.R. Routray, B.K. Sharma, S.K. Patra, Simple effective interaction: infinite nuclear matter and finite nuclei. J. Phys. G: Nucl. Part. Phys. **40**, 095105 (2013)
43. B. Behera, X. Viñas, T.R. Routray, L.M. Robledo, M. Centelles, S.P. Pattnaik, Deformation properties with a finite-range simple effective interaction. J. Phys. G: Nucl. Part. Phys. **43**, 045115 (2016)
44. B. Behera, X. Viñas, T.R. Routray, M. Centelles, Study of spin polarized nuclear matter and finite nuclei with finite range simple effective interaction. J. Phys. G: Nucl. Part. Phys. **42**, 045103 (2015)

Chapter 18
Effective Surface Properties of Light and Medium Mass Exotic-Nuclei

Abdul Quddus and S. K. Patra

Abstract In this work, we have calculated the effective surface properties such as the symmetry energy, neutron pressure, and symmetry energy curvature for the neutron-deficient/rich isotopes of light and medium mass nuclei. For this, we have considered the isotopic series of O, Ca, Ni, and Zr, which are proton magic isotopes. The ground-state properties of these isotopes along with the density distribution of nucleons are calculated within the spherically symmetric relativistic mean field model by using the recently developed IOPB-I parameter set, and the widely used NL3 set. The calculated densities of the isotopes are used in coherent density fluctuation model in order to calculate the effective surface properties. We found the surface properties as non-traditional way to predict the shell/sub-shell closure of nuclei. The present investigations are quite relevant for the synthesis of exotic-nuclei with high isospin asymmetry also to constrain an equation of state of nuclear matter.

18.1 Introduction

The nuclear symmetry energy is defined as the energy required to convert one neutron to proton or vice-versa. It is important for both the systems, finite nuclei and infinite nuclear matter, having significant role in different areas of nuclear physics. For example, in structure of ground state nuclei [1–3], dynamics of heavy-ion reactions [4, 5], physics of giant collective excitation [6], and physics of neutron star [7–10]. As stated in [11], the astrophysical observations and availability of exotic beam in a laboratory have raised an interest in the symmetry energy. Nuclei near the drip-line have been understood with the density dependence of nuclear symmetry energy

A. Quddus
Department of Physics, Aligarh Muslim University, Aligarh 202002, India
e-mail: abdulquddusphy@gmail.com

S. K. Patra (✉)
Institute of Physics, Bhubaneswar 751005, India
e-mail: patra@iopb.res.in

Homi Bhabha National Institute, Anushakti Nagar, Mumbai 400085, India

R. K. Puri et al. (eds.), *Advances in Nuclear Physics*, Springer Proceedings in Physics 257, https://doi.org/10.1007/978-981-15-9062-7_18

[12]. It is important to characterize the symmetry energy experimentally to interpret neutron-rich nuclei and neutron star matter. However, limitation is that symmetry energy is not directly measurable and thus extracted from the observables related to it. In [11, 13], it has been shown theoretically that the symmetry energy of nuclei can be used to indicate/determine magic nuclei.

The neutron pressure is an essential quantity in determining the equation of state (EoS) of nuclear matter [11, 13–15]. For finite nuclei, the neutron pressure depends on the interaction strength of nucleons and their distributions, while the symmetry energy curvature is important for the scattering phenomenon. The symmetry energy, neutron pressure, and symmetry energy curvature are collectively referred as effective surface properties [11, 13]. The physical importance of the surface properties and their sensitivity to density dependence have motivated us to pursue their systematic study for the isotopic series of O, Ca, Ni, and Zr. There are various ways of calculating the symmetry energy and the related quantities. Recently, the symmetry energy of finite nuclei has been studied by using various formulae of the liquid drop model [16–18], the random phase approximation based on the Hartree–Fock (HF) approach [19], the energy density functional of Skyrme force [20–22], the relativistic nucleon-nucleon interaction [23, 24], and the effective relativistic Lagrangian with density-dependent meson-nucleon vertex function [25]. In this work, we have calculated the effective surface properties of the nuclei within the coherent density fluctuation model (CDFM). The reason for choosing the CDFM approach is that it has the following advantages over other methods. It takes care of (i) the fluctuation arises in the nuclear density distribution via weight function $|f(x)|^2$, and (ii) the momentum distributions through the mixed density matrix (i.e., the Wigner distribution function) [11, 13, 26]. In other words, the CDFM approach is adopted to comprise the variation arise from the momentum and density distributions at the surface of finite nuclei. The input to CDFM approach is density of a nucleus which has been calculated within the relativistic mean field (RMF) approach with IOPB-I [27] and NL3 [28] parameter sets.

The paper is organized as follows: in Sect. 18.2.1, we outline the relativistic mean field model, which has been used to calculate the densities of the nuclei. Section 18.2.1.1 contains the general idea of calculating symmetry energy and relevant quantities and how they are calculated within CDFM. The calculated results are discussed in Sect. 18.3. Finally, the work is summarized in Sect. 18.4.

18.2 Formalism

The densities of the considered isotopes, as mentioned, are calculated within the RMF formalism. These calculated densities are further used in the coherent density fluctuation model to obtain the effective surface properties of finite nuclei from the corresponding quantities of infinite nuclear matter. The formalism adopted here to find the effective surface properties is briefed below.

18.2.1 Relativistic Mean Field Theory

Relativistic mean field (RMF) theory is one of the most successful microscopic approaches to solve the many body problem of nuclear system, where the nucleons are assumed to interact through the exchange of mesons. It predicts well the properties of finite nuclei and infinite nuclear matter as well. The effective field theory motivated relativistic mean field model (E-RMF) is just an extension of RMF in which, in principle, all possible types of self and cross-couplings of mesons are considered. To handle E-RMF numerically, the ratios of fields and the nucleon mass are used in the truncation scheme as a constrain of naturalness. The detailed formalism of E-RMF model and its various parameterizations can be found in [27–31]. Here for the sake of completeness, we express the energy density, obtained within the E-RMF Lagrangian by applying mean field approximation, as

$$
\begin{aligned}
\mathcal{E}(r) = \sum_i \varphi_i^\dagger(r) \Bigg\{ & -i\boldsymbol{\alpha}\cdot\nabla + \beta\left[M - \Phi(r) - \tau_3 D(r)\right] + W(r) + \frac{1}{2}\tau_3 R(r) \\
& + \frac{1+\tau_3}{2} A(r) - \frac{i\beta\boldsymbol{\alpha}}{2M}\cdot\left(f_\omega \nabla W(r) + \frac{1}{2} f_\rho \tau_3 \nabla R(r)\right)\Bigg\}\varphi_i(r) \\
& + \left(\frac{1}{2} + \frac{\kappa_3}{3!}\frac{\Phi(r)}{M} + \frac{\kappa_4}{4!}\frac{\Phi^2(r)}{M^2}\right)\frac{m_s^2}{g_s^2}\Phi^2(r) - \frac{\zeta_0}{4!}\frac{1}{g_\omega^2}W^4(r) \\
& + \frac{1}{2g_s^2}\left(1 + \alpha_1\frac{\Phi(r)}{M}\right)(\nabla\Phi(r))^2 - \frac{1}{2g_\omega^2}\left(1 + \alpha_2\frac{\Phi(r)}{M}\right)(\nabla W(r))^2 \\
& - \frac{1}{2}\left(1 + \eta_1\frac{\Phi(r)}{M} + \frac{\eta_2}{2}\frac{\Phi^2(r)}{M^2}\right)\frac{m_\omega^2}{g_\omega^2}W^2(r) - \frac{1}{2e^2}(\nabla A(r))^2 \\
& - \frac{1}{2g_\rho^2}(\nabla R(r))^2 - \frac{1}{2}\left(1 + \eta_\rho\frac{\Phi(r)}{M}\right)\frac{m_\rho^2}{g_\rho^2}R^2(r) - \Lambda_\omega\left(R^2(r) \times W^2(r)\right) \\
& + \frac{1}{2g_\delta^2}(\nabla D(r))^2 + \frac{1}{2}\frac{{m_\delta}^2}{g_\delta^2}\left(D^2(r)\right), \qquad (18.1)
\end{aligned}
$$

where Φ, W, R, D, and A are the fields which have been redefined as $\phi = g_\sigma\sigma$, $W = g_\omega\omega^0$, $R = g_\rho\rho^0$, and $A = eA^0$. The variables m_σ, m_ω, m_ρ, and m_δ are the masses and g_σ, g_ω, g_ρ, g_δ, $\frac{e^2}{4\pi}$ are the coupling constants for σ, ω, ρ, δ mesons, and photon, respectively. The total energy of a nucleus is given by following expression:

$$
E = \int \mathcal{E}(r) d^3r + E_{cm} + E_{pair}, \qquad (18.2)
$$

where the first term represents the total contributions of mesonic and nucleonic energy densities given by (18.1). While, the second and third terms are the center-of-mass correction energy and pairing energy, respectively. The expression for E_{cm} is given as

$$E_{cm} = -\frac{3}{4} \times 41A^{-1/3}. \tag{18.3}$$

To describe open-shell nuclei, certainly pairing plays a crucial role. Here, we have considered the quasi-BCS pairing by following the procedure of [30]. In our calculations, we take the bound-state contributions and the levels coming from the quasi-bound states at positive energies [32] and the expressions for E_{pair} are written as

$$E_{pair} = -\frac{\Delta_i^2}{G_i}, \tag{18.4}$$

where Δ_i and $G_i (= C_i/A)$ are, respectively, the pairing gap and strength with $i = n, p$. The C_i are chosen in a way to reproduce the binding energy of a nucleus with mass number A. For the IOPB-I set, $C_n = 19$ and $C_p = 21$ MeV.

18.2.1.1 Coherent Density Fluctuation Model (CDFM)

The energy per nucleon of nuclear matter $\mathcal{E}/A = e(\rho, \alpha)$ (where ρ is the baryon density) in terms of the isospin asymmetry parameter is $\alpha \left(= \frac{\rho_n - \rho_p}{\rho_n + \rho_p}\right)$:

$$e(\rho, \alpha) = \frac{\mathcal{E}}{\rho_B} - M = e(\rho) + S(\rho)\alpha^2 + \mathcal{O}(\alpha^4), \tag{18.5}$$

where $e(\rho)$, $S(\rho)$ and M are the energy density of symmetric nuclear matter (SNM) ($\alpha = 0$), the symmetry energy, and the mass of a nucleon, respectively. The odd powers of α are forbidden by the isospin symmetry and the terms proportional to α^4 and higher orders have a negligible contribution. The contribution of the symmetry energy $S(\rho)$ can not be neglected for a neutron-deficient/rich nucleus as it is explicitly clear from the expression for the total energy of a nucleus within the liquid drop model [7, 16]. The symmetry energy $S(\rho)$ is defined by

$$S(\rho) = \frac{1}{2}\left[\frac{\partial^2 e(\rho, \alpha)}{\partial \alpha^2}\right]_{\alpha=0}. \tag{18.6}$$

The symmetry energy can be expanded through the Taylor series expansion around the saturation density ρ_0 as

$$S(\rho) = J + L\mathcal{Y} + \frac{1}{2}K_{sym}\mathcal{Y}^2 + \frac{1}{6}Q_{sym}\mathcal{Y}^3 + \mathcal{O}[\mathcal{Y}^4], \tag{18.7}$$

where $J = S(\rho_0)$ is the symmetry energy at saturation and $\mathcal{Y} = \frac{\rho - \rho_0}{3\rho_0}$. The slope parameter ($L$-coefficient) and the symmetry energy curvature (K_{sym}) are defined as

$$L = 3\rho \frac{\partial S(\rho)}{\partial \rho}\bigg|_{\rho=\rho_0}, \quad (18.8)$$

$$K_{sym} = 9\rho^2 \frac{\partial^2 S(\rho)}{\partial \rho^2}\bigg|_{\rho=\rho_0}, \quad (18.9)$$

respectively.

In the present work, we have used the energy density for nuclear matter within the method of Brueckner et al. [33, 34], considering the pieces of nuclear matter with density ρ_0(x). In the Brueckner energy density functional method, the matrix element V(x) of the nuclear Hamiltonian is given by

$$V(x) = AV_0(x) + V_C + V_{CO}, \quad (18.10)$$

where

$$V_0(x) = 37.53[(1+\delta)^{5/3} + (1-\delta)^{5/3}]\rho_0^{2/3}(x) + b_1\rho_0(x) + b_2\rho_0^{4/3}(x) + b_3\rho_0^{5/3}(x) + \delta^2[b_4\rho_0(x) + b_5\rho_0^{4/3}(x) + b_6\rho_0^{5/3}], \quad (18.11)$$

with $b_1 = -741.28$, $b_2 = 1179.89$, $b_3 = -467.54$, $b_4 = 148.26$, $b_5 = 372.84$, and $b_6 = -769.57$. The V_0(x) in (18.10) is the energy per particle of nuclear matter (in MeV) which accounts for the neutron-proton asymmetry. V_C is the Coulomb energy of charge particle (proton) in a Flucton,

$$V_C = \frac{3}{5}\frac{Z^2e^2}{x}, \quad (18.12)$$

and V_{CO} is the Coulomb exchange energy given by

$$V_{CO} = 0.7386Ze^2(3Z/4\pi x^3)^{1/3}. \quad (18.13)$$

On substituting V_0(x) in (18.6) and taking its second order derivative, the symmetry energy S_0^{NM}(x) of nuclear matter with density ρ_0(x) is obtained

$$S_0^{NM}(x) = 41.7\rho_0^{2/3}(x) + b_4\rho_0(x) + b_5\rho_0^{4/3}(x) + b_6\rho_0^{5/3}(x). \quad (18.14)$$

The corresponding parameterized expressions for the pressure P_0^{NM}(x) and the symmetry energy curvature ΔK_0^{NM}(x) for such a system within Brueckner energy density functional method have the forms

$$P_0^{NM}(x) = 27.8\rho_0^{5/3}(x) + b_4\rho_0^2(x) + \frac{4}{3}b_5\rho_0^{7/3}(x) + \frac{5}{3}b_6\rho_0^{8/3}(x), \quad (18.15)$$

and

$$\Delta K_0^{NM}(x) = -83.4\rho_0^{2/3}(x) + 4b_5\rho_0^{4/3}(x) + 10b_6\rho_0^{5/3}(x), \tag{18.16}$$

respectively. These quantities are folded in the (18.23)–(18.25) with the weight function to find the corresponding quantities of finite nuclei within the CDFM.

In the CDFM, the one-body density matrix ρ ($\mathbf{r}, \mathbf{r}'$) of a nucleus is written as a coherent superposition of infinite number of one-body density matrices ρ_x ($\mathbf{r}, \mathbf{r}'$) for *Fluctons* (spherical pieces of the nuclear matter) [11, 35, 36],

$$\rho_x(\mathbf{r}) = \rho_0(x)\Theta(x - |\mathbf{r}|), \tag{18.17}$$

with $\rho_o(x) = \frac{3A}{4\pi x^3}$, where x is the spherical radius of a nucleus contained in a uniformly distributed spherical Fermi gas. The one-body density matrix for a finite nuclear system can be given as [11, 26, 36],

$$\rho(\mathbf{r}, \mathbf{r}') = \int_0^\infty dx|f(x)|^2\rho_x(\mathbf{r}, \mathbf{r}'), \tag{18.18}$$

where $|f(x)|^2$ is the weight function (18.22). The term $\rho_x(\mathbf{r}, \mathbf{r}')$ is the coherent superposition of the one-body density matrix and defined as

$$\rho_x(\mathbf{r}, \mathbf{r}') = 3\rho_0(x)\frac{J_1\left(k_f(x)|\mathbf{r} - \mathbf{r}'|\right)}{\left(k_f(x)|\mathbf{r} - \mathbf{r}'|\right)} \times \Theta\left(x - \frac{|\mathbf{r} + \mathbf{r}'|}{2}\right). \tag{18.19}$$

Here, J_1 is the first-order spherical Bessel function. The Fermi momentum of nucleons in the Fluctons with radius x is expressed as $k_f(x) = (3\pi^2/2\rho_0(x))^{1/3} = \gamma/x$, where $\gamma = (9\pi A/8)^{1/3} \approx 1.52A^{1/3}$. The Wigner distribution function of the one body density matrices in (18.19) is

$$W(\mathbf{r}, \mathbf{k}) = \int_0^\infty dx|f(x)|^2 W_x(\mathbf{r}, \mathbf{k}). \tag{18.20}$$

Here, $W_x(\mathbf{r}, \mathbf{k}) = \frac{4}{8\pi^3}\Theta(x - |\mathbf{r}|)\Theta(k_F(x) - |\mathbf{k}|)$. Similarly, the density ρ (r) within CDFM can express in terms of the same weight function as

$$\rho(r) = \int d\mathbf{k}W(\mathbf{r}, \mathbf{k}) = \int_0^\infty dx|f(x)|^2\frac{3A}{4\pi x^3}\Theta(x - |\mathbf{r}|) \tag{18.21}$$

and it is normalized to the nucleon numbers of the nucleus, $\int \rho(\mathbf{r})d\mathbf{r} = A$. The differential equation for the weight function can be obtained in the generator coordinate by taking the δ-function approximation to the Hill–Wheeler integral equation [36]. The weight function for a given density distribution ρ (r) can be expressed as,

$$|f(x)|^2 = -\left(\frac{1}{\rho_0(x)}\frac{d\rho(r)}{dr}\right)_{r=x}, \tag{18.22}$$

with $\int_0^\infty dx|f(x)|^2 = 1$. For a detailed analytical derivation, one can follow [36, 37]. The CDFM allows us to make a transition from the properties of nuclear matter to those of finite nuclei. The symmetry energy S, neutron pressure P, and symmetry energy curvature ΔK for a finite nucleus are defined below, within the CDFM, by weighting the corresponding quantities for infinite nuclear matter [26, 36–38],

$$S = \int_0^\infty dx|f(x)|^2 S_0^{NM}(\rho(x)), \tag{18.23}$$

$$P = \int_0^\infty dx|f(x)|^2 P_0^{NM}(\rho(x)), \tag{18.24}$$

$$\Delta K = \int_0^\infty dx|f(x)|^2 \Delta K_0^{NM}(\rho(x)). \tag{18.25}$$

Here, the quantities on the left-hand-side of (18.23)–(18.25) are the surface weighted average of the corresponding nuclear matter quantities with local density approximation, which have been determined within the method of Brueckner et al. [33, 34].

18.3 Results and Discussions

18.3.1 Densities and Weight Functions for the Nuclei

The total density and the weight function for the ^{40,52}Ca isotopes as representative cases corresponding to the NL3, and IOPB-I parameter sets are shown in Fig. 18.1. The black color with circles represents the curve for NL3 set while the red one with squares is the curve for the IOPB-I set. This representation is same throughout the work. It can be noticed from the figure that the density corresponding to IOPB-I set is larger as compared to that of NL3. These calculated densities of the isotopes are further used in (18.22) to obtain the weight functions for the corresponding nucleus. From the right panel of the figure, it can be noticed that the behavior of the density profile (left panel) is reflected in weight function curve. The lower value of the central density yields the lesser height of the weight function for an isotope of the particular nucleus. The NL3 parameter set predicts larger value of the weight function for all the nuclei as compared to the IOPB-I set. The weight function is used to make a transition from infinite nuclear matter to finite nuclei. That is, properties of infinite nuclear matter are folded with the weight function to find the corresponding properties for finite nuclei. The reason to call the symmetry energy, neutron pressure, and symmetry energy curvature as an effective surface properties is well illustrated in [36].

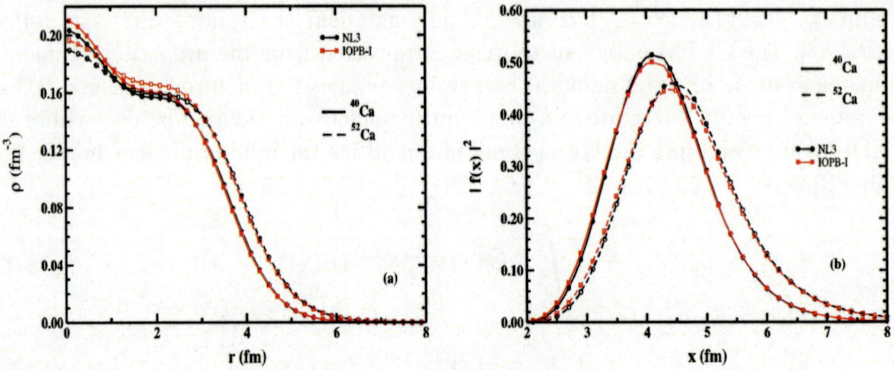

Fig. 18.1 **a** The total density profile and **b** the weight function, for ^{40,52}Ca as the representative case corresponding to IOPB-I [27], and NL3 [28] parameter sets

18.3.2 The Effective Surface Properties of the Nuclei

The symmetry energy S, neutron pressure P, and symmetry energy curvature ΔK for the isotopic series of O, Ca, Ni, and Zr nuclei are shown in Fig. 18.2. The first, second, and third row of each panel of the figure represent the S, P, and ΔK, respectively. The value of the S for finite nuclei lie in the range 23–30 MeV. It can be seen in the figure that the S is larger for the IOPB-I parameter set for all the cases except the isotopes of O and a few isotopes of Ca. The peaks/kinks observed from the S curves for each of the isotopic series represent the magic numbers and/or shell/sub-shell closures at the corresponding neutron numbers. The peaks in the S curve signify more stability of the corresponding nuclei as compared to the neighboring isotopes, which imply that more energy would be required to convert one neutron to proton or vice-versa. Apart from the peaks at the magic neutron numbers, a few small peaks are also evolved which may arise due to the shell structure on the density distribution of the nuclei, and can be referred as weak magic numbers.

However, the P and ΔK show the opposite nature to that of the S with respect to the force parameter sets. The opposite nature means that higher the S values of nuclei corresponding to the particular interaction, lower the P and ΔK are for the same parameter set and vice-versa. The isotopes of O have negative pressure for all the force parameters. The IOPB-I predicts negative pressure for Ca and Ni too, while NL3 predicts negative P values for some of the isotopes of Ca and neutron-rich isotopes of Ni. It is to note that negative values of P arise due to the behavior of the density distribution of these light nuclei in the surface part. It can be remarked after observing the figures that the symmetry energy decreases with the increase of neutron number while pressure and symmetry energy curvature increase with neutron number for an isotopic series.

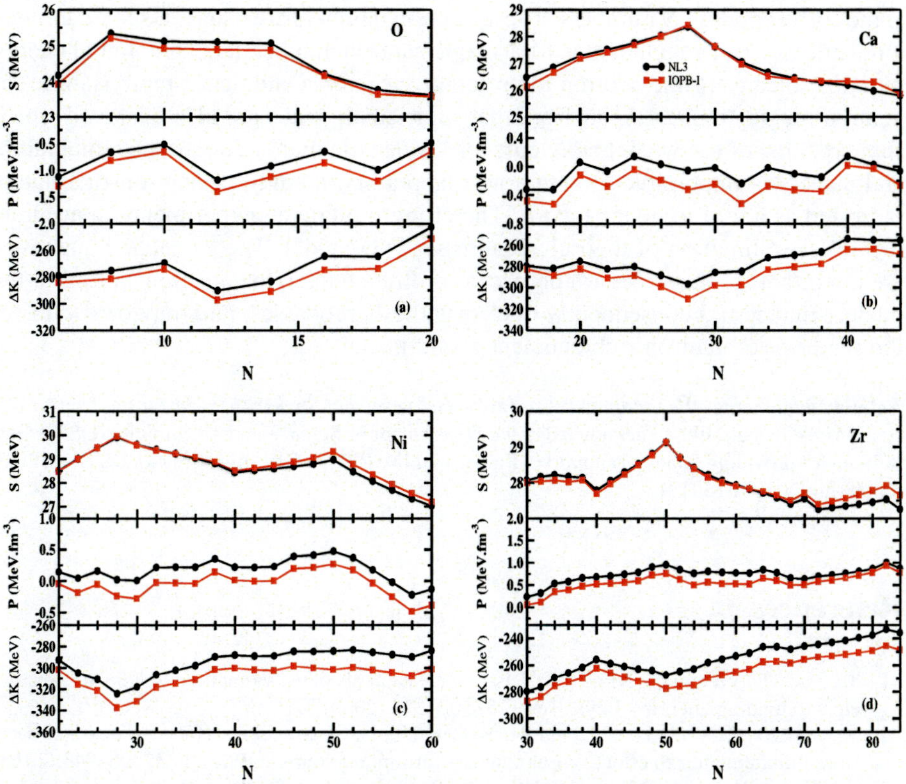

Fig. 18.2 The symmetry energy (S), pressure (P), and symmetry energy curvature (ΔK) for the isotopic series of O, Ca, Ni, and Zr nuclei corresponding to NL3 and IOPB-I parameter sets

18.4 Summary and Conclusions

In summary, the effective surface properties like the symmetry energy, neutron pressure, and the symmetry energy curvature have been studied for the isotopic series of O, Ca, Ni, and Zr nuclei within the coherent density fluctuation model by using the relativistic mean field densities of the isotopes solved in the spherical coordinate. We have used the recently developed IOPB-I parameter set along with the widely used NL3 set within E-RMF to calculate the densities. In CDFM, the effective surface properties for finite nuclei are obtained with folding the corresponding properties for infinite nuclear matter with the weight function; a function used to make transition from infinite nuclear matter to finite nuclei. The symmetry energy, neutron pressure, and symmetry energy curvature for infinite nuclear matter are calculated within Brueckner energy density functional model. The IOPB-I set predicts large S values for the isotopic series except O and few isotopes of Ca. While NL3 set predicts large values for the pressure and symmetry energy curvature as compared to the IOPB-I set. We observe some peaks/kinks in the S and minima in P and ΔK curves at the corresponding N numbers. These neutron numbers are usual close-shell/magic numbers. One can conclude that for a magic neutron number, large amount of energy required to convert one neutron to proton or vice-versa and hence more stability of neutron magic number. Apart from that some small peaks are also observed which support some of the weak magic numbers. Since the matter at extreme temperature and density is impossible to create in a laboratory, a study of neutron-rich nuclei is treated as a tool to understand it. Therefore to guide an experiment, an accurate theoretical estimation of their characteristics is essential. The calculated quantities are important for the structural properties of finite nuclei, to constrain an EoS of the nuclear matter and consequently nucleosynthesis processes, and may be useful for the synthesis of neutron-rich/deficient exotic-nuclei.

Acknowledgements We are thankful to Mitko Gaidarov and Shakeb Ahmad for the fruitful discussions. AQ would like to acknowledge the Department of Science and Technology (DST), Gov. of India for providing financial support in the form of INSPIRE fellowship with No. DST/INSPIRE fellowship/2016/IF160131.

References

1. T. Niksic, D. Vretenar, P. Ring, Relativistic nuclear energy density functionals: adjusting parameters to binding energies. Phys. Rev. C **78**, 034318 (2008)
2. N. Van Giai, B.V. Carlson, Z. Ma, H. Wolter, The Dirac-Brueckner-Hartree-Fock approach: from infinite matter to effective Lagrangians for finite systems. J. Phys. G **37**, 064043 (2010)
3. E.N.E. van Dalen, H. Müther, Relativistic effects in nuclear matter and nuclei. Int. J. Mod. Phys. E **19**, 2077–2122 (2010)
4. B.-A. Li, L.-W. Chen, C.M. Ko, Recent progress and new challenges in isospin physics with heavy-ion reactions. Phys. Rep. **464**, 113–281 (2008)
5. M. Colonna, Nuclear matter and nuclear dynamics. J. Phys. Conf. Ser. **168**, 012006 (2009)
6. V. Rodin, Neutron skin and giant resonances in nuclei. Prog. Part. Nucl. Phys. **59**, 268–276 (2007)
7. A.W. Steiner, M. Prakash, J.M. Lattimer, P.J. Ellis, Isospin asymmetry in nuclei and neutron stars. Phys. Rep. **411**, 325–375 (2005)
8. F.J. Fattoyev, W.G. Newton, J. Xu, B.-A. Li, Generic constraints on the relativistic mean-field and Skyrme-Hartree-Fock models from the pure neutron matter equation of state. Phys. Rev. C **86**, 025804 (2012)
9. M. Dutra, O. Lourenço, J.S.S. Martins, A. Delfino, J.R. Stone, P.D. Stevenson, Skyrme interaction and nuclear matter constraints. Phys. Rev. C **85**, 035201 (2012)
10. M. Dutra, O. Lourenço, S.S. Avancini, B.V. Carlson, A. Delfino, D.P. Menezes, C. Providència, S. Typel, J.R. Stone, Relativistic mean-field hadronic models under nuclear matter constraints. Phys. Rev. C **90**, 055203 (2014)
11. M.K. Gaidarov, A.N. Antonov, P. Sarriguren, E. Moya de Guerra, Surface properties of neutron-rich exotic nuclei: a source for studying the nuclear symmetry energy. Phys. Rev. C **84**, 034316 (2011)
12. W.D. Myers, J. Swiatecki, Droplet-model theory of the neutron skin. Nucl. Phys. A **336**, 267–278 (1980)
13. M. Bhuyan, B.V. Carlson, S.K. Patra, S.-G. Zhou, Surface properties of neutron-rich exotic nuclei within relativistic mean field formalisms. Phys. Rev. C **97**, 024322 (2018)
14. J.M. Lattimer, Symmetry energy in nuclei and neutron stars. Nucl. Phys. A **928**, 276–295 (2014)

15. A. Quddus, M. Bhuyan, S. Ahmad, B.V. Carlson, S.K. Patra, Temperature-dependent symmetry energy of neutron-rich thermally fissile nuclei. Phys. Rev. C **99**, 044314 (2019)
16. W.D. Myers, W.J. Swiatecki, Nuclear masses and deformations. Nucl. Phys. **81**, 1–60 (1966)
17. P. Möller et al., Nuclear ground-state masses and deformations. At. Data Nucl. Data Tables **59**, 185–381 (1995)
18. K. Pomorski, J. Dudek, Nuclear liquid-drop model and surface-curvature effects. Phys. Rev. C **67**, 044316 (2003)
19. A. Carbone, G. Coló, A. Bracco, L.-G. Cao, P.F. Bortignon, F. Camera, O. Wieland, Constraints on the symmetry energy and neutron skins from pygmy resonances in ^{68}Ni and ^{132}Sn. Phys. Rev. C **81**, 041301(R) (2010)
20. L.-W. Chen, C.M. Ko, B.-A. Li, Nuclear matter symmetry energy and the neutron skin thickness of heavy nuclei. Phys. Rev. C **72**, 064309 (2005)
21. S. Yoshida, H. Sagawa, Isovector nuclear matter properties and neutron skin thickness. Phys. Rev. C **73**, 044320 (2006)
22. L.-W. Chen, C.M. Ko, B.-A. Li, J. Xu, Density slope of the nuclear symmetry energy from the neutron skin thickness of heavy nuclei. Phys. Rev. C **82**, 024321 (2010)
23. C.-H. Lee, T.T.S. Kuo, G.Q. Li, G.E. Brown, Nuclear symmetry energy. Phys. Rev. C **57**, 3488–3491 (1998)
24. B.K. Agrawal, Asymmetric nuclear matter and neutron skin in an extended relativistic mean-field model. Phys. Rev. C **81**, 034323 (2010)
25. D. Vretenar, T. Niksic, P. Ring, A microscopic estimate of the nuclear matter compressibility and symmetry energy in relativistic mean-field models. Phys. Rev. C **68**, 024310 (2003)
26. M.K. Gaidarov, A.N. Antonov, P. Sarriguren, E.M. de Guerra, Symmetry energy of deformed neutron-rich nuclei. Phys. Rev. C **85**, 064319 (2012)
27. B. Kumar, S.K. Patra, B.K. Agrawal, New relativistic effective interaction for finite nuclei, infinite nuclear matter, and neutron stars. Phys. Rev. C **97**, 045806 (2018)
28. G.A. Lalazissis, J. König, P. Ring, New parametrization for the Lagrangian density of relativistic mean field theory. Phys. Rev. C **55**, 540–543 (1997)
29. R.J. Furnstahl, B.D. Serot, H.B. Tang, Analysis of chiral mean-field models for nuclei. Nucl. Phys. A **598**, 539–582 (1996); R.J. Furnstahl, B.D. Serot, H.B. Tang, A chiral effective Lagrangian for nuclei. Nucl. Phys. A **615**, 441–482 (1997)
30. M. Del Estal, M. Centelles, X. Viñas, S.K. Patra, Pairing properties in relativistic mean field models obtained from effective field theory. Phys. Rev. C **63**, 044321 (2001); M. Del Estal, M. Centelles, X. Viñas, S.K. Patra, Pairing properties in relativistic mean field models obtained from effective field theory. Phys. Rev. C **63**, 024314 (2001)
31. A. Quddus, K.C. Naik, S.K. Patra, Study of hot thermally fissile nuclei using relativistic mean field theory. J. Phys. G: Nucl. Part. Phys. **45**, 075102 (2018)
32. E. Chabanat, P. Bonche, P. Haensel, J. Meyer, R. Schaeffer, A Skyrme parametrization from subnuclear to neutron star densities Part II. Nuclei far from stabilities. Nucl. Phys. A **635**, 231–256 (1998)
33. K.A. Brueckner, J.R. Buchler, S. Jorna, R.J. Lombard, Statistical theory of nuclei. Phys. Rev. **171**, 1188–1195 (1968)
34. K.A. Brueckner, J.R. Buchler, R.C. Clark, R.J. Lombard, Statistical theory of nuclei. II. Medium and heavy nuclei. Phys. Rev. **181**, 1543–1551 (1969)
35. A.N. Antonov, P.E. Hodgson, I.Zh. Petkov, *Nucleon Momentum and Density Distributions in Nuclei* (Clarendon Press, Oxford); *Nucleon Correlations in Nuclei* (Springer, Berlin, 1993)
36. A.N. Antonov, D.N. Kadrev, P.E. Hodgson, Effect of nucleon correlations on natural orbitals. Phys. Rev. C **50**, 164–167 (1994)
37. C. Fuchs, H. Lenske, H.H. Wolter, Density dependent hadron field theory. Phys. Rev. C **52**, 3043–3060 (1995)
38. A.N. Antonov, D.N. Kadrev, M.K. Gaidarov, P. Sarriguren, E. Moya de Guerra, Temperature dependence of the symmetry energy and neutron skins in Ni, Sn, and Pb isotopic chains. Phys. Rev. C **95**, 024314 (2017)